OCT   2011

CAPTURING CARBON

ROBIN M. MILLS

Capturing Carbon

The New Weapon in the War against Climate Change?

Columbia University Press
New York

Columbia University Press
Publishers Since 1893
New York   Chichester, West Sussex
Copyright © Robin M. Mills, 2011
All rights reserved

Library of Congress Cataloging-in-Publication Data

Mills, Robin M., 1976–
 Capturing carbon : the new weapon in the war against climate
change / Robin M. Mills.
    p. cm.
 Includes bibliographical references and index.
 ISBN 978-0-231-70186-0 (alk. paper)
 ISBN 978-0-231-80047-1 (ebook)
 1. Geological carbon sequestration. 2. Carbon dioxide mitigation.
 3. Renewable energy sources. 4. Global warming. I. Title.

 TD885.5.C3M55 2010
 628.5'32—dc22

                        2010042798

∞

Columbia University Press books are printed on permanent and durable acid-
free paper. This book is printed on paper with recycled content.
Printed in India

c 10 9 8 7 6 5 4 3 2 1

Dedication

To Peter, who challenges my ideas, promotes my books, debates environmental issues, helps developing countries, and lives a low-carbon life.

'More than any other time in history, mankind faces a crossroads. One path leads to despair and utter hopelessness. The other, to total extinction. Let us pray we have the wisdom to choose correctly.'

Woody Allen

CONTENTS

CONTENTS

CONTENTS

CONTENTS

LIST OF ILLUSTRATIONS

PREFACE

I remember clearly the first time I became involved in carbon capture. In January 2005, my colleague Geir Vollsæter, a long-time expert and advocate for carbon capture, and I sat down in our office in the Norwegian oil town of Stavanger, where we were working for Shell. We were convinced that climate change was real, and that the petroleum industry had to play a major part in fighting it. Geir explained to me the concept for capturing carbon dioxide emissions from a power plant in a depleting offshore oilfield; I ran the numbers which showed that, although tricky, it could be economically feasible.

Later that year, I was in the Norwegian Arctic archipelago of Svalbard, which had only the week before experienced record temperatures, and where the recent retreat of glaciers was strikingly apparent in the landscape. This was just a reminder for me of what I already knew. I was taught at Cambridge University by, amongst others, Harry Elderfield, who demonstrated the dramatic global warming event in the Eocene period, 55 million years ago, and Nick Shackleton, the pioneering climatologist. I have, therefore, been aware of the dangers of climate change for at least fifteen years, longer than many who are now engaged with the issue. I am not wedded to fossil fuels—I am of an age, if I wish, to continue my career in alternative energy.

So although I work in the oil industry, I have tried to write an accurate appraisal of this key technology. It is not a panacea—a major role is required for increased energy efficiency, renewable energy and other mitigation options. I have not, except from my publisher, received any encouragement, instruction or payment to write this book. So I hope it can be judged on the merits of my arguments, not on preconceptions, some linked to hidden (or open) agendas.

PREFACE

Carbon capture and storage (CCS) is a fast-evolving field. Although existing projects are few, the literature is voluminous. New technologies emerge, some adaptations of existing methods, others promising but speculative. Economic volatility changes the prices of materials, of fossil fuels and of competing energy sources—and carbon permits, which are also subject to regulatory risk. Cost estimates span a wide range and are highly uncertain. To spare the reader, I am not aiming to provide a comprehensive survey of every opinion or estimate on CCS. But I do want to describe thoroughly the key aspects of technology, economics and policy.

ACKNOWLEDGEMENTS

I have benefited from many helpful discussions and comments while writing this book. Most thanks should go, of course, to my editor, Michael Dwyer of C. Hurst & Co. Publishers, and his staff, particularly Daisy Leitch.

I would also like to express my gratitude to Alex Bakir for his help on the policy side; Melissa Barrett of A.D. Little for insights into recent research; Kamel Bennaceur of Schlumberger for assisting with his deep expertise in the subject; Paul Bryant of Hydrogen Energy for his expertise and encouragement; Heleen de Coninck of the IPCC for her clear-headed view of the fundamentals; Professor Jon Gibbins for his assistance on the capture side, particularly in sharing the latest research; Professor Jon Gluyas of Durham University for his thorough review and valuable suggestions; Chris Hansen of CERA for his energy expertise; Neil Hirst of the IEA for his articulate views; David Harding for drawing my attention to a significant paper; Nader Jandaghi for his insights into solar power; David Johnston and Tony Rogers for discussions on petroleum fiscal systems; Bill Kennedy of Avicenna Partners for his advice; Jacques Merour of Statoil for discussions on the practical issues; Peter Mills for numerous debates on environmental issues; Richard and Jean Mills for perceptive reviews of the manuscript; the people of Ndop, Babungo, Bamenka and Bambalang, Cameroon, for their hospitality and practical demonstrations of climate challenges and renewable energy in the developing world; Emily Rochon of Greenpeace for her robust challenge to carbon capture; the late Professor Nick Shackleton for teaching me the fundamentals of climate science; Geir Vollsæter for his inspiration on Norwegian carbon capture; the editorial and journalistic team at *The National*, Abu Dhabi, par-

ACKNOWLEDGEMENTS

ticularly Rupert Wright, Chris Stanton and Tamsin Carlisle, for helping me develop some of these ideas in print.

INTRODUCTION

The global environment faces today a seemingly insoluble challenge. A system powered mostly by fossil fuels has, for more than two centuries, delivered generally reliable, cheap energy to a growing proportion of humanity, and fuelled unparalleled prosperity. Yet it has become increasingly clear that this same energy system emits unsustainable amounts of greenhouse gases, and that its continuation threatens catastrophic climate change.

Renewable energy and increased efficiency are the most widely proposed solutions to this conundrum, while nuclear power may also have a role. These options are promising, but form only an incomplete picture. We need to add another pillar of future energy: how to use carbon fuels without increasing atmospheric carbon dioxide. Relying only on renewables, efficiency and perhaps nuclear greatly magnifies the costs and risks of what will anyway be a perilous journey.

- Moving to alternative energy sources does not address the large and growing stock of fossil fuel-fired infrastructure, cheap to continue operating, vastly expensive to replace.
- Alternative energy is still generally more costly than carbon fuels, a concern for the health of the world economy, especially the great number of the world's poor who lack access to modern energy.
- Transforming energy systems at sufficient speed will inevitably spawn unintended consequences: bottlenecks, undeserved windfalls, rampant cost increases, perhaps further environmental damage and social impacts.
- The technical and commercial viability of running a whole major economy on non-carbon energy has not been demonstrated in practice.

1

- Most alternative energy technologies can do nothing about the carbon dioxide already in the atmosphere.

If the renewable energy industry's projections are wrong, it will be very costly for both people and the environment. The new mantra should be not 'sustainability' but 'resilience'.[1] As opposed to sustainability, which is essentially static, resilience emphasizes diversity, back-up plans and flexibility. Resilience aims to minimise risks and to maximise our chance to survive the unfathomable challenges of the next century and beyond.

The emerging technology of carbon capture and storage (CCS) is therefore a vital part of the journey to a resilient energy future. CCS covers a broad set of methods which remove carbon dioxide from the waste streams of power plants and other industrial facilities, or directly from the atmosphere, and lock it away—underground, or in plants and soils, or in minerals. Some CCS approaches have long, well-tested commercial histories; others are more speculative. But we have essentially all the components we need today.

What is still missing, although gradually emerging, is the right combination of financial support, a policy framework, and industry innovation, to create a major carbon capture industry. The recent success of some renewable energy technologies has shown the potency of such an alliance.

The idea of CCS, unfortunately, attracts significant opposition.[2] It has received far less publicity than, for instance, solar or wind power, and so is not well-known or comfortingly familiar. Some reject it out of hand, as perpetuating an 'outdated' reliance on fossil fuels, and distracting us from renewable power. Some fear it on safety grounds, or are hostile because of its links to 'big oil' and 'big coal', and feel it gives these unloved industries an undeserved lifeline. To be a little Machiavellian, those who are heavily invested in alternative energies may see it as a threat to their domination of the future. And governments have been slow to realise carbon capture's potential, and to give it the same support extended to other energy technologies.

Others, including many within the energy business, deny the reality of climate change, or see carbon capture as technically unproven, or too expensive or difficult to develop on the required scale. The two main players, the power and petroleum industries, have been slow to move ahead. In the US in particular, great resources have been devoted to lobbying against inevitable climate regulations, instead of trying to provide real solutions. This allows energy companies to be painted as the villain of the climate problem, and hampers their efforts to obtain the same government support for carbon capture that renewable energy has enjoyed. If the corporations do not mobilise soon to make CCS a reality, the opportunity may be lost.

In this book, I hope to show that carbon capture is a viable and vital part of the fight against climate change. I want to describe the key ideas behind CCS, and suggest what needs to happen to make it a large-scale reality. I also want to tackle some concerns and suspicions that have arisen about the idea. Whether we work on or research energy, invest money, participate in debates and vote on public policy, or simply pay our electricity and gas bills, understanding carbon capture is crucial to our future.

Chapter 1: The Need to Capture Carbon. Climate change, caused primarily by human emissions of carbon dioxide, is a reality and the consequences will be very damaging. All the main mitigation options have varying problems, and none is sufficient on its own. Carbon capture is more mature and less risky than often supposed, and can be a major part of the solution.

Chapter 2: Capture Technology. There are three main ways of capturing carbon dioxide from power stations, all in use today but not demonstrated at full scale. Industrial sources can also be suitable for carbon capture but the ease of doing so varies greatly. Carbon capture significantly reduces the efficiency of the plant. Some 'break-through' carbon capture methods, including capture from ambient air, are being researched.

Chapter 3: Transport and Storage. Carbon dioxide transport is mature and straightforward. Underground geological storage, in a variety of settings, is well-understood, and carbon dioxide can be used to extract additional oil and gas. Safety should be good and significant leakage is unlikely for thousands of years. Storage as solid minerals is at a very early stage.

Chapter 4: Bio-sequestration. Carbon can be captured in forests, soils, charcoal and biomass (plant material), mainly in the developing world. Initial costs are potentially low and technology required is simple, but implementation and verification are challenging. Bio-sequestration is complementary to technological carbon capture, not an alternative.

Chapter 5: Scale, Costs and Economics. Carbon capture is capable of being a large, perhaps the largest, element of emissions cuts. Current costs of carbon capture are relatively high and would add significantly to electricity prices. Costs should fall substantially with a large programme of implementation. Carbon capture can be cost-competitive with other low-carbon power generation.

Chapter 6: Policy. A short-term 'kick-start' programme can establish carbon capture commerciality by as early as 2020. After that, consistent carbon costs, and various legal reforms, can be sufficient for carbon capture to compete with other low-carbon technologies. Many projects are already under development and there are major opportunities for businesses and nations.

Chapter 7: Risks. Carbon capture faces significant risks, especially costs, public acceptability and institutional barriers in both industry and government. Bio-sequestration confronts problems of implementation, competition for land and social justice. These risks are not insuperable, and some groups exaggerate them for ideological reasons.

Chapter 8: Conclusion. Carbon capture is a realistic contender for a place in the future energy portfolio. Ideology and fear should not distract us from real solutions. What carbon capture means for you.

1

THE NEED TO CAPTURE CARBON

*'"Avoiding dangerous climate change" is impossible—dangerous climate change is already here. The question is, can we avoid **catastrophic** climate change?'*

David King, UK Chief Scientist[1]

'It [carbon capture] *might be the single most effective—but also controversial—contribution to fighting global warming.'*

Niels Peter Christensen, Chief Geologist, Vattenfall[2]

What is Carbon Capture?

The build-up of carbon dioxide (CO_2) in the atmosphere, due to human activities, is leading to increasingly dangerous and costly climate change. Carbon dioxide is a colourless gas produced by burning carbon-containing materials, such as coal, oil, natural gas and wood, and by living creatures' normal processes of respiration.

Carbon capture and storage (CCS) is a suite of techniques for preventing carbon dioxide entering the atmosphere, or for removing it once it is there. CCS, as generally understood, covers technological methods for capturing carbon at industrial plants, such as power stations. I, however, broaden the discussion to include direct removal of carbon dioxide from the air, and to biological methods. Enhancing the carbon in the biosphere (that part of the Earth which contains life, including plants, animals, marine life, soils and so on) is typically referred to as sequestration, to distinguish it from storage.

The technological approach involves using chemical or physical methods to separate carbon dioxide from a power plant or other industrial facility, such as an oil refinery or paper mill. This can be done either before or after combusting the fuel. The carbon dioxide is then piped to a suitable location where it can be stored permanently underground, for instance in a depleted oilfield. Alternatively, it may be possible to convert the carbon dioxide to a solid mineral. Technology can be extended to processing air directly, to tackle emissions where on-site capture is not practicable, such as transport.

The biological method involves using plants to soak up carbon dioxide, then sequestering the resultant carbon, either as the living organism, as wood, or in soils, potentially as partly combusted material, biochar. Alternatively, changes in land-use and farming practices can cause plants and soil to hold more carbon.

Climate Change is Real, and a Threat

I do not intend to repeat all the details of the global warming debate here, since excellent accounts can be found elsewhere.[3] A brief review, though, is helpful for setting the terms of the issue. The 'deniers', who do not believe in anthropogenic climate change, naturally do not see a need for carbon capture, perhaps viewing it as a boondoggle to extract money from government, or as caving in to environmentalist pressure.

Yet there is now overwhelming consensus[4] that human-caused climate change is happening and has negative consequences which will approach disastrous levels if we do not change course. The following account is generally accepted by most scientists.[5]

Various gases, of which carbon dioxide and water vapour are the most important, are transparent to visible light but absorb infrared light;[6] they are the 'greenhouse gases' (GHGs). They are only a small part, some 0.03%,[7] of the atmosphere, but they are the key to the warming effect, since the other main atmospheric gases—nitrogen, oxygen and argon—allow infrared to pass. Visible light from the sun penetrates the atmosphere, but re-radiated infrared light from the Earth's surface is trapped. Without the natural greenhouse effect, the Earth's average temperature would be around -15°C. This is a matter of basic physics.

In the millennium before the industrial age (starting in, say, 1750), the atmospheric concentration of carbon dioxide did not vary by more than 4%.[8] Since then, it has risen by nearly 40%, growing steadily from 280 parts per million (ppm) to 390 ppm today.[9] Atmospheric carbon dioxide has been measured systematically since 1958,[10] and its increase is now being tracked from space.[11] Giving a feeling for the magnitude of this change, it is at least as large as that separating us from the last Ice Age,[12] when ice sheets covered Manchester and mammoths roamed North Carolina. When carbon dioxide levels were last this high, about 20 million years ago, there was no Antarctic ice-cap and sea levels were 25–40 metres higher than today.[13]

This change is due mainly or entirely to human activities, the most significant of which are the burning of fossil fuels (coal, oil and natural gas), cement manufacture, and deforestation (releasing carbon from plant matter and soils). Other agricultural and industrial processes release the potent greenhouse gases methane, nitrous oxide and ozone, as well as rarer gases such as chlorofluorocarbons (CFCs) from refrigerants and aerosols,[14] and black soot which absorbs heat. This conclusion is supported by data from many sources, including direct measurements, computer models, ice cores and the unmistakeable isotopic 'fingerprint' of fossil carbon.

In 1970, emissions were about 20 gigatonnes[15] of carbon dioxide per year; by 2005, that had reached 36 Gt.[16] About half of our carbon dioxide emissions are taken up by sinks in the oceans and on land,[17] and it is true that human emissions of carbon dioxide are only a small fraction of the total flux. But although this point is made much of by some climate change deniers, the key issue is that most of this carbon dioxide is merely in transit, moving from atmosphere to land and oceans or back again. New net additions, for example from volcanoes,[18] are minor, only some 1% of human carbon dioxide. These additions are balanced by the weathering of silicate rocks to form carbonate minerals, mediated by living organisms,[19] but only on a geological timescale (thousands to millions of years[20]). The rapid anthropogenic addition of carbon dioxide therefore puts the whole carbon cycle out of balance.

A good analogy, made by MacKay,[21] is an airport arrival hall, with 1,000 passengers arriving every hour, and immigration officers who can check 1,000 people per hour. The system is in balance, and the queue does not grow. Now imagine that other flights are diverted to the same airport, bringing an additional fifty people per hour, without any increase in the capacity of the immigration staff. The increase in passengers is small, compared to the total flux. But by the end of the day, there will be 1,200 people standing in line.

In step with this increase in carbon dioxide and other greenhouse gases, global temperatures have risen. Warm air holds more moisture, and so a positive feedback effect operates—a relatively small initial increase in temperatures raises atmospheric levels of water vapour, a powerful greenhouse gas, so triggering further warming. The Intergovernmental Panel on Climate Change (IPCC) has written that 'Warming of the climate system is unequivocal...Most of the observed increase in global average temperatures since the mid twentieth century is very likely[22] due to the observed increase in anthropogenic GHG concentrations.'[23]

Warming is somewhat offset by another form of pollution: sulphates, mostly from burning coal, which form reflective 'aerosols' in the atmosphere and so screen out some of the sun's heat. However, increasingly strict pollution controls in the developed countries, which have had the welcome effect of reducing acid rain and improving air quality, are reducing this cooling effect.[24]

Warming manifests itself in melting glaciers, rising sea levels, changes in the range of many species, and other climatic disturbances such as shifting rainfall patterns. With better data, it has also become clear that changes in the atmosphere—warming in the lower part, cooling in the upper—fit the greenhouse explanation.

Climatic models that do not include these artificial inputs do not fit the observed history; indeed, without human activity, solar and volcanic effects[25] would probably have led to some cooling in the last fifty years, and particularly since the mid 1980s when solar activity has dropped.[26] Of course, natural climatic cycles operate on various timescales, and the rather cooler northern hemisphere weather of 2008–9 was probably part of one of these. But the warming in the last half-century greatly exceeds the natural background changes in speed and magnitude. Since neither the volcanic nor longer-scale solar variations are well understood, natural cyclicity may somewhat offset some artificial warming in the next few years, but equally it may amplify it.

This climate change will have some positive effects (such as longer growing seasons, greater biodiversity in the Arctic,[27] and reduced heating requirements in high latitudes), but the overall balance will be negative for ecosystems—including humankind—and will become increasingly damaging as global warming progresses. Negative effects include rising sea levels inundating cities and fertile, heavily populated land;[28] decreased rainfall in arid areas[29] (Figure 1.1) and heavy, unpredictable showers elsewhere; the loss of glaciers on which huge populations depend for water; increased tropical diseases such as malaria; higher heat stress in summer with increased mortality, especially amongst the elderly; more severe forest fires;[30] and possibly more intense hurricanes. It may be that as many as 150,000 deaths annually are already attributable to climate change.[31] Summer temperatures in the south-east of England may become as much as 12°C higher than today.[32] Higher levels of carbon dioxide in sea water increase its acidity, threatening marine life such as corals, and so indirectly important fishing grounds.

Even if carbon dioxide emissions were to stop today, the built-in inertia in the climate system would lead to temperatures increasing further. In addition to the 0.75°C rise since the nineteenth century, we are already committed to a further warming of 0.6°C. If emissions, and hence temperatures, continue to rise, warming may be as much as 4°C by 2050—and locally much more, 15°C hotter in the Arctic and 10°C in western and southern Africa. At this level, climate impacts will become more and more serious. Extinctions are likely to increase

Figure 1.1. A dust storm from drought-stricken land blankets Sydney, Australia (Photo: istockphoto.com)

sharply, while extreme heat-waves, forest die-offs, flooding of major river deltas, persistent severe droughts, mass migrations,[33] wars and famines are all possible. We may soon pass, or already have passed, the point at which, over the next few centuries, parts of the West Antarctic and Greenland ice sheets melt irreversibly, with potential sea level rises of 1.5 and 2–3 metres respectively.[34]

Due to feedback mechanisms and poorly understood components of global climate, there is even the possibility of a sudden, rapid catastrophic change. For example, open ocean absorbs more heat from the sun than ice. Melting permafrost[35] and warming ocean bottom waters[36] release carbon dioxide and the powerful greenhouse gas methane, driving further warming. Carbon sinks will become increasingly ineffective[37] as forests die off, soils dry out and warmer oceans dissolve less carbon dioxide, so that ecosystems may become net contributors of carbon dioxide to the atmosphere, rather than net absorbers as today. The shade of clouds may diminish over warming oceans,[38] while melting ice shelves may lead to sudden collapse of grounded ice, and hence rapid rises in sea level.[39] The picture is complicated further by some offsetting effects, due for instance to increased plant growth in a warmer, more CO_2-rich world. Changes in cloudiness, snowfall and albedo (reflectiveness) of vegetation may have warming or cooling effects.

Such positive feedbacks[40] may greatly accelerate warming. Unpredictable, non-linear effects can lead to prolonged droughts in the Mediterranean, California[41] or West Africa,[42] or to weakening of ocean circulation[43] with knock-on effects including a rise in North Atlantic sea levels of up to 1 metre, a collapse of fisheries, disruption of the South Asian monsoon,[44] and possibly (albeit unlikely) sharp cooling in Europe.[45] Similar rapid changes are documented from Earth history, as at the end of the Ice Ages. At one time, at the end of the so-called Younger Dryas event around 12,000 years ago, Europe warmed by some 5°C within two decades.[46] It seems increasingly clear, from geological studies, that the climate system is unstable and prone to abrupt transitions from one state to another, so further warming might trigger entirely unforeseen consequences.[47] We should not give in to alarmism, and such disastrous shifts are thought to be unlikely—but their consequences are serious enough to be worth guarding against.

This is about as far as the weight of consensus has reached.[48] Yet many individuals and corporations continue to deny the reality of anthropogenic climate change. The US petroleum and coal businesses, in particular certain commentators,[49] and many of the general public across the world,[50] continue to maintain that the climate is not warming, that elevated carbon dioxide does not cause warming, that rising carbon dioxide and temperatures are not caused by humans, that the consequences of climate change will be benign, or some combination of these positions.

Beyond this understanding, there remains great uncertainty and debate on how much warming will occur for given changes in atmospheric carbon dioxide, how serious the impacts of this warming will be, how the climate will change at regional and local levels, how much it is worth spending to reduce climate change,[51] exactly what types of action we should take, and how we should go about encouraging global action. Despite extensive and continuing research, these major uncertainties will persist for the foreseeable future. Some of the debate is a normative one, about the values of our civilisation, and therefore is not even capable of being solved by scientific inquiry. Such uncertainty and controversy, though, is not a reason for inaction. After all, we ban certain drugs suspected to be carcinogenic, without waiting for absolute proof, and we will only know the truth about some of these climate change disasters when they actually strike.

I will take as my starting point here, in this fast-evolving area of research, the view that we should attempt to keep total warming below 2–3°C.[52] The original goal of the EU, recommended by the International Climate Change Task Force, was for a maximum temperature rise of 2°C,[53] but given the delay in taking major action, and the latest science, this already seems to be very tough to achieve. Anything above 2°C is already dangerous,[54] but, with luck, avoiding rises over 3°C will prevent the most damaging effects of climate change. Otherwise, we will venture into uncharted territory, where the risk of abrupt climatic changes is high: 'Once the world has warmed by 4°C, conditions will be so different from anything we can observe today (and still more different from the last ice age) that it is inherently hard to say where the warming will stop.'[55]

This temperature range suggests long-term stabilisation of greenhouse gases at no more than 550 ppm equivalent.[56] Given that we are already at 430 ppm CO_2e, this is a very challenging goal. Furthermore, the current global warming effect is somewhat reduced by 'global dimming' due to pollution (particularly sulphur dioxide from low-tech coal plants, and forest burning). Cleaning up power plants in China and India, in particular, although a good thing for human health, is likely to accelerate warming.

To achieve climate stabilisation, global emissions will have to peak around the year 2023 and fall to half of today's by 2050. The longer that emissions continue growing, the sharper and faster any subsequent cuts will have to be. This is because the key thing is not so much the maximum level of emissions but the total amount of carbon emitted.[57] The most recent research suggests that cumulative future emissions[58] should not exceed some 1,600 billion tonnes of CO_2 (gigatonnes, Gt) if we are to have a reasonable chance of staying below 2°C of warming. Even if emissions do not grow at all from today's levels, we are currently on course for total release of 1,250 Gt by mid century.

The cost of making at least some cuts is almost certain to be less than the damage that uncontrolled global warming will inflict. There has been much debate on the expenses versus benefits of tackling climate change with varying amounts of urgency.[59] This is a very complex issue, as we have to judge how much warming there will be, and how much damage (offset by some benefits) it will cause, recognising that these are both highly uncertain impacts.

Then we have to decide how to weigh the costs of action, which will mostly be in the near future, against the benefits which will accrue to future generations, over many centuries. Our descendants will suffer the effects of climate change, but, especially if we make wise decisions now, they are also likely to be much richer and technologically more advanced than us. Being richer, future generations may place a greater weight on non-material welfare, such as biodiversity, natural parks, etc., precisely those goods that are threatened by climate change.[60] There is a similar issue of equity, in that it is fair for rich nations to help poorer countries avoid global warming damage. For instance, it seems justifiable for millionaires to spend a thousand dollars, which they hardly notice, to help labourers in a developing country avoid $500 of climate losses, which might be a year's wages.

We should be prepared to spend extra to reduce the risk of unlikely but disastrous events, a kind of insurance. It is also important to recognise that climate change is almost totally irreversible—by the time it becomes obvious that serious impacts are occurring, it will be too late to stop them. The risk of low-probability catastrophes may therefore turn out to dominate the economics.[61]

All these considerations are generally dealt with by economists in the choice of a 'discount rate', an annual percentage by which we reduce future benefits or costs relative to today.[62] For instance, at a discount rate of 2%, it would make sense to spend $45 in 2010 on preventing climate change that would cause $100 of damage in 2050. But if we used 4%, we would only be prepared to pay $21 in that year to avoid the same damage. The choice of discount rate was one of the most controversial aspects of the Stern Review, prepared by the leading economist Sir Nicholas Stern for the British government.[63] Stern chose (in the opinion of some) a low rate, in the range 1–3%, which suggests we should spend heavily now to prevent future damage from climate change. A high discount rate would imply a 'wait and see' approach.

The choice of discount rate raises issues of economic theory, but, more importantly, is highly value-laden. It is a moral issue whether we ought to weight human welfare in this way at all.[64] There is a balance between spending on reducing carbon emissions, adapting to their effects, or tackling other non-climate problems. As the 'sceptical environmentalist' Bjørn Lomborg stresses, we might help the poor more by increasing spending on basic issues such as health, education and water, rather than climate change.[65]

Adaptation is also part of the response—for instance, building sea defences, sowing drought-resistant crops, strengthening houses against hurricane damage. As the climate changes, adaptation is inevitable, and, if well-planned, is likely to be a vital component of future strategies.[66] Lomborg in particular argues that almost all the consequences of global warming can be dealt with more cheaply by adaptation than by reducing greenhouse gas output.[67] Existing models are very limited in capturing the full range of adaptation possibilities, particularly for agriculture,[68] let alone in capturing the innovations that may arise over the next hundred years. On the other hand, recent studies suggest that the cost of adaptation may be two to three times higher than the UN's figure of about $100 billion per year.[69]

Regardless of the theoretical optimum, it is also questionable whether the current global political system can choose and implement the best policies. Indeed, after many years of campaigns, we are still far from allocating the rather modest amounts that Lomborg calculates are required to solve a variety of non-climate problems. Sending aid to the poor in order to remedy the damage we are simultaneously inflicting on them via climate change might seem ethically suspect, and certainly runs the risk of instilling a dependency culture. Optimal adaptation might also be slowed or prevented by social and cultural barriers, as I have witnessed in Cameroon. Finally, it is reasonable to debate whether it is possible to draw any sensible conclusions with such forecasts extending over many centuries.[70]

Whatever the discount rate, though, the logic of a limit on cumulative emissions argues for action now—a steady reduction from today, even with less advanced technology, is likely to be less expensive and disruptive than trying to cut carbon dioxide very rapidly from a higher peak.[71] In an extreme case, we might have to reduce emissions by some 8% annually,[72] implying wrenching change for society and the economy. The Great Recession has brought falls in CO_2 emissions of 2.6%—do we want to experience something three times worse, year after year?

Stern argued that spending about 1% of global GDP (i.e. some $570 billion per year today) would save ultimate damage amounting to 5–20% of GDP.[73] If this burden were borne entirely by rich countries, it equates to 1.8% of their GDP (compared to the handy yardstick of military spending, which amounted to 4% of US GDP in 2005).[74]

This level of stabilisation implies that by 2050 we could emit no more than 2 tonnes of CO_2 per person per year. Compare this to worldwide emissions in 2000 of 6.8 tonnes per capita,[75] and to US levels of 22.9 tonnes. Consider that a single return flight from New York to London[76] emits some 1.6 tonnes of CO_2—and, after taking this flight, such a carbon budget would leave very little for heating, lighting, driving and all the other accoutrements of a modern lifestyle, nor for agriculture or any deforestation. Only a few of the poorest countries in the world manage on less than this carbon diet today.

From such economic calculations, we can derive the 'social cost of carbon'. This is the damage (present and future, discounted to today) caused by emitting an additional tonne of carbon dioxide at some date, typically this year. The social cost of carbon (SCC) generally rises with time—or, actually, with cumulative emissions. Consider that the first tonne of CO_2 emitted by James Watt's steam engine did very little, if any, damage to climate; the trillionth tonne that triggers the melting of the Greenland ice cap is highly damaging, and has a much higher SCC. Similarly, scenarios which forecast very high emissions of carbon dioxide in 2050 will have higher estimates for the social cost of additional emissions than those with lower carbon dioxide output.

The social cost of carbon should be compared to the abatement cost (i.e. how much it costs to save a tonne of carbon, for instance by installing solar panels to replace coal, or by cancelling an important business journey). The marginal abatement cost (the cost to save the last, most expensive tonne of CO_2), opposed to the social cost of carbon, tends to rise the lower the carbon emissions are. After all, turning off the power switches when leaving for holiday is a simple, low-cost action. Making major cuts in emissions, e.g. by cancelling a medical flight carrying a life-saving transplant, is very costly.

The point at which the social cost of carbon meets the marginal abatement cost is the price for emitting a tonne of CO_2, which would arise under a well-designed carbon tax or trading scheme (Figure 1.2). Estimates for its level vary, and it is generally expected to grow with time as we make the 'easy' cuts in emissions and as climate damage becomes more severe. Stern calculated a value of $85 per tonne;[77] the IPCC suggested $62/tonne,[78] while Greenpeace's value rises to $50/tonne by the year 2050.[79] So, on these figures, any method that can cut emissions for less than $50–85/tonne CO_2 can be a viable part of tackling climate change (Chapter 5).

It is worth noting, of course, that other authors have argued for a much lower cost of carbon, perhaps only $2 per tonne, rising to $27 per tonne at end-century.[80] At such a low price, not only carbon capture, but most low-carbon and renewable energy technologies[81] would not be viable for many decades.

Figure 1.2. The social cost of carbon, marginal mitigation cost
and optimal carbon tax/cap

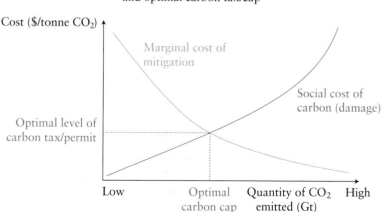

The choice of discount rate has an impact on how much emission reduction we require, but not perhaps so much on the choice of CCS versus alternatives, since most of the major options for tackling climate changes—CCS, solar, wind, nuclear and radical improvements in energy efficiency[82]—involve large capital investments today. There are some low-cost measures to tackle climate change (such as straightforward efficiency improvements, avoiding deforestation, solar water-heating, some onshore wind power and some biofuels[83]). But, with the current level of knowledge, between the current contenders for serious emission cuts—electric cars, nuclear, offshore wind, solar power, carbon capture—there are wide ranges of uncertainty and no clear winners on costs. We simply will not know the optimal solution until these contenders are tried out on a large scale. If there are any carbon cuts beyond a token amount, then, as I am about to discuss, carbon capture will have to play a major role.

Carbon Fuels are Abundant

Coal, gas and oil exist in great abundance and, despite a spike in prices in 2008, can be extracted at reasonable costs. In terms of conventional pollutants, such as acid rain and smog, they are being used increasingly cleanly. If it were not for the problem of carbon dioxide, they would probably continue to dominate energy production through most of the twenty-first century.

There has been much debate recently, spurred by record high oil prices, that we may be approaching 'peak oil', the point at which oil production begins to fall inexorably due to increasing exhaustion of the resource base. But, as I have argued in *The Myth of the Oil Crisis*,[84] peak oil is not imminent. Conventional oil resources, including potential for discovering more and increasing recovery from known fields, are very substantial, probably at least twice what has already been produced (Figure 1.3). Unconventional oil, such as the heavy bitumen from oil sands, and the oil that can be 'cooked' from the kerogen in oil shales, form a vast resource base, far exceeding the endowment of conventional oil. Long-run production costs for these resources are likely to be below the $100 per barrel oil prices encountered in 2008. Unconventional oil is often objected to on environmental grounds, particularly its carbon intensity—but that, of course, strengthens the case for requiring carbon capture.

Figure 1.3. Drilling rig offshore Saudi Arabia. Middle East oil reserves remain abundant (Photo: Tor Eigeland *Saudi Aramco World*/SAWDIA)

Beyond oil, gas is still in a relatively early stage of development. Numerous large gasfields around the world remain undeveloped for lack of markets, in East Siberia, Central Asia, Iran, Iraq, West Africa, Australia, Indonesia, Bolivia, Alaska and elsewhere. Recent strong production growth in the supposedly mature basins of North America has been led by technological breakthroughs in extracting gas from shales and coals. Such gas-bearing formations, also known in countries that include Australia, China and India, are likely to be ubiquitous around the world. Gas can be converted to oil, and a number of commercial plants are in operation or will start up soon.[85] Beyond that, although not yet commercially feasible to extract, there are poorly quantified but vast quantities of gas dissolved in underground aquifers and trapped in the strange ice-like hydrates of Arctic regions and the deep ocean.

Then there is coal. The least glamorous of the fossil fuels, despite having been exploited at least since Roman times,[86] remains in huge abundance. Extraction of some large known resources, such as those in Alaska, has not even begun. Reserves can be further increased by exploration and more intensive exploitation. New approaches, such as underground coal gasification, could exploit unminable seams (too deep or thin for conventional extraction), as well as the enormous coal beds found beneath the seabed. Finally, coal can be converted into oil (and gas), thus helping to make up for any shortfall in petroleum resources.

The volumes of available carbon are more than sufficient to cause dangerous climate change.[87] Table 1.1 shows potential ultimate recovery of fossil fuels, including reasonable assumptions on new discoveries and increasing use of currently uneconomic or 'unconventional' resources, but not some very large sources (such as oil shales and gas hydrates) that are not being commercially extracted today. This is compared to a scenario, from the Intergovernmental Panel on Climate

Table 1.1. Future recovery and consumption of carbon fuels

	Reserves + resources	Maximum consumption 2010–2100[88]
Coal (billion tonnes)	14,900[89]	2,242
Oil (billion barrels)	3,400[90]	2,984
Gas (trillion cubic feet)	22,500[91]	15,803
Potential emissions (Gt carbon dioxide)	49,000	8,640[92]

Change (IPCC), for fossil fuel consumption to the end of this century, in the absence of significant action against climate change. I have chosen the IPCC scenario with the highest fossil fuel consumption.

Even in this case, assuming extremely high fossil fuel use, the resource base (especially of coal) vastly exceeds the projected consumption. Burning less than a fifth of our carbon fuel resources, as demonstrated here, could take atmospheric CO_2 far over 1000 ppm, and temperature rise to very dangerous levels, well above 4°C. Even if we take an extremely pessimistic view on fossil fuel resources, it is clear that we have vastly more available than the ~1600 Gt CO_2 which would lead to dangerous warming.

If extraction continues to increase, then, of the three fossil fuels, oil seems likely to reach resource limits first. If conventional oil production were to begin to decline, then some of the key expedients would be to convert gas and coal into liquid fuels, and exploit oil sands and oil shales more intensively. Since such 'unconventional oil' processes typically emit large quantities of carbon dioxide, a belief in 'peak oil' implies a greater need for carbon capture.

The objection is sometimes made that, because resources of oil, gas and coal are clearly finite, their use is not 'sustainable' and therefore we should move away from them as fast as possible. This is a fallacy which ignores the generally accepted definition of 'sustainability': '[to meet] the needs of the present without compromising the ability of future generations to meet their own needs'.[93] The use of non-renewable resources is acceptable as long as we have replacements ready in good time. To demand a duration for 'sustainability' of 1,000 years is senseless.[94] Where would we be today if the Anglo-Saxons or Song Chinese had decided that they did not have a millennium's worth of iron and its use was therefore 'unsustainable'? A realistic maximum time frame for energy policy would be some forty to seventy years, the lifetime of major pieces of infrastructure.[95] Beyond that, the uncertainties and unknowables are too great, and there is sufficient time to change course if resource limits are becoming apparent.[96]

Carbon Fuels Will Continue to be Used

Carbon fuels retain many attractions. If they could be burned without releasing carbon dioxide to the atmosphere, they would probably continue to dominate global energy supply through most of the twenty-first century. Even with very high scenarios of fossil fuel use, substantial reserves of oil, and particularly coal and gas, will remain in the year 2100.

Costs are also low compared to the alternatives. The spike in oil prices in 2008 misled many observers into assuming that fossil fuel prices would be high and rising indefinitely. For instance, Greenpeace's scenario for a rapid rise in renewables relies on an oil price of $100 per barrel in 2010, rising to $140 in real terms by 2050,[97] yet 2009 saw prices well below $100. Contrasting with their forecast of a 400% increase in natural gas prices over 2005–2050, prices in the USA dropped due to breakthroughs in producing 'unconventional' gas.[98] By August 2009, gas prices were back to the levels of August 2002.[99] Predictions of simultaneously falling fossil fuel consumption with rising prices do not take account of the probability that, as it has been able to do many times before, the oil industry would respond to the challenge with significant improvements in efficiency, and perhaps with technological game-changers similar to the 1990s introduction of 3D seismic or horizontal drilling, radically reducing production costs.

It is now clear that many of the cost increases in finding and developing fossil fuels were caused by the global boom, pushing up prices of steel, cement, skilled workers, drilling rigs and so on. If action on climate change leads to a sustained fall in fossil fuel use, then fossil fuel prices will drop sharply. The marginal cost of managing a slow decline in output from an existing oil- or gasfield is generally low. Thus, unless carbon costs rise to very high levels, there is a natural limit to the cost-competitiveness of alternative energy.

At least as pressing as the fight against climate change is the need to bring affordable energy to massive numbers of the world's poor. Recent price falls reinforce the belief that carbon-based energy will often be not just the most economic option, but also the most humane and environmentally friendly. For instance, an investment of $8 billion in modern LPG stoves for the developing world, replacing unsustainable harvesting of wood, could yield benefits of $90 billion in time-saving, avoidance of land degradation and improved air quality.[100]

Power generation systems (of all types) tend to have very long lifetimes, typically forty years for a coal-fired plant. Even if we cease building new fossil fuel capacity today, the existing units will continue to emit for decades. This legacy electricity is extremely cheap, since a coal plant's fuel is only a small part of its costs. If a power station is decommissioned, it will be relatively cheap to build a new one on the same site, since many facilities, power lines and so on will already be in place, and the site is unlikely to be attractive for many alternative uses. This is not some 'unfair advantage' of fossil fuels, as some commentators have rather naïvely suggested;[101] the situation is exactly the same for a hydroelectric dam. It is merely an indication that it is not efficient to abandon capital goods before the end of their useful life.[102]

Similarly, the traditional energy industry has tremendous resources of skilled people, institutional knowledge, political relationships, physical assets and financial strength. To some extent, that can be turned to developing renewable energy. But, just as coal-miners from northern England mostly did not find new employment as North Sea roughnecks, a rapid abandonment of fossil fuels would waste many of these strengths. Continuing use of carbon energy, combined with carbon capture, though, can continue to use this intangible capital. Environmental organisations might reflect that the fossil fuel industry is unlikely to cooperate in its own destruction. However, it can be a very powerful ally in realising a new energy future, as long as the will for cooperation is there, from all parties.

Fossil fuels also have the supreme virtue in energy supply, that of reliability. They have been used for well over a century, with a high level of safety and predictability. Suggesting they should be phased out as 'old energy', as some environmental organisations have advocated, is just dogma—ships and trains are 'old technology' too, but that does not make them obsolete. If we are concerned about the environmental or social impacts of fossil fuel extraction, then those issues should be tackled directly—for example, by mandating double-hulled oil tankers, banning open-cast coal mining in sensitive areas, or prosecuting oil companies complicit in human rights abuses. Banning carbon capture would be a completely ineffective, indeed counter-productive, way of tackling such unrelated problems.

The use of fossil fuels is becoming increasingly clean.[103] The appalling pollution in Chinese cities, due to heavy coal use, might seem to contradict this, but modern particulate filters and scrubbers on power plants,[104] and catalytic converters and low-sulphur diesel for vehicles, have dramatically improved air quality in the developed world.[105] This is evident when thinking back to the 'pea souper' smogs that persisted in London from Victorian times to the 1950s. I have breathed the dreadful air in cities in India and Nepal due to open-air burning of wood and rubbish, which could be cleaned up by switching to modern fuels. Gas, virtually pure methane, is very clean in any case. Upstream extraction, despite its bad public image, has low environmental impact with modern methods (two admitted exceptions being oil sands mining and some open-cast coal mining), and, despite the Macondo disaster in the Gulf of Mexico, spills from oil platforms are extremely rare.

Notwithstanding such improvement, we should continue to ameliorate the environmental and social effects of fossil fuel extraction and use. Even ignoring its carbon dioxide emissions and considering only the developed world, truly 'clean coal' remains elusive—best-in-class pollution control would have major health and economic benefits.[106]

Those countries with competitive advantages in fossil fuels will continue to wish to use them. The former Soviet Union is obviously rich in coal, oil and gas. Canada and Venezuela have major deposits of bitumen. Major coal users and exporters include the USA, China, India,[107] Germany, Australia, South Africa and Indonesia. Although the Middle East has great solar potential, this is far outweighed by its remaining oil and gas, exceptionally low-cost to extract.[108]

It will be very hard to reach a global political consensus on tackling climate change, if it requires such countries to abandon their major source of natural wealth. Signs of this are already apparent, as the OPEC[109] states suggest they should be compensated for any limits on oil use, while Poland, a heavy coal user, argued for less stringent limits on EU emissions. Altogether twenty-five US states produce coal, giving an almost unassailable Senate bloc against curbs on the fuel.[110]

This deadlock will be exacerbated if certain countries dominate renewable energy technologies. For instance, if wind turbines are mostly manufactured in China or India, then some of the fervour for 'green jobs' in the USA may evaporate. For coal-rich countries to support carbon capture is not cynical; it is just practical. Otherwise, we might accuse Scotland or Spain of cynicism for promoting wind or solar power.

Even in the scenario of a rapid shift to renewable energy, fossil fuels remain important.

- Firstly, as I'll discuss, it is very difficult for renewable energy, starting from a low base, to move to dominating energy supply in the time we have to tackle climate change.
- Secondly, one of the key short-term changes to reducing emissions is to replace coal with gas, since a modern gas-fired plant produces less

than half the carbon dioxide of an equivalent coal generator. Coal can even be converted into gas by underground gasification.

- Thirdly, fossil fuels and renewable energy can have valuable synergies. For instance, hybrid solar thermal/gas-fired power plants provide reliable electricity even when the sun is not shining. 'Biogas' from organic matter can be blended with fossil fuel gas; coal is co-fired with biomass; and biofuels are added to petroleum-derived petrol and diesel. Hot water produced at oilfields is a possible source of geothermal energy, while solar collectors can be used to generate steam for heavy oil recovery projects and to boost coal power plants.[111] The largest user of solar electricity in Canada is the oil business, for powering remote, off-grid locations.[112] Offshore wind turbines can be combined with oil platforms, as at the Beatrice field in the North Sea, and in general, the engineering and construction of offshore wind turbines lean heavily on experience from the oil business. Some of the technological advances, such as long-distance direct current (DC) electricity transmission, that are important for renewable energy could also favour carbon capture, allowing power plants to be next to a good storage site.

- Fourthly, renewable electricity sources are mostly intermittent. It is a big challenge to maintain a stable electricity grid, and provide power at peak times, even with the current system, a fantastically complicated entity which has grown up, in developed countries, over more than a century. This balancing task becomes increasingly difficult with large quantities of wind, which is rather unpredictable, and solar, which does not necessarily match peak load, especially in winter in northern climates. A given wind turbine, for instance, produces no power at all between a sixth and a third of the time. In cloud, solar panels generate only 10% of the electricity they do in direct sunlight,[113] particularly problematic for northern Europe and parts of the tropics during the rainy season. To some extent, this can be managed by intelligent management of demand, redundant generation capacity, including a geographically dispersed range of sites and plugging in other renewable sources such as geothermal and ocean power, employing 'dispatchable' renewable power such as biomass-fired plants, electricity storage (including 'pumped storage' behind dams and perhaps the batteries of a future fleet of electric vehicles), and long-distance imports.

But these solutions become increasingly expensive as the proportion of renewables in the energy mix increases. Views on required back-up capacity vary, but might be as high as 60–95% for wind power. In fact, wind power may end up mainly displacing (low-carbon) gas-fired generation, and lowering electricity prices during windy spells, rather than replacing high-carbon coal.[114] It has yet to be demonstrated, beyond theoretical calculations, that a national-scale, mainly renewables-powered grid can achieve reliability at acceptable costs.[115] The rapid growth in alternative energy could easily be dented by a couple of high-profile black-outs and consequent loss of public faith. The most economic solution is therefore likely to involve a mix of generation options, including coal-fired baseload fitted with CCS, and gas-fired capacity, some of which is held in 'spinning reserve', operating at low levels, ready for demand surges.

Carbon Capture is Feasible Today

It is often claimed that carbon capture and storage is 'not proven'. For example, the US's oldest environmental organisation, the Sierra Club, has said, 'We don't have any idea whether or when this [carbon storage] will be possible...it's pie in the sky.'[116] Andrew McKillop dismisses the idea as 'exotic technological fantasies',[117] while Greenpeace commented that the Swedish utility Vattenfall was attempting to deceive ecologists with its CCS plans.[118]

Yet, as we will see, the individual elements of CCS are all technologically proven. Four industrial-size carbon storage projects are operating, in various parts of the world, and numerous pilots are investigating all the aspects of carbon capture, transportation and storage. Long-distance carbon dioxide transport and storage for enhanced oil recovery is commercially proven, and operating on a large scale. Experience from these projects suggests that they are safe and that leakage will be minimal.

It is also true that carbon capture is a less mature technology than, say, wind, nuclear or some energy efficiency options, and that it will take time to scale up. It may still encounter technical problems that render it if not impossible, at least undesirable. Carbon capture from power stations is not yet commercially proven, and is likely to be expensive, above current costs of emitting carbon.[119] That is not an argument for moving slowly—quite the opposite! As I will discuss (Chapter 6), carbon capture may advance to a major mitigation option much faster than many critics imagine.

All that is needed is a sufficiently high price of carbon, bold moves by industry, and supportive policies. Technological advances and economies of scale will bring down the costs significantly. In these respects, carbon capture is no different from solar power, which has enjoyed lavish subsidies to reach its current state of development, and which, even including carbon pricing, is still not commercially cost-effective. The technology for CCS is not particularly complicated, not when compared to the breakthroughs still required for competitive hydrogen fuel cells, 'second generation' biofuels, electric cars and advanced solar photovoltaics (PV).

CCS is also capable of working at the required scale. Currently, 8,100 large point sources (mostly big power stations and industrial plants) emit more than 60% of all anthropogenic carbon dioxide.[120] Nearly all these facilities could be candidates for carbon capture, as can their successors to be built over the next century. Implementing some 8,000 carbon capture projects seems a large but not insurmountable task.

Six countries[121] (China, the USA, EU, India, Japan, Russia) account for more than four-fifths of all coal use; adding South Africa, Australia, Canada and South Korea brings this to nine-tenths.[122] So a relatively small grouping, more than half of them wealthy nations, could in principle agree largely to eliminate coal as a global source of greenhouse gases, with carbon capture as a key part of the solution. The key, of course, is China, still a relatively poor nation, but one that represents not far short of half the world's coal use.

Reasonable challenges to CCS are entirely valid. New coal plants, certainly in the developed countries, should not be built with vague assurances that carbon capture will ultimately be available. The technique has to be tested and scrutinised thoroughly. If major objections to its use arise, then they have to be solved or else the use of CCS will have to be abandoned. But it is much better that we find out fast whether or not carbon capture can make a big contribution to tackling emissions. If it is not going to be successful, we need to know as soon as possible, so that we can move on to other solutions. It makes no sense to say that CCS is useless because it will not be available 'before 2030',[123] and then to oppose researching and developing it now. As Kennedy said, 'The great French Marshall Lyautey once asked his gardener to plant a tree. The gardener objected that the tree was slow growing and would not reach maturity for 100 years. The Marshall replied, "In that case, there is no time to lose; plant it this afternoon!"'[124]

Opposition to carbon capture has to be informed and rational, not knee-jerk or based on ideological hostility to fossil fuels. The battle against climate change is often compared to the Apollo Project, or the recruitment of American industry to fight the Second World War. It is worth reflecting that neither of those crash programmes would have achieved much had they been held up by legal challenges requiring that we go to Mars instead, or by protesters chaining themselves to the gates of aircraft factories and demanding that they be built in someone else's city.

Carbon Capture is not only about Fossil Fuel Electricity

Carbon capture is often attacked on the grounds that it perpetuates a reliance on fossil fuelled electricity. Yet many industrial processes, other than power generation, emit carbon dioxide: hydrogen production (from coal or natural gas), cement, fertilisers, ammonia, chemicals, iron and steel. To some extent, we might be able to create substitutes, or reduce the demand for these products, or the carbon intensity of their production. But in general, such vital industries would be expensive, difficult or technically unfeasible to run on renewable sources. For cement in particular, which accounts for some 5% of global carbon dioxide emissions[125] (more than aviation), the production of carbon dioxide is a fundamental part of the process. And even wind turbines and nuclear plants require large amounts of concrete for construction. Anti-CCS activists have the duty to explain how they would decarbonise these basic industrial processes at acceptable cost. There are some limited alternatives, but they are mostly expensive and/ or at considerably earlier stages of development than carbon capture.

Some industrial processes, in particular ammonia and hydrogen manufacturing, produce concentrated streams of carbon dioxide which are ideal for capture. Carbon capture is also an enabling or enhancing technology for several future energy sources. Hydrogen, in particular, is often touted as a key part of a zero-carbon economy, as a fuel for home heating and possibly vehicles. But making hydrogen from gas is far cheaper than the cumbersome and inefficient route of generating renewable electricity and using that to electrolyse water. Centralised co-generation of hydrogen and power, with capture of the produced carbon dioxide, is an attractive option. By keeping hydrogen and electricity cheap, we greatly improve the prospects for achieving zero-carbon transport and residential sectors.[126]

Carbon capture can also be applied to combined heat and power (CHP) facilities, power plants running on biomass, landfill gas or waste; biofuel refineries; and possibly geothermal projects with high carbon dioxide emissions, to boost further their climate credentials.

Therefore, carbon capture remains important even if we are shifting away from coal- and gas-fired power. And it is not an enemy of renewable energy—it is a potentially valuable addition to the arsenal of future energy techniques.

No Single Solution to Climate Change is Perfect or Sufficient

Writers and researchers on the climate change issue often fall into what I call the 'panacea fallacy'. This is the idea that a single technology or approach is the way to solve climate change—whether the panacea is nuclear power,[127] renewable energy, reforestation, carbon capture, energy efficiency, geo-engineering or some other concept. The converse is to pick on a particular technology, show that it cannot solve the climate change problem on its own, and therefore dismiss it as useless. In fact, given the scale of the problem, even a 5% solution is highly worthwhile.

Another similar mistake is to attempt to design the perfect world energy system from the ground up.[128] If such commentators are trying to demonstrate that at least one possible way exists of meeting climate change targets, that is perfectly reasonable. But our current energy system has grown up over more than a century, meeting varying challenges, advancing by trial and error, adapting to new technologies and social trends as they arise. Trying to build, in the space of a couple of decades, a centrally planned replacement seems a piece of dirigiste folly, unlikely to be cheap, efficient, effective or popular, and perhaps not even very environmentally friendly.

Of course, government has an essential role to play in distributing information, setting policies, facilitating new projects and initiatives, nursing infant technologies and above all in reaching international agreement on enforcing rigorous limits and costs for greenhouse emissions. But beyond this, we have to allow the ingenuity of entrepreneurs, the consensus-building of NGOs and civil society, and the practical capability of industry sufficient freedom to meet our energy and climate needs.

A very rapid shift to renewables and higher energy efficiency is often propounded as the climate solution par excellence. Of course, both these approaches are key parts of twenty-first-century energy.[129] But there are major reasons to believe that relying solely on them, while phasing out fossil fuels (and nuclear power, as Greenpeace and others advocate), is likely to fall far short of achieving both economic and environmental goals. Environmental groups have to get serious about tackling climate change—that is the number one priority, and should not be distracted by attempts to achieve other agendas. Throwing away some of our best options is as though Churchill were to announce that defeating Hitler was his top priority, but that he intended to do it without using planes or tanks.

Part of this fallacy is to say that carbon capture (or nuclear, or some other energy option) is not required because the author's favourite plan is already 'enough'. This assumes, firstly, that energy demand is a fixed quantum. In reality, of course, if we had abundant, cheap, green energy, we would find ways to use more—and that is not a bad thing, since it would enhance human welfare.

Secondly, such talk gives the impression that 2°C of global temperature rise (or 3°C, or 450 ppm atmospheric CO_2, whatever our climate target is), is a magic number: below it, nothing bad happens; above it, hell is unleashed. In fact, if keeping the temperature rise below 2°C is a good thing, limiting it (at reasonable cost) to less than 1.8°C, or 1.6°C, would be even better (it is only at very low levels of warming that we might argue there is a net global benefit). And there is only a hazy knowledge of how much carbon dioxide will cause 2°C warming (and, indeed, how much carbon dioxide a given plan will really emit over the next half-century)—so every tonne of carbon dioxide saved reduces the risk of rapid, catastrophic climate change. In short, any plan that rules out *ab initio* some valid options is bound to be sub-optimal.

The following discussion is not meant to be an attack on renewable energy; it is simply a realistic discussion of some of the issues that will hamper a rapid transition away from fossil fuels.

The first issue is the fantastically rapid rate of change required. Renewables growth has been very impressive in recent years (15–30% annually for various technologies during 2002–6, and a 30% increase in global wind capacity in 2008).[130] But this was, firstly, at a time of unusually high oil and gas prices and, secondly, covered early-stage industries in the take-off phase. Sustaining such growth over long periods in increasingly mature businesses, which will come to account for a large fraction of the world economy (and hence of demand for materials, capital, land, personnel, etc.), is likely to be very challenging.

Greenpeace, for instance, forecast 18% compound annual increase in solar photovoltaic capacity during 2005–50. Even by the time solar PV will have become a relatively mature industry, in the decade 2020–30, the projected growth rate is 13% per year. By comparison, in its best years, in the late 1960s, oil production grew at about 7% annually. Global PV capacity today, 16 gigawatts (GW),[131] covers about 0.2% of global consumption, but, even with vast projected gains in energy efficiency, has to increase to 25% in Greenpeace's scenario.[132] Consider that a single very large coal-fired plant can produce as much electricity as all the photovoltaic cells in the world today.[133] Wind and solar together produce about 1% of US electricity, despite several years of strong growth.[134] Estimates of current renewable energy capacity are often misleading, because they include large hydropower (with limited capacity for further growth) and biomass which, particularly in the developing world, are frequently not renewable at all, but derived from the unsustainable destruction of forests.[135] I have witnessed at first hand the widespread deforestation in Nepal caused by wood-gathering.

We have seen in recent times, for wind turbines, solar cells, skilled engineers and oil industry equipment, that a rapid growth in energy demand leads to soaring costs. Bottlenecks emerge at various points of the supply chain. Such rapid, sustained growth is likely to hamper or even overwhelm cost savings from technological improvements. For instance, the cost of wind turbines rose 17% during 2006.[136] Polysilicon, a vital component of most solar panels, could be bought for less than $10 per kilogram after the dot-com bust, but reached $400–450 in 2008. The economic crisis drove prices down to $110, but this good news was accompanied by the bad, a fall in solar panel demand of a fifth.[137]

Costs will also tend to be driven up as the proportion of renewable energy in the grid increases (as mentioned earlier), and as the best renewable sites are used and new projects have to be placed in increasingly unfavourable or remote areas. The best locations for hydroelectric dams, particularly in the developed world, are already taken. Achieving grid connections is proving a major obstacle to renewable energy expansion, for instance in the UK. The history of cross-border gas and oil pipelines, which tend to be long-delayed and encounter many thorny political and legal issues, is not very encouraging for the prospect of rapidly installing long-distance DC cables from the Sahara to Europe.

The second major issue is the cost of renewables, which remains, in most cases, well above that of conventional energy. Some renewable sources, notably large hydro-power, utility-scale onshore wind, and solar water heating, are often highly competitive with fossil fuels. But others remain very expensive. For example, Germany has spent/will spend 68 billion euros on feed-in tariffs for solar power in the period 2000–12.[138] If we assume this solar power was replacing coal,[139] then the cost of carbon dioxide avoided was about \$650/tonne.[140] This is ten times the estimates for cost of CCS and for typical values for the social cost of carbon; it is one hundred times the cost of carbon avoidance at the In Salah CCS project in Algeria.[141] Of course, that solar subsidy also helped to advance the science and lower the cost of solar power, but it would have done that for CCS too; we might now already have a mature means of carbon capture. And Germany has plenty of domestic coal, whereas its solar resource is limited—about the same as Alaska's. This subsidy programme drove up the price of panels that might otherwise have been installed in sunnier climes.

The combination of public and political pressure to move to renewables, very rapid industry growth, and the continuing need for subsidies[142] creates significant risks of showering undeserved windfalls on certain players. Profits may go to those who play the subsidy game most effectively, not those with the best technology.[143] As many hydrocarbon exporters have already discovered, it is much easier to implement subsidies than to wean users off them.

Within the current range of uncertainty, it is close to impossible to 'pick winners' on cost grounds. Onshore wind in good locations certainly seems to offer the cheapest carbon mitigation, and solar photovoltaic the most expensive. Tidal power is perhaps the next most costly, but all the other leading options (carbon capture, nuclear, biomass, offshore wind) are broadly comparable, and fall within the \$50–100/tonne CO_2 range for likely long-term carbon costs.[144]

The only reasonable approach, as with any investment portfolio, is to be diversified, and to aspire to resilience. It is always possible that a breakthrough in a certain technology—solar, ocean power, electricity storage, small-scale nuclear power, carbon capture, or perhaps something more speculative such as algal biofuels or nuclear fusion—might rapidly advance to dominate our energy future. Since we don't know the chances of such a 'positive surprise', we should place as many bets as possible. But at the moment, we do not appear to be close to a 'silver bullet' for future climate-friendly energy. Instead, we need 'silver buckshot'.

Therefore it is more likely, especially given the scale of the problem, that we will need most or all of the available tools to tackle climate change.[145] As the World Wildlife Fund says, 'It's clear that the energy system as a whole has to undergo a revolution to become low-carbon or zero-carbon by 2050. We don't think that can be done in time with only efficiency measures or renewables.'[146] Given the size, variety and long-lived capital stock of the global economy, even a transformational technology would probably take three to five decades to achieve dominance. As with today's energy system, a whole range of different technologies will be established, each with its own niche.

Renewable energy is also exposed to the risks that all new technologies face. In the 1950s, we were assured that nuclear electricity would be 'too cheap to meter'. But high-profile accidents and rising costs have led to disillusionment and a very poor public image for nuclear power. Something similar may happen if major renewable energy projects, for instance offshore wind farms, run into problems of budget, schedule or reliability. At Horns Rev in Denmark, the first large-scale offshore wind project, there were 75,000 maintenance trips by helicopter in the first year and a half of operation, working out at two per turbine per day.[147] In that period, the eighty turbines only operated together for half an hour. All had to be removed to shore and replaced. Manufacturing faults and the rough marine environment were blamed.[148]

There is also often significant local opposition to wind turbines: the UK has 2 gigawatts of wind capacity, but 9 GW of potential is mired in planning permission.[149] Technological progress in solar photovoltaics, batteries and hydrogen fuel cells may stagnate. Energy crops seem particularly vulnerable to protests by competing land-users.

Nor are renewable energy sources flawless environmental performers.

- The recent rush to biofuels has highlighted their environmental limitations. It quickly became apparent that some biofuels were contributing to deforestation (destroying habitats and, ironically, accelerating global warming), driving up food prices, and reducing carbon emissions only marginally, if at all. The prospect of increasing biomass use by a factor of ten, as proposed by Greenpeace, would probably lead to the establishment of vast monoculture plantations, with low biodiversity, vulnerable to disease, and probably with negative social impacts. Fast-growing energy crops also tend to be thirsty for water and fertilisers—and nitrous oxide from agriculture is a serious greenhouse gas in its own right. Widespread drought, increasingly likely as climate change advances, could cause not only famine, but also an energy crisis. Really effective biofuels seem likely to require genetic modification, a prospect which, no doubt, many environmental groups would oppose.

- The burning of biomass (plant material and organic wastes) can be dirtier than coal, in its emissions of air pollutants, especially particulates[150] responsible for lung disease. Biomass combustion also releases the greenhouse gas methane. Burning the biofuel ethanol in car engines creates smog.[151]
- Hydroelectric dams and tidal barrages are damaging to the local environment; imagine the public opposition if a plan with impacts similar to that for the Severn Tidal Barrage[152] in the UK were proposed by the oil business! Dams produce surprising amounts of methane from rotting vegetation: Indian dams yield a fifth of that country's greenhouse gas emissions.[153] They are not necessarily safe either: a major explosion at a Siberian hydroelectric power station in August 2009 killed seventy-six people and also polluted the Yenisei River.[154]
- Wind farms built on peat—as are more than half of those in windy Scotland—also generate large amounts of carbon dioxide, while damaging unique ecosystems.[155]
- Geothermal energy can release carbon dioxide from underground waters, and has been blamed for triggering earthquakes.[156] These were minor but did trigger local concern.
- Proponents of covering the Sahara Desert with solar panels should recall that the desert is a unique and diverse ecosystem in its own right, something I am vividly aware of through trips into the fringes of the famous Rub' Al Khali (Empty Quarter) in the UAE. The Sierra Club and other environmentalists held up construction of a solar plant in California's Mojave Desert, as it threatened the habitat of a rare tortoise.[157]
- As another example, both cadmium and tellurium, used in some solar panels, are toxic, cadmium highly so. Solar panel manufacturing yields 4 tonnes of poisonous silicon tetrachloride for every tonne of silicon, and there are instances of this being dumped by Chinese manufacturers. Environmentally responsible manufacturing might increase costs by 50–300%.[158] Mining rare earth metals has been blamed for pollution and damaging landscapes in south-eastern China.[159]

- The energy payback period for solar panels is about four years, and some manufacturing processes release the powerful greenhouse gas nitrogen trifluoride, only just recognised as an environmental problem and so not even included in the Kyoto process. Marine and wind power also require substantial amounts of carbon-intensive concrete and steel. The life-cycle carbon emissions from renewable energy are certainly low, but not zero—averaging typically 10% of those for gas-fired power generation. Solar electricity is responsible for about two-thirds the amount of greenhouse gases of a capture-equipped coal power plant.
- Solar thermal power plants require large amounts of water.[160] Water availability is increasingly a concern for new solar power in California, and water-efficient facilities are considerably more costly.[161]

Furthermore, a number of renewable energy technologies, notably biomass and to some extent hydropower, are threatened by the very climate change they are trying to prevent, if crop yields fall due to drought and higher temperatures, and river and wind flows decrease or become less predictable.[162] The destruction by floods of the Namche Bazaar hydropower facility in Nepal is one example;[163] slowing winds over the USA and globally, reducing wind turbine output, perhaps another.[164]

Some renewable sources also have high demands for land,[165] a problem for crowded countries in Western Europe and parts of Asia. Renewable energy sources are usually very diffuse. A complete transition to renewables for a wealthy, densely populated country such as Japan, the Netherlands or the UK implies an almost inconceivable level of industrialisation of the countryside, for example an area of UK wind farms the size of Wales. This blight of 'energy sprawl'[166] seems directly contrary to the ethos of the green movement.

For instance, a Concentrating Solar Power (CSP) plant requires 3 km^2 for 30 MW of capacity (compared to a gas power plant of the same size, which covers about 0.006 km^2, or about twenty-five tennis courts; see Figure 1.4). Over its lifetime, this solar plant would avoid carbon dioxide emissions of about 1.8 million tonnes,[167] while a carbon capture scheme fitted to a coal plant could, over its operating period, easily store 30 million tonnes of CO_2 using the same area of above-ground facilities.[168] A similar expanse of typical suburban woodland in the USA contains about 0.2 Mt of CO_2. Because of the large extent and often remote locations of renewable energy sites, there is additional land disruption for long-distance power lines to customers. Carbon capture is, therefore, a vastly more efficient use of land for avoiding carbon emissions than reforestation or some renewable energy technologies.

For offshore developments, a large 300 MW wind farm might require about 100 turbines, with associated disturbance to the seabed, visual intrusiveness (if near to shore) and so on. A single gas platform with minimal footprint—indeed, a handful of sub-sea wells, invisible from above the waves, with a submarine pipeline to shore—can produce an equal amount of energy. Yet in the USA, offshore renewable energy leasing is forging ahead while petroleum leasing remains under moratoria.[170] None of this discussion is intended to suggest that renewable energy is a bad thing—it is a great opportunity. But it should be judged on the same standards as fossil fuel use with carbon capture, not given some kind of free pass on its environmental impacts.

It is sometimes suggested that reforestation is superior to industrial carbon capture schemes:

It is a typical modern human response by the [UK] Government to excess carbon—implement a complicated, technological industrial system that is expensive, requiring carbon to be transported over great distances and much energy use, thereby creating as much of a problem as a solution....Nature, of course, has a much more elegant carbon capture system: plant life. So, Ed Miliband, if you want to capture carbon, plant more trees.[171]

However, reforestation raises several problematic issues, including competing land use, the vulnerability of forests to natural or human-

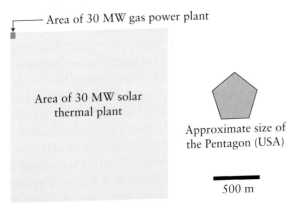

Figure 1.4. Comparison of the area of gas-fired and solar thermal power plants, with the Pentagon for scale[169]

Area of 30 MW gas power plant

Area of 30 MW solar thermal plant

Approximate size of the Pentagon (USA)

500 m

caused destruction, the social challenges of establishing and maintaining them, and the long periods they require to reach maximum carbon absorption. I explore some of the other issues in Chapter 4.

I am personally very sceptical of the idea of imminent, fundamental resource limits for most minerals,[172] due to careful study corroborated by the repeated failures of predictions of 'peak oil', 'peak gas', minerals shortages, and so on.[173] But to be intellectually consistent, 'peak oil' believers should admit the possibility of resource limits for other minerals as well.

For example, cadmium telluride is a promising material for making solar cells, but tellurium is a very rare element in the Earth's crust, rarer than platinum. Current tellurium production is about 215 tonnes per year, all as a by-product of copper refining,[174] and prospects for major increases are dim.[175] If just 1% of Greenpeace's proposed solar panels were made of cadmium telluride, we would need some 4,000 tonnes annually by the year 2050. Lithium, used in the popular lithium ion batteries for electric cars, may also, on some estimates, face resource constraints,[176] as platinum, a component of fuel cells, possibly will.[177]

Even if we don't believe in imminent resource limits, it would at least appear that scaling up mining for these substances will take time, increase their price, and may have environmental side-effects. Prices can be even more volatile than those of oil and gas.[178] And the world might remain dependent on a small number of countries producing key minerals—witness the recent concerns about Chinese control of rare earth metals, used in wind turbines, efficient light-bulbs and hybrid car batteries.[179]

The idea that carbon capture 'wastes' energy[180] is also a fallacy. Indeed, as Chapter 2 discusses, a CCS plant uses more fuel than one without CCS. But, firstly, primary energy (in the form of coal, oil or gas) is not the only scarce resource; we are also constrained by capital, labour and land. It is not sensible to talk about 'efficiency' as if it only applied to energy; a fantastically costly scheme that uses very little energy is wasteful too, just in a different way. Secondly, the energy is not being 'wasted': it is being used to provide a useful service, i.e. cleaning up the atmosphere. Otherwise we might also consider refuse disposal trucks, and catalytic converters on car exhausts, as wastes of energy. Thirdly, environmentalists who propound this view propose to leave remaining fossil fuels in the ground. Essentially coal will then become a worthless black rock. Abandoning 100% of a fossil fuel appears considerably more 'wasteful' than burning 20–30% of it to capture carbon, the remainder giving usable energy.[181]

Renewable energy and efficiency have some energy security benefits. In general, a diversity of sources is always better for energy security than dependence on a single supplier. Some countries may feel safer relying on domestic wind and solar power rather than imported hydrocarbons. But again, these benefits should not be overstated. After all, gas and oil can easily be stockpiled for months, coal and uranium for years. Electricity, though, is very hard and expensive to store in quantity.

Some schemes being proposed, for example powering Europe via long-distance cables from solar plants in North Africa, expose the consuming nations to severe risk of disruptions, whether from accidents, sabotage or political action. This is particularly so if such a halt were timed to coincide with a period in which domestic renewable energy was operating at low ebb, for instance because of storms, lulls in wind or during the short daylight hours of winter. Similarly, it is not clear that a transport system dependent on biofuels from Africa or Brazil would be much more secure than one relying on Russian and Middle Eastern oil.

A more insidious risk would be that a group of 'solar exporters',[182] acting as monopolists, could capture the 'rents', that is the price difference between their own abundant solar power and the much more limited resource in Europe. They might persistently under-invest in capacity, to ensure higher prices, or use their market power to win political concessions from the EU in other areas. And subsidising uncompetitive domestic energy in the name of 'security' or 'energy independence' is, at best, a dubious policy. As the Soviet Union found, short-term energy self-sufficiency is useless when the national economy itself is moribund.

Similar caution applies to the fervour about 'green jobs'.[183] That a given energy technology 'creates' more jobs than another is not a point in its favour.[184] Material wellbeing has increased due to growing productivity. This should be obvious when we consider that global GDP grew about 4% annually from 1950 to 2006, while population rose at only 1.7% annually in the same period.[185] Subsidising employment-heavy 'green' technologies might appear to create some jobs in energy, but it will destroy many more jobs amongst businesses that consume energy. Capital is cheap, labour is expensive, as anyone who has tried to secure a plumber in a developed country will know. The ideal energy system uses as few employees as possible, not as many—freeing up people to do more creative, productive and fulfilling work. After all, powering our civilization by using millions of slaves to turn treadmills would create a lot of jobs. It might, though, not be very efficient, popular or respectful of human rights.

A renewable energy future would not necessarily usher in an era of 'energy democracy' based around small-scale, sustainable local communities. It seems more likely to be dominated by massive solar technology companies (such as Microsoft and Google); big-scale utilities (offshore wind farms, tidal barrages, etc.) able to manage large, complicated projects and to balance risk and demand across a large generating portfolio; major mining companies extracting important rare minerals; and large manufacturers relentlessly trying to drive down costs, probably outsourcing work to developing countries.[186] Intermittent power generation will be hard to manage within a decentralised system; as conceded in some environmental studies,[187] more long-distance power interconnections will be needed, rather than fewer.

CCS may be criticised as promoted by vested interests, but the renewable energy business is becoming a vested interest too. It does not make sense to argue that CCS is an 'easy way out' for the fossil fuel business; does that mean we should prefer difficult, costly solutions? It is also questionable whether most people really want responsibility for their own electricity generation. Some local generation for district heating and grid support from micro-generation is certainly worthwhile, and householders might appreciate receiving a cheque in the mail for selling some local power to the utility company. But in general, life is complicated enough; people are content to receive power as a service, without having to worry about where it comes from, or actively balancing their own load and bills.

Carbon capture also has advantages over the other major option for low-carbon baseload generation: nuclear fission. I personally consider that nuclear energy probably has a role to play in fighting climate change,[188] but many environmentalists and members of the public are determinedly opposed to it. In the developed world, at least, the approval of new nuclear plants is likely to be a long and difficult process, substantially raising its costs.[189] If concerns about the links of civilian atomic energy to nuclear proliferation and terrorism are perhaps overstated, there is still the issue of long-lived radioactive by-products. Many experts maintain that the technical issues have largely been solved, and that new plants produce much less waste than previous generations, but waste storage has, as yet, no politically agreed solutions. And CCS installations, though taking longer to build than a wind turbine, are likely to come onstream faster than new nuclear capacity.

The other key pillar of a 'clean energy' future is generally proposed to be energy efficiency. As experience has shown, high energy costs and the right policies can certainly lead to sustained improvements in the energy efficiency of specific activities. Heating and cooling buildings, for instance, is one area where it ought to be possible to drive energy use to very low levels, by 'passive' designs, heat pumps and combined heat and power (CHP) power stations. Reducing stand-by power on electronics,[190] and employing more efficient transport, particularly cars, are other areas with potential for inexpensive savings.[191]

Greenpeace's vision has global energy consumption barely increasing from 2005 to 2050, despite a world economy predicted to grow more than four times larger than today. Yet although this may be technically feasible, it is far from clear that such heroic improvements in efficiency are politically or economically likely.[192] The 'business-as-usual'[193] forecasts already include strongly increasing energy efficiency.[194] If, as is suggested, a shift to renewable energy will ultimately lower energy costs, then there will be even less incentive for further efficiency gains.

It is often claimed that energy efficiency improvements 'save money', even without considering carbon cuts.[195] The question then arises as to why intelligent consumers and profit-oriented businesses have not already made these free savings. The reasons,[196] still a matter of academic study, include:

- Lack of capital (can we afford a Toyota Prius, even though we calculate it would save money in the long term? Or possibly we are a government department or school that has a fixed investment budget regardless of profitability).
- Competing investment opportunities (an energy efficiency project that has a ten-year payback might appear to 'save' money, but not if investors can put their money into buying distressed real estate that pays back within five years).
- Imperfect awareness (do we know where we use most energy, and what technology there is to reduce energy use?).
- The 'hassle factor' (energy efficiency opportunities are often small-scale and it takes a lot of effort and expertise to detect and implement all of them).

- Subsidies for energy use (particularly prevalent in the developing world, and in oil exporting states).
- Misaligned incentives (the owner of a building does not fit insulation when the tenant will gain the savings on the gas bill).
- The risk of new technologies (unforeseen problems with our new hybrid car may mean more breakdowns and maintenance bills).
- New products being imperfect substitutes for older ones (will that LED give as good quality light as the old incandescent bulb?).

Many of these obstacles can be tackled by improved technology (e.g. smart meters), policies (regulation, information, removing energy subsidies) and 'energy service companies', who improve the efficiency of others in return for a share of the benefits. Yet because of these barriers, the full potential for efficiency is unlikely to be realised. More fundamentally, attempts to make radical increases in energy efficiency are likely to fall foul of two phenomena: the 'rebound' effect and the Khazzoom-Brookes Postulate (or Jevons Paradox).

The rebound effect occurs because, at the level of an individual consumer, making energy use more efficient encourages us to use more.[197] If we are driving a super-efficient hybrid, we may make that cross-country trip to see Grandma that we would not have done in our fuel-hungry Hummer. Levels of rebound vary depending on the context, and are hard to estimate accurately, but for transport are probably in the range 10–30%: i.e. improving car efficiency by 50% would cause us to drive 5–15% more (or use heavier vehicles, cars with more accessories, etc.).

The Khazzoom-Brookes Postulate operates at the level of an entire economy. It states that improvements in energy efficiency spur economic growth, and so lead ultimately to higher energy consumption. When washing machines become more efficient, people buy dishwashers; when heating becomes cheaper, they build bigger houses.[198] For example, as Len Brookes comments, 'It is inconceivable that we should have had the high levels of economic output triggered by the industrial revolution if energy conversion had stayed where it was at the beginning of the nineteenth century',[199] i.e. if we were still reliant on Thomas Newcomen's 1712 steam engine, with its efficiency of 1%. From 1750 to 2000, the efficiency of lighting increased by 860 times; the efficiency of transport has increased by a factor of 2,000 since 1800.[200] Yet the global middle class of today use much more, not less, energy for light and mobility than King George II or President Jefferson.

The Soviet Union's Communist system did not allow energy efficiency to improve, and it therefore collapsed partly under the weight of its gross wastage of energy.[201] Energy use only dropped when the Soviet economy fell apart. Conversely, growing energy efficiency in Europe and Japan, particularly in oil use, enabled these economies to recover and continue growing after the 1970s oil shocks.

Related to this postulate is the observation that many estimates of future energy efficiency assume that the devices we use in 2050 will be largely the same as today, only more efficient. Yet in the last decade or two we have seen the introduction of a variety of energy-hungry gadgets which were hardly imagined before: mobile phones (now web-enabled), laser printers, video-conferencing systems, plasma TVs, set-top digital boxes, patio heaters, internet servers. Some of these are arguably wasteful luxuries, but others have become key parts of the modern economy and society. No doubt many more will be invented over the next half-century. We might imagine that if we were able to develop low-cost, low-carbon transport, we might see the emergence of long-distance commuting in high-speed robotic automobiles, widespread use of personal helicopters, intercontinental hypersonic flight and space holidays. Substantial growth in hydrogen supply for vehicles and homes, proposed by many as a clean fuel, would add to energy demand, due to the need to manufacture and store the hydrogen.

Such radical energy efficiency gains, then, are only compatible with low energy demand if the consumer cost of energy remains high, whether through taxes, carbon prices, efficiency mandates, rationing or another mechanism. But if the carbon price is already set efficiently to reflect the costs of climate damage, then there is no economic rationale for further charges. Energy efficiency is a false economy if it wastes other resources, such as capital or labour (people's time). And relentless increases in energy taxation will eventually exhaust the patience of businesses and citizens.

Much stress has also been put on tackling climate change as a 'moral imperative', and seeing changes in behaviour as key: for example, a voluntary shift to public transport, a reduction in consumerism and materialism, perhaps even a return to a pre-industrial way of life. I might agree that stopping climate change is a moral issue, given that it is largely damage inflicted by the rich upon the poor, and by the people of today on unborn generations. But I am not sure that framing the problem as a moral one is helpful in finding solutions. We have to solve climate change with the human beings who exist, not with hypothetical, tremendously far-sighted and rational saints.

I feel that methods of mitigating climate change are more likely to succeed if they work with human nature and society, rather than against it. Providing attractive public transport and safe bicycle lanes are appealing examples. Japan, an unusually homogenous and conformist society, has been fairly successful in reducing energy use by social change, for instance encouraging office workers not to wear ties during the summer, and so to turn down air conditioning. Yet Japan's reduction in carbon intensity has been much more predicated on its expansion of nuclear and gas-fired power. And it seems unlikely that many countries could reach Japanese levels of consensus (perhaps the Scandinavians are one other candidate). For one thing, the problem of 'free riders', who refuse to change their carbon-intensive habits, will remain. There are big doubts about whether exhortations for energy efficiency are really effective.[202]

Capture Can Tackle Carbon Dioxide Already in the Atmosphere

We will quite possibly discover, in the next ten, twenty or thirty years, that our emissions abatements have been insufficient, and that the climate is much more sensitive than we had imagined. In any case, the allowable annual carbon emissions by 2050 are going to be very small, around 2 tonnes per person, less than a third of current levels—and that in the context of a much richer world. A single aeroplane journey can eat up most of this budget.

In this case, only carbon capture can help us. We will have to reduce the carbon dioxide concentration in the atmosphere rapidly—not merely reducing our net emissions, but actually taking them below zero. This can also help reduce ocean acidification, a non-greenhouse but serious impact of the build-up of carbon dioxide.

In order to be ready for this eventuality, we need to develop carbon capture techniques today on the easier opportunities—coal-fired power stations and so on—and have a network of carbon dioxide pipelines and storage sites ready. I, for one, don't wish to discover in 2050 that disaster is upon us, and regret that it's too late by then for a realistic 'Plan B'.

There are several possible techniques for 'sequestration plus' or going 'carbon negative'. One is to add biomass to the feedstock of ordinary fossil fuelled plants with carbon capture. Another is the whole suite of biological techniques: reforestation, land-use changes, biochar and so on. A third is to process air directly, to remove its carbon dioxide, a method that, perhaps surprisingly, does not appear unfeasibly expensive.

A Portfolio of Solutions is Required

For all these reasons, we need a broad range of possibilities, rather than focussing on a single path. If one solution fails to deliver because of reasons such as costs, technical disappointments or unanticipated environmental drawbacks, we will have other options. This is far safer than committing ourselves to a single path which, if it proves unworkable after ten or twenty years, will leave us an unpalatable choice between freezing in the dark[203] or enduring irreversible climate change. Indeed, given the vast uncertainties in future technology, economics and climate, it is remarkable that any observer can be so confident as to assert dogmatically that any one of a number of plausible mitigation measures will or will not work.

The costs of a balanced set of options are likely to be far lower than relying on a single one.[204] Compare this kind of a portfolio to a tool-kit. A hammer is good at driving in nails. It can also be used, if we have nothing better, to knock in screws. But if we relied on it also to cut timber, it would be highly expensive (in wasted wood). Now add a screwdriver, and then a saw, and the effectiveness of our tool-kit is greatly increased. This illustrates the folly of trying to rely on a single tool, a metaphorical hammer, to solve all of our energy and climate challenges.

Consider that wind power, for instance, may be highly effective in some localities, but uneconomic if used in a low-wind area, such as the Middle East or the south-eastern USA. It would be highly expensive to meet our energy needs from wind alone. If we now add, for instance, solar power in sunny areas, we have less need for the expensive subset of wind sites, and so the overall cost comes down. Now, instead of also trying to install solar power in Canada or northern Russia, include instead carbon capture in these coal-reliant countries with good storage sites. Again the total cost is reduced.

The portfolio concept has been well conceptualised by the Princeton researchers Stephen Pacala and Robert Socolow, in their famous 'wedges'.[205] They realised that, as with any large and difficult task, the climate problem can be made more manageable by breaking it down into smaller pieces. As Socolow says, their method 'decomposes a heroic challenge...into a limited set of merely monumental tasks'.[206]

In their original paper, they considered keeping emissions stable at 2004 levels until 2050. To do this, we have to save a total of 26 Gt CO_2 in 2050 (or 7 Gt of carbon, in the original). This can be divided into seven 'wedges', each of 3.7 Gt CO_2. It has been argued that we will need more wedges, perhaps eleven initially, depending on the climate stabilisation goal and when we begin, but the concept remains valid[207] (Figure 1.5).

We can then define a suite of solutions, each of which tackles one 'wedge'. For instance:

- Doubling car efficiency.
- Pursuing known methods of energy efficiency in houses.
- Increasing average coal power plant efficiency from 40% to 60%.
- Doubling nuclear capacity.
- Increasing wind power by fifty times.
- Increasing solar photovoltaic power by 700 times.
- Decreasing deforestation to zero while doubling new plantations.
- Carbon capture and storage at 800 GW of coal power stations.

Figure 1.5. Pacala and Socolow 'wedges' of emissions reductions

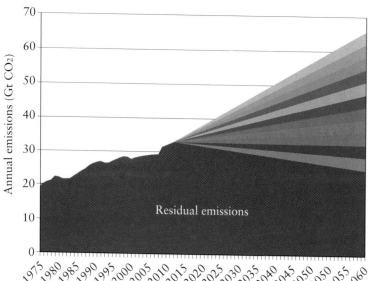

43

Pacala and Socolow suggested in total fifteen of these wedges, but many more possibilities exist. Some are mutually exclusive, or at least partly overlap, but in general, we can pick seven to achieve our climate change target. Importantly for the current discussion, three of the suggested fifteen wedges involve carbon capture: at coal and/or gas-fired plants; at hydrogen plants; and at synthetic fuels facilities. The scale of any single wedge is very imposing (for example, halving car travel, creating 3500 CCS projects,[208] or building 2 million wind turbines), so emphasising even more the impracticability of relying on a single solution. Of course, we are not limited to a single wedge of carbon capture on coal power stations: emissions in this case are large enough to support at least three wedges.

There is an idea that investing in carbon capture will take money away from renewable energy projects. But, except at the margin, this appears unlikely. $100 billion was invested in renewable energy R&D and manufacturing during 2007,[209] which hardly suggests a shortage of finance. The economic crisis is being tackled with heavy government-led spending focussed on 'green energy'. Adding more realistic options for tackling climate change will, if anything, increase the finance flowing to the sector; for one thing, it would open up the participation of cash-rich petroleum and utility companies. Only the more marginal and costly renewable energy projects are likely to be displaced by CCS.

Halving carbon emissions by 2050 will cost, by one estimate, $2 trillion without carbon capture, but 'only' $0.75 trillion with it.[210] This is due to the role of CCS in replacing these very high-cost and marginal carbon abatement options. The same study found, disappointingly for many environmentalists but not surprisingly, that ruling out CCS would probably lead to more use of nuclear (the other big low-carbon generation option) and higher overall carbon emissions. CCS also has the scale that is required. A single CCS scheme would be comparable in carbon dioxide abatement to 40% of all the wind turbines developed in the UK.[211] These big schemes also have the advantage of much lower transaction costs than many small ones: only a single management team, single set of government approvals, single financing package, and so on.

Risks to CCS

Of course, there remain significant barriers in the path of carbon capture. It is quite possible that it will not take off. The often self-interested denials and obfuscatory campaigns waged by the petroleum, electricity and coal businesses have clouded the climate debate and delayed action.[212] The fossil fuel industry has now succeeded in painting itself as the villain of climate change, and badly damaged its credibility when speaking about the environment.[213] It has thus become much harder to make the case for carbon capture than it might otherwise have been.

Quite apart from the over-riding risk—that the world does not manage to take effective action against climate change at all—such CCS-specific issues include its technological immaturity, high costs, residual environmental problems, safety concerns over CO_2 leakage, possible public opposition and institutional barriers within government and business. These are all described in more detail in Chapter 7.

Despite these risks, there is a strong case for looking seriously at carbon capture as a major part of the climate change puzzle. We will only know the feasibility of CCS, its costs, the best technology options, the required legal framework and its acceptability to the public, once several large-scale projects are in place. We now need to answer: how can carbon dioxide be captured, and how and where should it be stored?

2

CAPTURE TECHNOLOGY

'It is very disappointing when we read that "CO_2 capture is a long way off" and "is too expensive to implement". There are ways to bring down these costs. We are working very hard to build a commercial plant that will demonstrate to the world that carbon capture can be done cost effectively now, and not 10 to 15 years from now as indicated by others.'

Dr Raphael Idem, Associate Dean of Engineering,
University of Regina, Canada[1]

'The technology will function. I'm sure. I'm absolutely sure.'

Reinhardt Hassa, Chief Executive, Vattenfall Europe Generation[2]

Capture Techniques

The most expensive and least mature part of CCS is the carbon capture itself. This is the point at which a relatively pure stream of carbon dioxide, ready for transport and storage, has to be isolated from other waste gases.

Carbon capture can, most realistically, be applied to large, stationary sources of pollution.[3] About 40% of global emissions are from electricity, heat plants and 'fuel transformation'.[4] A further 26%, about two-thirds of which is suitable for capture, is produced by industry, notably iron, steel, cement, paper, oil refining and chemicals. The remainder comes from residences, commercial properties, agriculture and trans-

Figure 2.1. Cooling towers at a coal-fired power station
(Photo: istockphoto.com)

port, typically from many small, dispersed and often mobile sources which will be very hard or impossible to capture economically.

There are more than 8,100 such large[56] point sources of carbon dioxide around the world today, accounting for 15 Gt CO_2 per year, nearly two-thirds of anthropogenic emissions.[7] Many more are under construction.

The large stationary sources are quite geographically concentrated: 41% are in Asia, mostly eastern China and Japan with some in India; 20% in North America; and 13% in Europe. To be successful, CCS will have to take off in at least two of these areas. The proximity of so many sources should make it easier to establish integrated networks to capture, transport and store carbon dioxide. It is likely that over the next 50 years the weight of emissions will shift further towards Asia, especially China and the Indian subcontinent, and to some extent also to South America and the Middle East.

Power Station Capture

Amongst power plants, coal produces the highest concentration of carbon dioxide in waste gases,[8] oil slightly less and gas the lowest. Carbon capture is therefore most applicable to big coal-fired stations. Large oil power plants are less than a tenth of total generation and CCS for them has been very little discussed, but the same capture options would be applicable to oil as to coal power. Similarly, petroleum coke, a by-product of oil refining, is very similar to coal. Biomass power stations produce even more carbon dioxide per unit of energy than coal,[9] but tend to be of smaller scale and hence not so economic for carbon capture. Co-firing coal and biomass is therefore attractive.

It is worth first briefly discussing the operation of conventional (non-capture) power stations burning carbon-based fuels.

Coal is a fossil fuel formed primarily from the remains of land plants from the geological past, which decayed in swampy conditions, and were converted to rock by the gradual increase of heat and pressure. The first step is peat, which is a fossil fuel but is not considered as a type of coal. True coals are classified from low-rank to high-rank, based on their content of energy, moisture and ash. Lignite (brown coal), common in Eastern Europe, is the lowest rank, followed by sub-bituminous, bituminous, and then the very high-grade anthracite (which is normally used in specialist applications rather than power generation).

A typical coal-fired power station works by pulverising coal into a fine powder (like flour). This is blown into a boiler and burned with air to produce high-pressure steam, which drives a turbine to produce electricity. The hot flue gases (mostly carbon dioxide, nitrogen and water) from the boiler are used to heat up the inlet air to improve efficiency. Older 'sub-critical' plants could turn about 33–37% of the energy in the coal into useful electricity; newer units, 'super-critical' and 'ultra-supercritical', running at higher temperatures and pressures, can reach around 43% efficiency.[10]

Low-rank coals burn less efficiently and produce more ash by-product, but are cheaper. A slightly different design from pulverised coal, a circulating fluidised bed (CFB) plant, crushes the coal into larger pieces, releases fewer pollutants, and is better-suited to burning low-grade coals. Biomass, biological matter such as wood and waste products including straw, rice husks, chaff and so on, can also be burned in this way.

Coal combustion is highly polluting, releasing sulphur oxides and nitrogen oxides (responsible for acid rain), mercury (a toxic metal which affects the nervous system) and fine particulates (implicated in respiratory diseases). Ironically, coal plants also produce more radioactive pollution than nuclear power, via release of uranium, thorium and other radio-nuclides found in coal. Coal plants also cause water pollution and produce large quantities of 'fly ash', which has to be disposed of. In more modern plants, a desulphurisation unit, and possibly other pollution controls, are used to remove acidic gases and particulates from the waste gases before they are vented to the atmosphere, also trapping most of the radioactive material.

Natural gas power stations are somewhat different. Older designs also burned natural gas to drive a steam turbine. But most new gas stations are 'combined cycle': the expansion of the flue gases generates electricity in a gas turbine (similar to a jet aircraft engine), before the hot gases make steam to run a steam turbine. The combination of these two units makes a combined-cycle plant much more efficient, 50% or more as compared to 40% for the single-cycle. Gas power plants tend to be much quicker and cheaper to build than coal (or nuclear) power, and come in smaller, more modular sizes. Gas produces very little pollution, other than carbon dioxide and some nitrogen oxides, the plants are smaller and less visually obtrusive, and there are none of the safety worries associated with nuclear power. Gas power stations are also easy and quick to fire up and turn off again. For these reasons, and others related to industry deregulation, there has been a boom in gas-fired generation in North America and Europe from the 1990s onwards.

On the other hand, natural gas is a much more expensive fuel than coal, and so running costs are greater, especially when, as in recent years, gas prices have risen sharply. For this reason, coal plants are generally run as 'baseload', operating all the time, while natural gas is used in intermediate and peak load, being turned on and off as needed to meet demand. Gas is therefore also important for balancing intermittent renewable sources, such as when the wind is not blowing. Natural gas is not produced locally in sufficient quantities to cover demand in some important regions, such as Europe, South Africa, China, India and parts of the USA. Here abundant domestic coal offers improved security of supply, as well as local employment.

Coal plants have long lifetimes, forty years or more.[11] A typical (500 megawatt) coal power station without CCS will produce about 3.4 million tonnes of carbon dioxide per year, depending on coal quality, plant type and load factor (how much of the year the plant is running).

A natural gas station of the same capacity, on the other hand, will emit some 1.2 Mt CO_2. This is partly because of the greater efficiency of gas power plants, but mainly because coal consists primarily of carbon, while methane, the main constituent of natural gas, contains only one carbon atom for every four hydrogen atoms. The emissions from an oil-fired plant lie between these extremes.

Carbon dioxide is a relatively small part of flue gases, typically 12–17% for a coal-fired plant, 11–13% for oil and 4–8% for a gas facility. Most of the rest is harmless, non-greenhouse nitrogen (which makes up almost four-fifths of the atmosphere), and water, which is easily removed by condensation. We cannot just capture all the waste gases from a power plant, for the following reasons:

- Compressing, transporting and injecting all of the flue gases would require much more capital investment and consume more energy than dealing only with the carbon dioxide.
- A mix of gases would not reach the 'supercritical'[12] state that carbon dioxide achieves under pressure, in which its volume is greatly reduced.
- Much available underground storage space would be wasted in storing the other gases.
- The mix of flue gases would not be so effective an agent for enhanced oil recovery as pure carbon dioxide.

Carbon dioxide separation is an energy-intensive process, and the energy required increases as the carbon dioxide concentration in the waste gases decreases. Because of this energy penalty, the most modern and energy-efficient plants, particularly industrial facilities that produce concentrated carbon dioxide, will be the most effective users of carbon capture. Indeed, efficiency has advanced so much over time that a new plant with CCS may be as efficient as an older one without. No process captures all the carbon dioxide, practical proportions being 90–99%, so there will still be residual emissions. Coal mining, in particular, also has 'upstream' greenhouse emissions,[13] due to the release of methane and carbon dioxide from the coal as it is mined;[14] this adds about 2% to the overall greenhouse emissions.[15]

Energy is also required to compress the carbon dioxide into a supercritical liquid suitable for transportation to a storage site. The exact proportions vary greatly with the technology used, but typically two-thirds of the energy penalty comes from the separation step, and one-third from compression.

The extra energy required for capture means that a power plant with CCS is less efficient than one without, and therefore has to burn more fuel for the same electrical output. This in turn means that the plant has to be larger and more expensive. The cost of carbon capture therefore has three main components (excluding transportation and storage, which are covered in Chapter 3):

- The capital cost (and other incidental operating costs) of installing and running the CCS equipment.
- The extra fuel (and other substances used in pollution control) required to make up for the energy penalty.
- The incremental cost of building a larger plant to produce the same electrical output.

These costs, though, may be partly offset by the higher efficiency and inherently better pollution performance of some CCS plant designs versus traditional plants. How these costs stack up is discussed in Chapter 5.

There are three main groups of techniques for capturing carbon dioxide. The first, 'post-combustion', is to strip carbon dioxide from the waste gases. The second, 'pre-combustion', removes the carbon from the fuel before it is burned. And the third, 'oxyfuel', burns the fuel in oxygen so that the waste gases consist mostly of carbon dioxide and no further removal is required. Some other possibilities are in the R&D stage. The existence of three distinctly different capture methods increases the chance that at least one will prove technically and commercially feasible, although it also complicates the task of researching and demonstrating such facilities.

Post-Combustion

Post-combustion capture is, of the three, the best-understood technique today. Post-combustion capture can be understood as a simple addition at the smokestack of a CO_2-emitting facility (Figure 2.2: the components that are found in a conventional non-capture plant are shown in grey). An advantage of this is that minimal modification of the plant design is required. The system can also be retro-fitted to existing facilities.

Post-combustion capture works as follows. The waste gases from fuel combustion are cleaned to remove pollutants such as particulates, nitrogen oxides and sulphur dioxide, a step which should be performed anyway for environmental reasons. The gases are made to react with a chemical which removes the carbon dioxide. The harmless remainder (typically, water and nitrogen, with some uncaptured carbon dioxide, about 10% of the total CO_2) is released into the atmosphere.[16] The proportion of captured carbon dioxide can be increased but at higher costs and energy penalty.

Figure 2.2. Schematic of post-combustion capture[17]

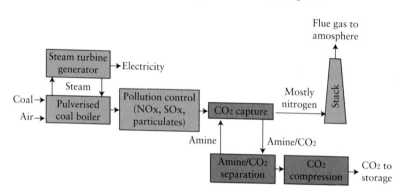

The most mature method of post-combustion capture uses chemical absorption by solvents, usually amines,[18] widely employed for decades to remove carbon dioxide from natural gas. All commercial solvents can work on gas-fired plants, but only some are suitable for coal, because of the pollutants present in the flue gas. Carbon dioxide will dissolve in the cold solvent, which is then heated to release a stream of almost pure carbon dioxide, after which the regenerated solvent can be re-used. The higher the concentration of carbon dioxide in the flue gases, the lower the heat energy required. The largest modern amine plants would need to be scaled up by a factor of ten for a typical coal plant.[19] However, this presents no obvious technical hurdles.

The energy required for solvent regeneration and carbon dioxide compression is substantial, thus creating a substantial energy penalty for post-combustion. A typical coal-fired plant with CCS might burn some 35% more coal for the same power output as a non-CCS plant.

One study has suggested the energy penalty may be considerably worse.[20] Improvements in power plant technology may partly offset this. For instance, today's cutting-edge plants convert 47% of the energy in the coal into electrical energy. Future (non-CCS) plants may reach 55% efficiency.[21] This is achieved primarily by increasing the temperature and pressure of combustion, from the older subcritical plants to newer supercritical and ultra-supercritical stations. Testing of the prototype amine unit at the Longannet power station in Scotland has reduced energy demands by about a third via new solvents and process improvements.[22]

Due to low coal prices, high thermal efficiency has not been a priority, but this can be expected to change as carbon costs increase. Alternatively, another source of heat could be used for solvent regeneration, such as waste heat from another power plant or industrial facility, or from a solar thermal system. The capture process itself is also advancing, energy efficiency having improved by a third since 1995, and expected to do so again in the next decade. Some amines show lower energy requirements but react more slowly, and research is investigating the use of additives to speed up the reaction while maintaining energy efficiency.

Further separation options are being investigated, and may also be used in industrial carbon dioxide separation or in pre-combustion capture. These include:

- High-pressure capture
- Physical capture
- Membranes
- Cryogenic separation
- Biochar scrubbing
- Biological and biomimetic approaches

High-pressure capture relies on integration with modern pressurised fluidised bed (PFB) power plants. Such plants are not very common (fewer than twenty operating worldwide), but are attractive regardless of CCS, because of their high efficiency, compactness, ability to fire waste coal and biomass, and low emissions of other pollutants. At a coal plant at Värtan in Sweden, the Norwegian company Sargas is testing a high-pressure system using cheap, non-toxic potassium carbonate.[23] The process, which relies mostly on proven technologies, has reported thermal efficiency of around 44%, with potential to improve above 50%, and captures up to 99% of the carbon dioxide. Its energy demand is less than half that of amine-based systems. It could also be applied to gas-fired plants.[24] An additional benefit is that it delivers high-pressure carbon dioxide, which therefore does not require additional compression for transport and storage.

Physical separation is more effective than chemical for relatively concentrated carbon dioxide. It is therefore more likely to be applied to industrial facilities rather than power plants. For physical absorption, the carbon dioxide dissolves in a solvent under high pressure, and is then liberated by reducing the pressure. In the future, hybrid chemical-physical solvents may be preferred.[25]

Other physical separation methods offer long-range research prospects. Some zeolites, porous artificial minerals composed mainly of aluminium silicates, show very high capacity to filter out and store carbon dioxide.[26] It may be possible to engineer structured materials which could trap and release carbon dioxide in response to an electric field rather than temperature. A number of ionic liquids show remarkably high solubility for carbon dioxide,[27] while numerous solid sorbents (activated carbon, sodium carbonate, metal-organic frameworks, molecular sponges and others) are under development.[28]

Membranes are barriers which are permeable to one fluid (say, carbon dioxide) but not to others (say, nitrogen).[29] They therefore effectively act as a sieve. They have higher capital costs than chemical or physical separation, but advanced membranes may achieve a much lower energy penalty. Many different types of membranes are under development, and the best option may be a combination of membranes with chemical solvents, which could greatly reduce capital costs and avoid some common operational problems.[30]

Cryogenic separation relies on cooling the flue gases so that carbon dioxide drops out as a solid. This might be an option for large, concentrated streams of carbon dioxide from some industrial facilities and is used on CO_2-contaminated natural gas fields. It is very energy-intensive and probably not applicable to power plants, but a cryogenic method has been proposed for coal plant flue gases, with reportedly high efficiency, capture rates and elimination of other pollutants.[31]

Biochar scrubbing has been proposed by the company Eprida, although progress since 2005 appears to have been slow. The principle here is to use biochar (a form of charcoal derived from plant matter, described in Chapter 4) together with by-product hydrogen to make ammonia, which then reacts with the carbon dioxide, and other pollutants such as sulphur and nitrogen oxides, to produce fertiliser.[32]

Biological processes include the use of large ponds of algae, which use carbon dioxide to grow, and possibly waste heat to warm them in the winter.[33] The algae can be used for biofuels, and perhaps the non-fuel residue could be burned in the power station. Algae do not need potable water; indeed, many strains thrive in salt water. They grow fast and have a very high productivity compared to other biofuels. The approach is being trialled at the Redhawk power plant in Arizona.[34]

However, this approach suffers from the rather slow take-up of carbon dioxide by the algae, which would require enormous ponds to process all the gas from a large power station. A 500 MW coal plant would need ponds as large as Manhattan or Vancouver, comparable to the area required by a solar thermal plant of the same capacity.[35] Costs of algal biofuels are still high, and projects have encountered various problems, including the growth of invasive weeds. The biofuel will, of course, be burned and its carbon dioxide released in a rather short period, so the greenhouse gas benefit is largely limited to displacing petroleum-based fuels. This would reduce, but not eliminate, the carbon footprint of coal-fired power, unless the storage of waste from processing the algae is enough to make the whole process carbon-neutral.

Biomimetic processes take their inspiration from living creatures. For instance, the enzyme (biological catalyst) carbonic anhydrase[36] speeds the (normally rather slow) reaction of carbon dioxide with water to form the bicarbonate ion HCO_3. This can then be separated to yield purified carbon dioxide, or made to precipitate limestone with modest heating (a kind of mineralisation, covered in Chapter 3).[37] However, scaling up such a process, and generating sufficient quantities of enzyme, are significant challenges.

Of the large-scale operating carbon capture projects worldwide, three use the solvent capture process for high-CO_2 natural gas: Sleipner (1 Mt CO_2 captured annually; Figure 6.7) and Snøhvit (0.7 Mt per year; Figure 2.3) in Norway and In Salah in Algeria (1 Mt per year; Figure 3.6). A similar project, the Gorgon Liquefied Natural Gas (LNG) project in Australia, will capture 4–5 Mt per year when it starts operations around the year 2015. Physical capture is also commonly employed to produce carbon dioxide for enhanced oil recovery, as at ExxonMobil's La Barge plant in Wyoming, USA, which will capture 6 Mt annually after a planned 2010 expansion.

In the US, the Shady Point coal power plant in Oklahoma, Warrior Run coal power station in Maryland, and Bellingham gas-fired co-

Figure 2.3. Snøhvit LNG plant in northern Norway; CO_2 is captured here for re-injection (Photo: Eiliv Leren/Statoil)

generation plant in Massachusetts all use amines to process small amounts of their flue gases,[38] extracting carbon dioxide for sale to industrial customers;[39] Alstom's Mountaineer demonstration power plant in West Virginia captures 0.1 Mt per year, equivalent to 30 MW output, from a coal-fired power station, using chilled ammonia,[40] a solvent that may be more effective than existing substances.

Post-combustion capture has primarily been considered for retrofitting of existing coal plants. Its low efficiency makes it less attractive for use in new-build plants but, if gas prices are low and carbon prices high, it could also be an economic option for gas-fired power.

Pre-Combustion

By contrast with post-combustion capture, pre-combustion aims to avoid carbon dioxide emissions by removing the CO_2 from the fuel before it is burned. It therefore requires a completely different kind of facility. For coal, the Integrated Gasification Combined Cycle (IGCC) plant is the leading example (Figure 2.4).

The first step is to separate oxygen from the air in an air separation unit (ASU). This oxygen is then used to gasify the coal to a 'syngas'

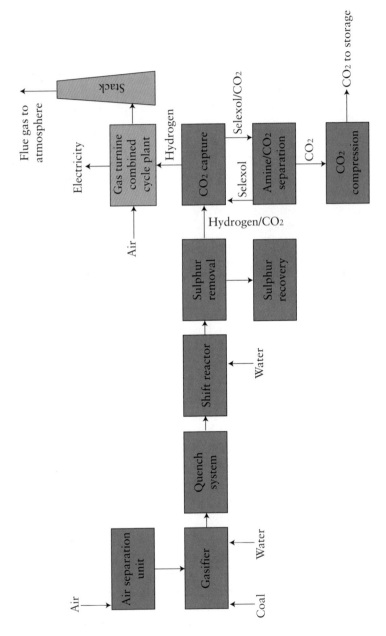

Figure 2.4. Schematic of IGCC plant[41]

(synthesis gas), consisting of carbon monoxide, carbon dioxide and hydrogen. This syngas has to be cleaned, particularly to remove sulphur compounds which would corrode the turbine. It then undergoes the 'water gas shift reaction' with steam to form carbon dioxide and more hydrogen. There are a variety of gasifiers available, of which the GE and Shell versions are well-tested; that of ConocoPhillips has less operational history.

The resulting gases are at high pressure, have high concentrations of carbon dioxide (15–40%) and are much smaller in volume than the flue gases in post-combustion capture. This is thus highly favourable for physical or membrane separation of carbon dioxide, as described above. About 90% of the carbon dioxide can be removed, its high pressure saving on compression for transport and storage. The syngas or, depending on the process, hydrogen, is then burned in a combined-cycle mode, where combustion drives first a gas turbine and then a conventional steam turbine. This is very similar to the highly efficient configuration of modern natural gas power plants. If syngas is the fuel, the carbon monoxide will burn to carbon dioxide which, at these high concentrations, is straightforward to remove. Alternatively, produced hydrogen can be used in a wide range of applications, possibly including, in the future, as a vehicle fuel. If the 'hydrogen economy' takes off, then generation from gasification and steam reforming plants may become even more important, being much cheaper than the alternative low-carbon method of creating hydrogen by electrolysing water with renewable or nuclear electricity.

IGCCs use much less water than standard power stations. They also have the advantage of removing nearly all[42] nitrogen oxides and sulphur oxides (responsible for smog and acid rain) and harmful mercury much more cheaply than conventional pollution controls. The Wabash River plant in Indiana, USA, is one of the cleanest coal plants in the world, with less than a tenth of the major (non-CO_2) pollutants of a conventional plant.[43] It should be noted, though, that Wabash River had some problems initially with excessive amounts of toxic selenium, cyanide and sometimes arsenic in its wastewater discharge.[44] The proposed IGCC at Edwardsport, Indiana would produce ten times the power of the old (1940s) plant it replaces, but emit less sulphur dioxide, nitrogen oxides and mercury, and also have barely half the carbon emissions per unit of electricity generated.[45] The plant will also begin capturing a fraction of its carbon dioxide emissions in 2013.[46] Even in a 'non-capture' IGCC, about a quarter of the carbon dioxide is removed by the solvents used for sulphur recovery, and this could be captured at low cost.[47]

New IGCCs today can reach 50% efficiency, and 56% is possible in the future. IGCCs running brown coal (lignite) are less efficient, but a straightforward refinement is to use waste heat from the gasifier to dry the coal (Integrated Drying Gasification Combined Cycle, IDGCC), bringing efficiency close to that of high-rank coals, with much cheaper fuel.[48] IGCCs' most energy-hungry step is the ASU, and efficiency here could improve by half with the use of membranes for oxygen separation. Probably the least mature part of the system is the hydrogen-powered gas turbines, which need further research to improve efficiency and reduce pollution,[49] since they produce considerable quantities of nitrogen oxides which have to be removed with scrubbers.

A variant of the process is to combine the gasification or water gas shift reaction with the carbon dioxide capture step, using a mix of catalysts and absorbents, so that pure carbon dioxide and hydrogen are produced in a single unit, reducing capital costs and increasing efficiency.[50] Alternatively, high-temperature membranes may allow for generation and separation of hydrogen in a single unit, leaving behind

Figure 2.5. Tampa Electric IGCC, Florida, USA. (Photo: Tampa Electric)

high-pressure, rather pure carbon dioxide. These processes are still in research or pilot stages.

Pre-combustion carbon capture can also be implemented at natural gas (and oil) power plants, in a similar fashion. The main difference is, of course, that no gasification is required. Instead, the gas reacts either with water ('steam reforming') or with oxygen ('partial oxidation') to form a syngas of carbon monoxide and hydrogen, which, in the presence of steam, forms carbon dioxide and more hydrogen. The carbon dioxide can be captured, typically by physical separation or solid adsorbents, while the hydrogen is burnt for power. The use of membranes for separation could significantly reduce energy demands,[51] as could novel reforming concepts such as 'HyGenSys'.[52] BP are especially interested in this route, proposing it for their (now cancelled) Peterhead project in Scotland, Bakersfield power plant in California (burning petroleum coke) and joint venture with Masdar in Abu Dhabi.

Gasification is the core technology of synthetic fuels, the production of petrol, diesel, jet kerosene and other oil products from gas and coal. IGCCs and other gasification systems can be integrated into 'poly-generation' plants, which produce a combination of electricity, hydrogen, synfuels, potable water, chemicals and heat.[53] Their output can be tuned to meet market needs: for instance, making more electricity at peak times and more synthetic fuels in low-demand periods. Such 'energyplexes' can be highly efficient, and their carbon capture system enjoys large economies of scale. Many such poly-generation facilities have been proposed and designed; signs of a move towards this concept include the integration of IGCCs burning petroleum coke into refinery sites.

IGCC plants are often described as 'capture-ready' (see below), because they can be fitted with CCS systems with relatively low cost and efficiency penalty. However, despite this advantage and their higher efficiency, they have a much higher capital cost than a pulverised coal plant, and have experienced problems with reliability. This is a key area that will require improvement, since electricity generators tend to prize consistent performance above almost all other criteria. They are harder to stop and restart, so are preferred for baseload power generation. Pre-combustion capture cannot be retro-fitted to existing facilities; an entirely new plant has to be built. However, it may be possible to salvage some components of an existing plant, such as the steam turbine, and of course the re-use of land and grid connection would save costs.

About 1,500 megawatts (MW) of coal-fired IGCCs are currently operating,[54] i.e. equivalent to about one large-sized conventional coal plant. However, the principles are well-understood, and there are about 130 gasification facilities of all types operating worldwide, with total capacity equivalent to about 1% of global electricity generation.[55] Despite some concerns about the reliability of the first generation of IGCCs, the Wabash River plant won an award[56] and maintained high uptime.[57] There has also been good experience with European IGCCs, mostly burning petroleum coke, a coal-like residue from oil processing. These plants have been more successful because they are integrated with refineries and run like chemical plants, essential for minimising downtime in the oxygen separator and gasifier.

The main operating pre-combustion capture project is that at the Beulah Synfuels plant in North Dakota, USA. This facility generates synthetic natural gas from coal, and captures about 1.2 Mt per year of CO_2 for enhanced oil recovery in Canada. FutureGen, the centrepiece of the US Department of Energy's 'clean coal' programme, will also use IGCC technology, if it goes ahead.[58]

Carbon dioxide removal at gasification facilities is likely to be a key technology for unconventional oil. For example, Nexen/OPTI's Long Lake facility extracts bitumen from the famous oil sands in Alberta, Canada. It generates steam for the extraction process by gasifying the heavy, low-value components of the oil, and thus avoids exposure to the volatile price of natural gas. This gives the plant a higher carbon footprint than other oil sands operations, which are already CO_2-intensive (a tenth to a fifth more than conventional crude[59]), but it would be relatively straightforward to capture the pure stream of carbon dioxide produced by the gasifier.[60] This could reduce the emissions from oil sands projects to no more than that of conventional oil. Long Lake was not built to be 'capture-ready', and it would save costs for future plants to envisage carbon capture from the outset.

Blending biomass (plant material) with coal in IGCCs can lead to completely carbon-neutral or even 'carbon-negative' power, since the vegetation actively removes carbon dioxide from the air as it grows. The Buggenum plant in the Netherlands burns coal with about one-third biomass, and will begin a CCS trial by late 2010.[61] The Kędzierzyn IGCC in Poland, starting operations in 2014, will burn about one-tenth biomass.[62] GreatPoint Energy is building a pilot plant at Brayton Point, Massachusetts, which produces methane from biomass and coal, by reacting them with steam in the presence of a catalyst.[63] About half the carbon in the feedstock is captured as carbon dioxide.

Oxyfuel

The major problem of processing the flue gases from conventional plants is that they consist of large amounts of nitrogen with only a small content of carbon dioxide. Oxyfuel technology is designed to solve this issue.

An oxyfuel plant is similar to a standard coal-fired plant, burning pulverised coal to drive a steam turbine (Figure 2.6). Circulating fluidised bed (CFB) plants could also use oxyfuel.[64] The main difference is that rather than burning the coal in air, an air separation unit supplies pure oxygen to the boiler. At the moment, ASUs are cryogenic (removing nitrogen from the air by cooling it to a liquid), but in the future, membranes or a bed of oxygen-separating minerals[65] may be more energy-efficient and cost-effective. The ASU required for a big (say 1 GW) coal plant would be very large, but not unprecedented.[66] An oxyfuel system needs about three times as much oxygen as an IGCC of the same electricity output.

Combustion with pure oxygen leads to very high temperatures, requiring special materials, so most of the flue gas is recycled. From

Figure 2.6. Schematic of oxyfuel combustion[67]

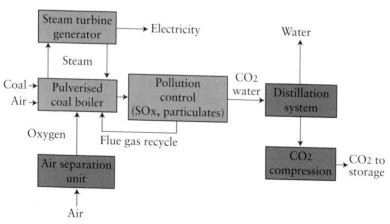

the resulting waste gases, sulphur oxides and particulate matter are removed in a separate step, water can be easily knocked out by condensation, and the remainder is nearly all carbon dioxide. There is also some nitrogen, depending on air leaks in the plant, which, if above a few percent, might make the carbon dioxide stream unsuitable for Enhanced Oil Recovery, though it would not affect its storage in other sites.

An additional advantage is the intrinsically higher efficiency of the oxyfuel boiler, about 5–8% more than air-firing[68]—although this has only been calculated theoretically, not yet proved in practice,[69] and may require new boilers to take full advantage. A boiler able to run on pure oxygen, without requiring flue gas recycling, would be much smaller and more efficient, and would save some equipment, but its design is a challenging R&D task. Oxyfuel also benefits from re-use of some of the waste heat in the flue gas, and removal of other pollutants, particularly nitrogen oxides, one of the causes of acid rain. Oxyfuel is a good candidate for retro-fitting to existing installations,[70] although new turbines would be required.[71]

Compared to pre- and post-combustion capture, oxyfuel has a higher carbon dioxide capture rate of up to 99%. It is less complicated and therefore less risky than the other two capture options. Since it has fewer variants (partly due, perhaps, to being a newer idea), a research and demonstration programme to investigate it thoroughly should be cheaper than for pre- and post-combustion capture.[72]

The main disadvantage of oxyfuel is the high electricity consumption of the oxygen unit,[73] though this is falling, and can be cut further by integration with the power cycle. The air separation unit itself might be vulnerable to reliability problems. The most interesting area of research is into new membrane systems, which might reduce the parasitic power requirement.[74] But these have to be highly optimised with the power generation equipment, and have problems with chemical and mechanical stability.[75]

The efficiency of a pulverised coal plant would probably be similar whether retro-fitted with oxyfuel or post-combustion capture. Oxyfuel is also less technically mature than the other possibilities, with only three pilot units in operation for CCS:[76] Vattenfall's Schwarze Pumpe coal plant in Germany, the Callide A plant in Queensland, Australia and Total's Lacq heavy oil power station in France.[77] These are only a tenth of the size of subsequent demonstration plants. Kimberlina in

California is testing a kind of hybrid pre-combustion/oxyfuel system based on rocket technology, although its carbon dioxide is not being captured as yet.[78]

'Second Generation' Techniques

'Second generation' capture techniques involve fundamentally different ways of generating power, or at least novel configurations of existing equipment. Some new systems do not burn fuel in air; others do not even involve combustion at all. The key objective is to reduce the capital costs and energy penalty of capture. Many such techniques have been proposed; some are being researched. Leading contenders include indirect coal-firing; chemical looping combustion; membrane combustion; ultra-high pressure; and different varieties of fuel cells.

The **Indirectly Fired Combined Cycle (IFCC)** plant is the first of these advanced techniques.[79] Here, pulverised coal is burned, as in a conventional boiler, but instead of using the heat to drive a steam turbine, high-temperature heat exchangers are used to heat air, which then runs a gas turbine. This is then followed by a standard steam turbine using the waste heat, giving an overall configuration similar to the conventional Natural Gas Combined Cycle (NGCC) plant. Efficiency of this system is up to 55%.[80] It could conceivably be retro-fitted to pulverised coal plants and, by making combustion more efficient, would improve the economics of post-combustion capture.

A better approach, though, may be to combine IFCC with oxyfuel. If the plant is run on oxygen, not only is it more efficient (due to better heat transfer to the air), but the flue gas is almost entirely water and carbon dioxide, and can therefore easily be captured. Again, the intrinsically high efficiency of the system makes capture more attractive than for conventional plants. It is also easier to run than an IGCC. The main challenge is developing materials robust enough for the high temperature heat-exchangers.

Chemical Looping Combustion (CLC), a kind of oxyfuel process, is one of the most promising avenues for research.[81] In CLC, the fuel is not burned in oxygen or air at all. Instead, a metal is used to carry oxygen. In the first step, the metal reacts with oxygen to form a metal oxide. In the second step, the metal oxide then reacts with the fuel (say, coal) to form carbon dioxide and to regenerate the metal. Various common metals, such as iron, nickel and cobalt, can be used.[82]

This method has two great advantages. Firstly, even in non-capture systems, it has potential for higher efficiencies than conventional power plants, without resorting to high temperatures and exotic materials. For a gas-fired plant, the efficiency might be only slightly lower[83] than a modern combined cycle gas turbine without carbon capture, while CLC can improve IGCC efficiency.[84] Secondly, since the fuel never comes into contact with air, the waste gases consist almost entirely of water and carbon dioxide. After straightforward condensation to remove the water, no further separation is required. CLC therefore achieves carbon capture without any energy penalty (other than that required for compressing the carbon dioxide for transport).

A variation on CLC is to use calcium oxide. This reacts with carbon dioxide to form calcium carbonate, the main constituent of limestone. This reaction can be used to capture carbon dioxide to encourage the production of hydrogen in a gasification plant. Alternatively, it can be integrated with a cement plant to avoid the emissions when limestone is calcined (heated to convert it to calcium oxide).[85] Chemical looping can also be used to make hydrogen from carbon fuels.

CLC has been tested at 120 kW scale (about 1,000 times smaller than a practical power plant), and a 10 MW demonstration plant is planned.[86] However, it remains an early-stage technique and there are significant practical difficulties, concerning circulating the metal particles, increasing the lifetime over which they remain effective, and making sure they do not damage the turbine. CLC might be ready for commercial demonstration around the year 2020.[87]

Advanced Zero Emission Power (AZEP) is another method, applicable to natural gas-fired plants, and similar to oxyfuel. It has no separate air separation unit. Instead, the gas is burned in a reaction chamber divided in two by a membrane. This relies on the recent development of membranes which, at high temperatures, allow oxygen to pass while excluding other gases.[88] As with oxyfuel, the waste gases amount to carbon dioxide and water with minor contaminants, but the energy consumption for oxygen separation is only a tenth of that needed by a conventional cryogenic ASU. AZEP is potentially very attractive, being highly efficient, with the potential to release no carbon dioxide at all, negligible nitrogen oxides, low capture costs and a minimal efficiency penalty.[89]

ZEPP (Zero Emission Power Plant) is one of a number of radically new designs which could come into existence from 2030 onwards, as yet only at the conceptual stage and requiring major advances in materials science. It is a natural-gas burning, super-compact and high-speed turbine operating at very high pressure. Potentially a system the size of a locomotive could produce 5 GW of power, as much as the largest power stations today, with efficiency of 70%. Under such ultra-high pressures, carbon dioxide would emerge as a liquid, ready for storage.[90]

ZECA (Zero Emission Coal Alliance) is an entirely different approach which avoids combustion entirely. It is rather complicated and has not yet been tested experimentally, but appears theoretically viable and efficient.[91] It relies on the gasification of coal with hydrogen to form methane, which then reacts with steam to give hydrogen. Half of the hydrogen is recycled for the gasification of more coal, while the remainder is sent to a high-efficiency fuel cell to generate electricity (or can be used in other applications). The carbon dioxide is removed by forming calcium carbonate, which then liberates a pure stream of carbon dioxide when heated with the waste heat of the fuel cell. The attractions of ZECA are its high efficiency (around 65–70%[92]) and integration with carbon capture.

Fuel cells are the subject of research and development for many small-scale power applications. Some varieties can use carbon or carbon-based fuels, including schemes from DirectCarbon,[93] SRI International[94] and the Lawrence Livermore National Laboratory.[95]

The principle is the same as for other fuel cells, which work rather like batteries, but are continuously supplied with fuel. Essentially carbon (say, coal), typically in the form of fine particles in a molten salt, reacts with oxygen. Or syngas, created by a gasification unit as in an IGCC plant, can react in an Integrated Gasification Fuel Cell (IGFC). A barrier between the carbon fuel and the oxygen prevents them from reacting directly. Instead, electrons travel from the carbon to the oxygen along wires, hence creating an electric current. The carbon dioxide created is of high purity, without nitrogen contamination, and is therefore suitable for capture without further treatment.

Even with CCS, fuel cells can achieve efficiency of 60%, some one and a half times that of most conventional coal power stations. They operate at high temperatures (650–900°C, hot enough to melt aluminium), allowing them to dispense with the expensive platinum or palladium catalysts of other fuel cells. They can run on a great variety of fuels, including natural gas, petrol, diesel, petroleum coke, coal and even biomass.

Their main drawback, at the moment, is durability, since they are degraded by the high temperatures and (for some versions) the corrosive electrolyte. They are also vulnerable to being 'poisoned' by the impurities, such as sulphur, found in coal, and have much higher capital cost than conventional power plants. They are still of small size, the largest units today being around 1 MW, enough to provide power for about 600 average Americans. Within a decade, they may reach a stage suitable for commercial power generation.[96]

If these problems can be overcome, fuel cells are extremely promising for clean, distributed generation, for example at the level of a city block. They appear to be the main contender for realistic carbon capture of such small sources,[97] although the cost of such a scheme is likely to be high due to the need for an extensive collection network. Fuel cells also benefit from substantial research unrelated to carbon capture. Their modularity and individual small size is promising for testing a variety of approaches, rather than the billion-dollar commitments needed for new central power stations.

Better compression methods are a final part of improving the capture process. As we have seen, compression of captured carbon dioxide accounts for about one-third of the energy penalty. Gas compression is a fairly mature technology, but there is still the possibility for new approaches. For instance, supersonic shock-waves, adapted from jet engines, could compress carbon dioxide in two stages rather than the six of conventional compressors, with considerably greater efficiency and less maintenance.[98]

Of all these advanced approaches, we do not know which, if any, will come to fruition. But there is clearly room for major improvements in existing capture methods, particularly once the first generation of units have been constructed and tested. Figure 2.7 gives a conceptual picture of the relative maturity of the various possibilities. Key areas for improvement include better oxygen separation and carbon dioxide absorption, improved solvents and membranes, and possibly cryogenic (low-temperature) separation, solid sorbents and biotechnology. Advanced materials, allowing combustion at higher temperatures and pressures, can greatly increase efficiency. With relatively modest investment in R&D, there is also the possibility of a breakthrough that would rapidly propel CCS up the league table of carbon mitigation. This might come, for instance, in some of the post-combustion capture techniques with enzymes or structured materials,

Figure 2.7. Illustrative technological maturity and cost for various capture options

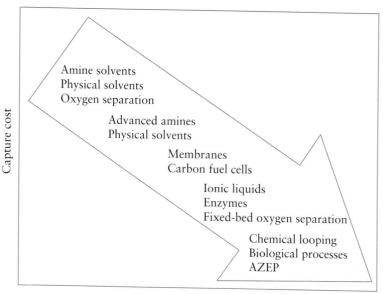

Amine solvents
Physical solvents
Oxygen separation

Advanced amines
Physical solvents

Membranes
Carbon fuel cells

Ionic liquids
Enzymes
Fixed-bed oxygen separation

Chemical looping
Biological processes
AZEP

Time to commercialisation

Chemical Looping Combustion, carbon fuel cells, or one of the radically new power plant designs.

Comparison of Capture Options

The various capture techniques can be compared in several ways. Cost is obviously key, and is considered in Chapter 5. On carbon mitigation, there is not much difference between the various mature techniques today (post-combustion capture on coal and gas, and IGCCs): all reduce plant emissions by about 85%. This percentage can be increased, at greater cost.

Conversely, on the 'efficiency penalty' for present-day facilities there is a significant difference. Figure 2.8 shows how carbon capture reduces the efficiency of three types of power plants, using today's technology. Natural gas plants are clearly more efficient; even a gas facility with capture is more efficient than a coal plant without. The efficiency penalty is also relatively small. A conventional pulverised coal plant,

Figure 2.8. Efficiency of power plants with and without capture, using today's technology

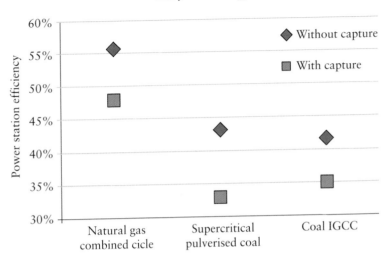

though, suffers a heavy penalty, the efficiency dropping by more than 10%. On this basis, IGCCs are far superior, with an energy penalty a bit less than gas (although they are intrinsically less efficient to begin with).

Figure 2.9 compares the efficiency penalty for some present-day plants and for some advanced designs. The first five bars show present-day coal plant types, four of which are new and one a retro-fit (of post-combustion capture to a subcritical coal plant). The next five bars are advanced coal plant designs. Three of these are only represented by a single study, and so there is no range given for efficiency; this does not imply, of course, that there is no uncertainty. The last three bars compare natural gas plants, of which the last two reflect relatively new technology.

The first observation is the very low efficiency when an existing plant is retro-fitted. This makes retro-fits less attractive than new-build plants, and emphasises the urgency of beginning the roll-out of CCS.

The second important point is the higher efficiency of natural gas pre-combustion capture versus post-combustion. It is probably no coincidence that pre-combustion capture has been advocated for planned projects in Scotland[99] and the UAE.[100]

Figure 2.9. Efficiency of power plants with and without capture, including present-day and advanced designs[101]

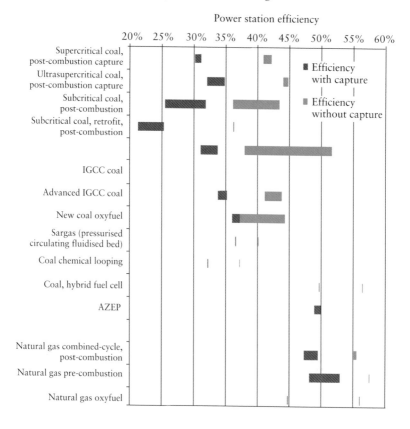

The third observation is the considerable efficiency gains offered by technological progress. Advanced IGCC and oxyfuel offer modest improvements.[102] The Sargas technique appears promising,[103] since it is about as efficient as oxyfuel, is technically straightforward and already being tested at pilot plants. Chemical looping combustion comes out less favourably, but the IEA's estimate (not shown here) is as good as for gas-fired plants.[104] Finally, two advanced possibilities in the research stage, fuel cells and the Advanced Zero Emission Power plant, and one concept, the ultra-compact Zero Emission Power Plant (ZEPP), have potential for breakthroughs in efficiency.

However, it should be recognised that, particularly in a conservative business such as electricity, which values reliability above all (and often has to satisfy regulatory demands), new, especially radically different, power plant designs tend to be adopted slowly. What works well in the laboratory may prove hard to scale up: a bench experiment has to be transformed into a system perhaps ten million times larger, the difference between building your own house and constructing New York. Unforeseen problems may arise when systems are operated for long periods under tough conditions of heat, pressure and corrosion. Complex, highly engineered systems can lack resilience and robustness, only approaching their theoretical efficiency when operated at optimum conditions. There is often a trade-off between the speed of a carbon dioxide capture process and its efficiency; efficient but slow capture processes require large and expensive facilities. Operating procedures have to be greatly revised for different types of plant, an approach that proved fruitful in improving the uptime of IGCCs. And in general, new technologies require many large-scale replications to bring their costs down.

A definite disadvantage of CCS against many renewable technologies is that (like present-day nuclear) it has to be on a large scale to be economic. This means that new projects take time, at least three to four years, and possibly up to ten. This acts to slow technological advance and learning. By contrast, wind and solar photovoltaics, for instance, can be easily implemented in small applications, giving a low risk for trying new approaches and allowing small companies or individual entrepreneurs to play a role.

For these reasons, we should be prepared that the first generation of carbon capture plants, indeed probably some generations beyond that, will rely on the most straightforward technologies. Once the basic concept of CCS is accepted, and integration with other components (transport and storage) is more mature, then it may be easier to try new technologies at demonstration scale.

There is increasing pressure, especially in Europe, for no new coal power plants to be built without CCS (see Chapter 6). This leads to the concept of 'capture-ready' plants, which can be more easily and cheaply re-configured to capture their emissions when this is legally or financially required. Some coal plants, the pre-combustion capture IGCCs (discussed above), are inherently largely capture-ready, though they may require modifications.

But for most designs, especially the basic pulverised coal (PC) plant which forms the bulk of the generating fleet, relatively little can be done ahead of time. The main items are to make some pre-investments for retro-fit, to identify a suitable nearby storage site, and to leave space for the additional facilities that will be required. Two new plants built by the big German utility E.ON, at Wilhelmshaven and Antwerp, have been certified as capture-ready by the verification firm TÜV Nord. Also, high-efficiency plants (supercritical and ultra-supercritical) are more suitable for later retro-fits. However, the loss of efficiency when adding capture will be greater than for a custom-built capture plant.

Industrial Capture

The capture of carbon dioxide is easiest when the waste gases contain a high proportion of carbon dioxide. There is therefore an excellent opportunity to begin carbon capture on a large scale by using industrial high-concentration sources. This would allow us to develop confidence in the transportation and storage of carbon dioxide, and to build the necessary infrastructure, without some of the risks of carbon capture at power plants. Alternatively, small but high-concentration sources, which would not be able to support their own carbon dioxide gathering and storage system, can piggyback on systems serving large power stations. Developing CCS for industrial sources is particularly important, since many of them are impractical to run on renewable energy, or produce carbon dioxide as an essential part of their process.

The highest-concentration industrial sources represent about 0.7% of global emissions, some 200 million tonnes (Mt) of CO_2 per year. However, this proportion may go up with increases in the production of hydrogen and synthetic hydrocarbon fuels. Capturing this would already be a worthwhile contribution to fighting climate change. Total industrial emissions suitable for capture are about 4.9 Gt, about 17% of total CO_2.

The world's largest operating CCS projects all target high-concentration industrial sources: three of them use carbon dioxide from processing 'contaminated' (high-CO_2) natural gas; the fourth is linked to a coal-to-gas facility. A third of carbon dioxide used for enhanced oil recovery (EOR) in the USA comes from artificial sources: mainly synthetic fuels and ammonia plants.

The major high-CO_2 concentration industrial facilities include:

- 'Contaminated' natural gas and 'acid gas'
- Ammonia and fertilisers

- Ethylene oxide
- Glass-making
- Ethanol
- 'Black liquor' gasifiers and 'carbon black'
- Hydrogen
- Oil refineries and heavy oil upgraders
- Synthetic fuels plants
- Iron blast furnaces and steel mills, including DRI (Direct Reduced Iron)
- Cement
- Aluminium and industrial solid wastes

Contaminated gas has to be processed to meet pipeline specifications, which typically specify no more than 2% CO_2. Ironically, ExxonMobil, long pilloried as climate change villains,[105] operate the world's largest carbon capture operation, at La Barge in Wyoming, USA. This facility processes natural gas containing 65% CO_2 and only 22% methane,[106] while an even larger plant, in West Texas, is due to start in 2010. This carbon dioxide is used in enhanced oil recovery (EOR) operations, as discussed in Chapter 3. At the moment it is therefore not optimised for maximum storage.

'Acid gas' is contaminated gas containing both carbon dioxide and hydrogen sulphide. Today, these are removed by amines. There are already numerous projects in Canada disposing of acid gas underground. A new process, 'Stenger-Wasas', may be able to make the two gases react together using a secret catalyst to form solid carbon, solid sulphur and water.[107] This would eliminate two problems in one: indeed, since this reaction is a source of energy, it could even be used for power. The solid carbon could be burnt for energy in a CCS-equipped coal power station, or simply sequestered directly (see Chapter 4), while the sulphur can be sold to the chemical industry. However, available volumes of hydrogen sulphide are probably too small for this to be more than a niche opportunity.

Ammonia production yields a highly concentrated carbon dioxide stream that can be captured directly. The same is true for manufacture of ethylene oxide, an important basic chemical. About 50% of the CO_2 from ammonia plants is used to produce the fertiliser urea, so is not available for capture.[108] However, new fertilisers incorporating ammonium carbonate and bicarbonate can be made from industrial carbon dioxide, and used to sequester carbon in the soil.[109] Some 0.7 Mt per year is already captured from ammonia plants in the USA and used for EOR; Dow Chemicals recently signed a deal to sell carbon dioxide from an ethylene oxide plant in Louisiana to Denbury Resources, an enhanced oil recovery specialist.[110]

Two interesting examples of industries that are essentially 'capture-ready' are glass-making and ethanol production.

Glass-making requires high temperatures and so, conveniently, furnaces already use oxyfuelling. Emissions are probably on the order of 75 Mt CO_2 per year,[111] which could be captured directly due to their high concentration.

Ethanol production as a gasoline substitute has increased significantly with the growing interest in biofuels. Fermentation of sugars yields cool, concentrated carbon dioxide, some 48 Mt in 2008. If this were captured, then the most efficient ethanol feedstocks, notably Brazilian sugarcane, could become 'carbon negative',[112] not only displacing fossil fuels but actively reducing atmospheric levels of carbon dioxide. Capture costs would be relatively modest, but transportation and storage might be more difficult as individual facilities are relatively small and dispersed. One interesting project in this connection is the Blue Flint ethanol plant in North Dakota, which is integrated with a coal-fired power station,[113] using its waste steam in the fermentation process. There is no carbon capture in this case, but co-location of small, low capture cost industries with large, more expensive coal plants can be the nucleus of CCS 'clusters' (Chapter 6).

'Black liquor' is a by-product of the paper and pulp industry, created from wood, and is burned as an energy source. Carbon dioxide could be captured from this, either by chemical absorption from conventional boilers, or from an IGCC-type boiler. Volumes of carbon dioxide available are very substantial, and, like ethanol, this is a 'carbon negative' option, since it is based on biomass.

'Carbon black', used in ink, rubber and plastics, is currently made by incomplete burning of fuel oil, with high carbon dioxide emissions, but could be generated by decomposing natural gas to carbon and hydrogen at high temperatures,[114] thus partly decarbonising the original fuel.

Hydrogen may become an important fuel for vehicles and possibly homes. It is already used in large quantities for chemicals and oil refining, and is most cost-effectively manufactured from natural gas or coal. Co-generation of hydrogen and electricity in an IGCC is economically attractive, and several pilot projects are in progress. BP's Bakersfield power plant in California will use petroleum coke, a refinery by-product, to generate hydrogen, with 4 Mt CO_2 per year captured. Existing older hydrogen plants (thirty years old and more) produce a concentrated stream of carbon dioxide which could be captured directly; newer plants yield a mix of carbon dioxide, nitrogen and water which would require post-combustion capture, or a redesign of the hydrogen separation system.[115]

Upgraders convert heavy oil to light synthetic crude which can be refined to useful products such as petrol (gasoline), kerosene and diesel. Upgraders are major customers for hydrogen, and heavy oil production is a major polluter via both the extraction and upgrading processes. Producing a tonne of bitumen from the Canadian oil sands via steam injection releases about 0.4 tonnes of carbon dioxide from gas that is burned for heat. A further 0.25–4 tonnes are emitted in the upgrading process, mostly when generating hydrogen. The carbon dioxide from upgrading plants is of high purity and could be easily separated. Volumes are in the order of 25 Mt per year from Canada alone.[116] Emissions from steam generation could be captured in the same way as for a power plant. By 2020, most[117] oil sands carbon dioxide could be suitable for capture, around 80 Mt per year.[118]

Large **oil refineries** produce about 800 Mt of moderately concentrated carbon dioxide annually. Some 30–40% of this could be captured, but at a high energy cost. Capture is being studied at the Mongstad refinery in Norway, operated by Statoil[119] (Figure 2.10).

Synthetic fuels plants, namely gas- and coal-to-liquids (GtL and CtL), have similarities to IGCCs.[120] These facilities produce synthetic oil from gas and coal (and, in the future, potentially from biomass). The recent period of high oil prices spurred interest in 'XtL'[121] plants. As well as older plants in Malaysia and South Africa, new facilities are being constructed in Qatar and are planned in Nigeria, China, Japan and elsewhere. The Secunda CtL plant in South Africa is often said to be the largest single source of anthropogenic carbon dioxide in the world, at some 50 Mt per year, as much as the whole of Sweden. Fortunately, these plants yield high-purity streams of carbon dioxide which should be straightforward to capture. A Japanese consortium is working on a GtL process which can use high-CO_2 gas.[122] The loss of efficiency from adding carbon capture to such plants is minimal.[123]

Figure 2.10. Mongstad Refinery in Norway, for which carbon capture is being considered (Photo: Harald M. Valderhaug/Statoil)

Iron and **steel** manufacture produces a huge quantity of global emissions, some 10% of the worldwide total in 2005. About 75% of this could be captured from the core process—it would be much more expensive to separate carbon dioxide from auxiliary units. Capture could use solvents regenerated by waste heat, which could perhaps be supplied by a nearby power plant. More promising would be to use oxyfuelling, which also reduces coal use. A demonstration plant is operating in South Korea, and could be fitted with carbon capture without any efficiency penalty. The ULCOS research project, funded by the EU, targets 50–70% emissions reduction from steel-making.

Alternatively, coal and coke could be replaced by hydrogen or electricity (with CCS at the generation plant), and more use could be made of the capture-ready DRI process, rather than conventional blast furnaces. DRI uses syngas, a mix of hydrogen and carbon monoxide, produced from coal or natural gas, to reduce iron ore to iron. DRI has several other attractions over conventional blast furnaces, such as its higher efficiency, higher-quality product, and potential for smaller-scale facilities.[124]

Cement is the other large contributor to industrial emissions, providing about 5% of the total. Two-thirds of this comes from limestone decomposition, which is an essential step in the process; the other third from fuel combustion. Some 95% of emissions could be captured by back-end chemical absorption. The concentration of carbon dioxide in waste gases is higher than for a coal power station, but the capture step would still be expensive and would require significant process adjustments.[125] Oxyfuelling could reduce costs and increase kiln productivity; it is cheaper than for power stations, since only part of the carbon dioxide is produced by combustion, and therefore relatively less oxygen is required. An alternative, mentioned previously, would be to use calcium oxide chemical looping, but this is only speculative at present. Cement, iron and steel plants are smaller sources than a typical coal plant, so economies of scale would be less, unless several facilities are grouped at a single site.

The manufacture of **aluminium** is a highly energy-intensive process, which also yields carbon dioxide from the carbon anodes used in the electrolysis process.[126] The carbon dioxide can be captured in the same way as for power plants, and Emirates Aluminium (EMAL)'s new smelter in the UAE is considering this process.[127]

Carbonation of industrial solid wastes, a kind of mineralisation technique, as described in Chapter 3, is applicable to alumina, steel, cement and mining of metals and oil shale. Processing bauxite into alumina leaves a waste product. When this reacts with carbon dioxide from the aluminium smelting process, it forms a solid residue which can be used as a building material or to improve soils.[128] Carbon dioxide removal could be around 2 Mt per year globally. Similarly, coal ash, steel slag, waste concrete, mine tailings and ash from oil shale burning can react with carbon dioxide, with global potential for carbon dioxide removal of about 450 Mt per year.[129] To this can be added unquantified amounts of fly ash from municipal waste incineration, and de-inking ash from paper recycling. Collectively, this is a low-cost (or even negative cost) opportunity, and of reasonable size: equivalent to 135 typical coal-fired power stations.

Environmental Impacts

As mentioned, CCS plants incur an energy penalty over comparable facilities that do not capture carbon dioxide emissions. This does not result in a higher emission of non-greenhouse pollutants, such as those responsible for acid rain, since CCS plants are also intrinsically much cleaner due to a variety of differences in the process. Pre-combustion plants have to remove sulphur and other impurities to avoid 'poisoning' the catalysts. The solvents used in post-combustion capture trap most acidic gases such as sulphur dioxide and nitrogen oxides. Oxyfuel plants have very low nitrogen oxide emissions (since nitrogen is removed from the air prior to combustion), but sulphur dioxide release is similar to a conventional plant. Post-combustion plants may also emit traces of ammonia from solvent decomposition.

All coal plants produce ash and slag, which can be used in construction. Carbonation of coal ash, use in cement-making, or conversion to 'green bricks'[130] are possible environmentally friendly alternatives. Sulphur (in various compounds or as the solid element) has some markets (for instance, in making sulphuric acid), but it is usually difficult to find a worthwhile use.

As with any power plant, additional pollution control systems can be installed, albeit at additional capital cost and energy penalty. And polluting combustion products are primarily an issue for coal (and biomass and oil), since gas-fired power stations are already very clean.

The increased fuel requirement leads to a commensurate increase in 'upstream' environmental impacts, i.e. those at the point of extraction of coal, oil or gas. For gas, the environmental footprint is generally very small.[131] The impact of oil production is also fairly minor, certainly with well-managed modern methods, but again with exceptions such as oil-sands mining and oil-shale retorting. Coal extraction can, however, have severe consequences, particularly open-cast mining in scenic areas. Mining also has a poor safety record, although fatality rates can be lowered dramatically with better safety standards, as we see if we compare, for instance, Australia with Ukraine or China.

The increased environmental burden from upstream activities is, though, not a reason to rule out carbon capture. Extraction-related emissions are relatively minor compared with the direct effects of the power plant.[132] New CCS plants may actually be more efficient than older non-CCS facilities. Incentives and mandates for carbon capture can accelerate the replacement of such inefficient old plants, and efficiency can also be raised by a major overhaul as part of retro-fitting CCS, especially for oxyfuel systems.

In any case, environmental protection should be tied as closely as possible to the problem it seeks to eliminate. Rather than preventing carbon capture, appropriate policies to reduce upstream environmental damage include moratoria on fossil fuel extraction from sensitive areas, unless the footprint can be reduced to acceptable levels; restrictions on pollutant emissions to water (e.g. oil droplets) or air; banning flaring of gas; laying strict conditions for land restoration of mining areas; discouraging coal-mining techniques such as mountain-top removal; reducing the use of surface water in oil sands projects; ensuring that local communities have a say in, and benefit from, the use of their natural resources. Similar stringent oversight needs to be applied to renewable energy projects.

Air Capture

Many carbon dioxide sources are not suitable for capture, typically because they are small and dispersed, such as residential gas- or oil-fired boilers. Emissions from mobile transport are, as discussed below, probably impossible to capture economically. Furthermore, if climate change proceeds rapidly, we may realise that, even with complete decarbonisation of the economy, it is not sufficient to wait for slow natural processes to reduce the carbon dioxide concentration of the air to acceptable levels.

To tackle this issue, interest is growing in indirect capture of carbon dioxide from the air. A number of methods have been proposed. Air, as we have discussed, is now about 390 ppm (0.039%) CO_2. Various chemicals can remove some of this carbon dioxide, which can then be compressed, transported to a storage site, and stored by any of the methods discussed in Chapter 3.

The energy required to capture this carbon dioxide is more than for a concentrated source, but not hugely so. To produce high-purity carbon dioxide from air requires about 30% more energy than to do so from a coal-plant's flue gas.[133] Air capture of other gases for industrial uses is routine, oxygen, nitrogen and argon all being extracted in this way. And, unlike for a power station, it is not necessary to capture all, or even much, of the carbon dioxide in the air moving through the unit. After all, there is plenty of air. The proportion of carbon dioxide captured is just an engineering trade-off; capturing a lower proportion of the air's carbon dioxide saves energy in the capture stage, but requires more energy to move larger volumes of air through the machine.

Professor Klaus Lackner of Columbia University has been the pioneer of air capture.[134] His company, Global Research Technologies, has produced a number of 'artificial trees'. These are the size of a shipping container (6 x 2.4 metres, 15 m³) and use a polymer-based resin which responds to changes in humidity, absorbing carbon dioxide when dry and releasing it when wet. A gentle breeze moves air past the device fast enough for it to be effective, and sea water can be used for 'washing' to release the trapped carbon dioxide.[135] Some low-grade heat is required for regeneration, and electrical power for compressing captured carbon dioxide. The low-grade heat could be supplied by a solar thermal plant, or by waste heat from a power station. Carbon dioxide is well-mixed in the atmosphere, and therefore this system would not produce areas of CO_2-depleted air, harmful for plants.

Lackner's is not the only machine available. Professor David Keith from the University of Calgary has devised a system which sprays a solvent into the air inside a tower, several metres tall.[136] The solvent dissolves carbon dioxide and drains to the bottom of the tower, where electricity or chemicals are used to reconstitute it. The device is simple and uses commonly available sodium hydroxide, capturing about half of the carbon dioxide that passes through it. According to Keith, it is intended to demonstrate that air capture is technically feasible and

Figure 2.11. Artist's impression of an 'air capture' machine (© Institute of Mechanical Engineers)

to set an upper limit for the possible cost. One disadvantage is the rather high water loss from evaporation, though this can be reduced in cold, humid climates,[137] or possibly made less consequential by using sea water.[138]

Air capture is, particularly at the moment, considerably more expensive than capture from large-point sources such as power stations. But it has several advantages over 'conventional' CCS.[139] The devices:

- Can be placed next to a suitable storage site, so avoiding transport costs.
- Can be located in a remote region, to avoid community objections.
- Do not have to be in any special location (unlike, say, wind turbines), other than being close to a suitable storage site, and supplied with a (fairly modest) quantity of energy.
- Are much less visually intrusive than power plants or wind turbines; indeed, they could probably be made completely unnoticeable.
- Are several thousand times more effective at removing carbon dioxide than natural trees.
- Could be run from a suitable cheap energy source, such as solar power in a desert, flared gas in a remote oilfield, or hydropower, lacking nearby consumers. Another possibility would be to take their power from intermittent sources (such as wind), during off-peak times, when output would otherwise be wasted, or would have to be stored.
- Can address one of the main objections to carbon capture, the risk of leakage. As I discuss in Chapter 3, I believe that leakage from good storage sites is likely to be negligible. But if we did discover that there was significant leakage, we could simply add additional air capture machines to compensate.

Indirect capture is therefore the ultimate backstop for climate policy. Storage capacity permitting, we can, at a cost in money and energy, remove any quantity of carbon dioxide from the atmosphere.[140] This may be crucial if we discover that we are on the path to sudden, catastrophic climate change. Even if we were to halt all emissions immediately, it would take millennia for the elevated concentration of atmospheric carbon dioxide to be fully absorbed. By contrast, air capture might be able to take us back to pre-industrial levels within some decades. As a 'geo-engineering' solution, it addresses the problem directly, rather than reducing global warming indirectly.[141] Undesirable side-effects are, as far as we can tell now, minimal compared with other geo-engineering techniques, and it also addresses the other key issue of ocean acidification.

Some major studies have dismissed air capture without serious consideration,[142] mainly on cost grounds. It is, indeed, likely to be one of the more expensive carbon mitigation options, but it does not have to compete with CCS on large centralised sources, nor with major low-carbon power solutions such as wind or nuclear. It is intended to address otherwise intractable polluters such as flying, and to provide a way of returning rapidly to a pre-industrial atmosphere. In contrast to other 'carbon offset' schemes such as forestry (see Chapter 4), which have been heavily criticised,[143] it offers completely verifiable, and undeniably 'additional', reductions. I will return to this issue in Chapter 6.

Decarbonising Transport

Transport represents the most difficult major class of carbon dioxide emissions to address. Modern transport is overwhelmingly dependent on fossil fuels, which are ideal because of their high energy density, good performance in vehicle engines, relatively low cost, ease of handling and well-developed infrastructure. As mentioned, about a fifth of global greenhouse emissions come from transport.

It may be possible to reduce transport emissions significantly by improved efficiency and by a shift to public transport. But, as travel correlates strongly with wealth, it is likely that economic growth will overwhelm these savings, unless carbon limits are very stringent.

Nor is it realistic, as some have suggested,[144] to abandon aviation. If some flights are unnecessary luxuries, many more are an indispensable part of the modern, globalised world.[145] Giving up flying is likely to deal a devastating blow to the fight against poverty; to the economy of many countries (particularly poor ones dependent on tourism); to scientific research; to international business; to the advance of technologies, including those that benefit the environment; and to mutual understanding between cultures and people. Shipping, too, is reliant on oil; in turn, growing global trade, which has underpinned the prosperity of the last sixty years, depends on an ever larger and more efficient shipping industry.

We must, therefore, find ways to reduce the carbon footprint of transport significantly, at reasonable cost, by direct or indirect means. Such attempts fall into four main classes.

- Capture the carbon dioxide produced on-board.
- Use zero-carbon transportation, notably electric or hydrogen-powered,[146] or, failing this, low-carbon fuels.
- Continue to use fossil fuels, but offset their emissions by removing carbon dioxide from the atmosphere elsewhere.
- Use synthetic fuels, either biofuels or others, that do not represent a net addition of carbon to the atmosphere.

It turns out that carbon capture is highly relevant to all these possibilities.

On-board capture can be dealt with first. This appears to have been first proposed in 1993,[147] and crops up again on occasion.[148] The idea, essentially, is to carry around a mobile carbon dioxide capture device, either as 'end of pipe' or by modifying the combustion process (similar to Chemical Looping Combustion).

These schemes might be just about feasible for ground transport, but it seems very unlikely they would be practical or economic. A huge network of carbon dioxide collection points would be needed, assuming that drivers would empty their carbon dioxide tank when filling up with fuel. Burning 1 kilogram of petrol[149] produces more than 3 kg of carbon dioxide, which would have to be stored, either combined chemically with some other substance, or in a pressurised tank. Its volume would be 30–50% more than the original fuel. Add the mass and size of the tank and capture device, and it seems that such vehicles would be very bulky and fuel-inefficient,[150] and for a weight-critical application like flying, it appears completely impossible. One possible exception might be the use of direct carbon fuel cells in large trucks. It might also work for trains, but it is far easier to drive them electrically.

Ships may be a different matter. Nuclear-powered ships are entirely feasible, but currently largely unacceptable on security grounds; use of auxiliary modern sails, or natural gas as fuel, can cut carbon dioxide emissions but not eliminate them. And running a ship on batteries while it sails across the Pacific also seems impractical. What about on-board CCS?

A number of carbon capture devices are compact enough to be placed on large vessels, and it seems less problematic for ships to carry around large quantities of waste carbon dioxide until it can be discharged. Consider an oil tanker, whose carbon dioxide generated from burning fuel on a twenty-day voyage is only about 5% of the weight and volume of the cargo.[151] A membrane separation unit capable of processing this much carbon dioxide has a volume of about 600 m^3 (about fifteen standard shipping containers[152]) and weighs 130 tonnes,[153] considerably less than the weight of the ship's engine. Molten carbonate fuel cells, which effectively integrate engine and CCS, would be even more compact and, with their high efficiencies, would produce less carbon dioxide (and require less fuel) to begin with. These simple calculations suggest that CCS might be applicable to at least larger ships. Since international shipping accounts for about 3% of emissions,[154] reducing its carbon footprint is highly worthwhile. Carbon capture on ships is, though, probably a long-term target, and might be pre-empted by the use of zero-carbon engines.

The next possibility is to use **zero- or low-carbon power sources** for transport. The leading contenders are electricity (in batteries, or powering electric trains) and hydrogen. A halfway house is the 'plug-in hybrid', which runs on batteries over shorter distances and switches to an internal combustion engine for longer trips. An electric motor or hydrogen fuel cell is more efficient than an internal combustion engine, but this has to be set against the energy lost in generation and transmission.[155]

In all these technologies, though, the electricity (or hydrogen) has to be generated somehow. Even with a rapid increase in renewables (and perhaps nuclear), fossil fuels are going to continue to be major components of power generation. And by far the most economical way of producing hydrogen is by steam reforming of natural gas. Therefore, for the time being, zero-carbon transport fuels will only deliver cuts in emissions if their electricity or hydrogen source is capturing its emissions.

At least in the medium term, it may prove more convenient to make partially-decarbonised fuels that can be used in the existing vehicle fleet. Synthetic diesel from natural gas has slightly lower carbon emissions than petro-diesel.[156] Methanol, butanol and dimethyl ether (DME) are compatible with modified car engines (DME being a diesel substitute), and, if made from natural gas with carbon capture, emit significantly[157] less carbon dioxide than conventional gasoline.[158] Compressed Natural Gas (CNG), already widely used in e.g. India, Pakistan and Egypt, is particularly low-carbon, about two-thirds of diesel's figure.[159]

The third option is to seek **'carbon offsets'** for transport emissions. Every tonne of carbon dioxide released to the atmosphere can be matched by an equivalent withdrawal elsewhere. This could be achieved by air capture, as mentioned above. Alternatively, as is currently popular for 'carbon neutral' flights, the offset can be a reforestation scheme (Chapter 4), at the moment rather cheap but, as we will see, carrying significant complications. The global warming impact of flying is greater than its carbon dioxide emissions suggests—perhaps by a factor of three, for long-distance flights—because aeroplane condensation trails (contrails) trap reflected heat from the Earth. Various techniques can be used to minimise contrail formation,[160] and the residual impact could be mopped up. For instance, a high-altitude flight could be required to finance the capture from air of carbon dioxide equal to three times its direct emissions.

The fourth possibility is to use **synthetic hydrocarbons** from non-fossil sources. They are, in several ways, ideal fuels: high energy density, easy to store and handle, and compatible with existing pipelines, filling stations and vehicle engines. This avoids the 'chicken and egg' problem that bedevils hydrogen: do you build a network of filling stations before the cars come, or wait for many people to buy cars that can't easily be refuelled? Synthetic liquids can also be much cleaner—zero-sulphur, for instance—than those made from fossil fuels.

The best-known approach is biofuels—using plant feedstocks to make biodiesel and ethanol that can be burned in conventional engines. As we have seen, fermentation for ethanol can be combined with CCS to give a carbon-negative fuel. A slightly different approach is to gasify biomass, in the same way as a coal- or gas-to-liquids plant, to produce synthetic diesel, petrol or jet fuel. In the same way as for GtL and CtL facilities, the concentrated stream of carbon dioxide is ideal for capture, and would also lead to a net removal of carbon from the atmosphere. A small gasification plant (without CCS) at Freiburg in Germany makes synthetic diesel from waste plant matter.[161] BtL, though, is unlikely to be competitive unless oil and/or carbon prices are very high. To maintain economies of scale, a CtL plant could be co-fired with some biomass.

Widespread use of biofuels, though, raises significant concerns, as has become apparent over the boom of the last few years (Chapter 4). Allowing for the energy used in growing and processing the crops, some biofuels, such as corn ethanol, may actually emit more carbon than fossil fuels. Converting large swathes of land for biofuels can displace local farmers, drive up food prices, compete for water, destroy virgin habitats and release carbon from the soil.[162] 'Second generation' biofuels, made from non-food crops or waste material such as stems and leaves, solve some of these problems but are still commercially unproven.

A relatively new idea, then, is to make 'carbon-neutral hydrocarbons'. Rather than using energy to make electricity or hydrogen for transport, we can employ it to synthesise fuels from the air. This is done by combining carbon dioxide with hydrogen under high pressure in the presence of a catalyst. The hydrogen is generated from fossil fuels with carbon capture (likely to be the cheapest option), or by electrolysing water using zero-carbon electricity from renewable or nuclear sources. The carbon dioxide is extracted from the air, as described above. A wide range of potential fuels can be produced, including octane (a major constituent of gasoline), methanol,[163] and DME. This carbon mitigation option is preferable at high oil prices and carbon costs, and has a good chance of being competitive against hydrogen vehicles.[164]

Conclusion

Capture is the most expensive and least mature part of the carbon capture, transport and storage chain. Three main methods exist: removing carbon dioxide post-combustion, decarbonising the fuel before combustion, and burning the fuel in oxygen rather than air. All of these methods have been demonstrated in pilot plants, and post-combustion capture is already widely used in gas processing. It is not yet clear which of these technologies will be most successful. Retro-fits require post-combustion capture or oxyfuel; new plants will probably mostly use pre-combustion capture or oxyfuel. Only a few capture projects are operating today, but many are under development (Figure 2.12).

Carbon capture leads to a significant reduction in power plant efficiency. There remains large scope for advances in capture efficiency and cost reductions, while some innovative break-through capture methods are being researched.

Carbon dioxide can also be captured from industrial sources. Some industries yield concentrated carbon dioxide emissions which can be easily captured. Others, such as iron and steel and cement, are major emitters but capture is more costly.

Carbon dioxide can also be captured from air, either directly or via biomass plants with carbon capture. This is a vital backstop to climate policy, giving the ability—at a cost—to reduce atmospheric carbon dioxide to any desired level within a relatively short time.

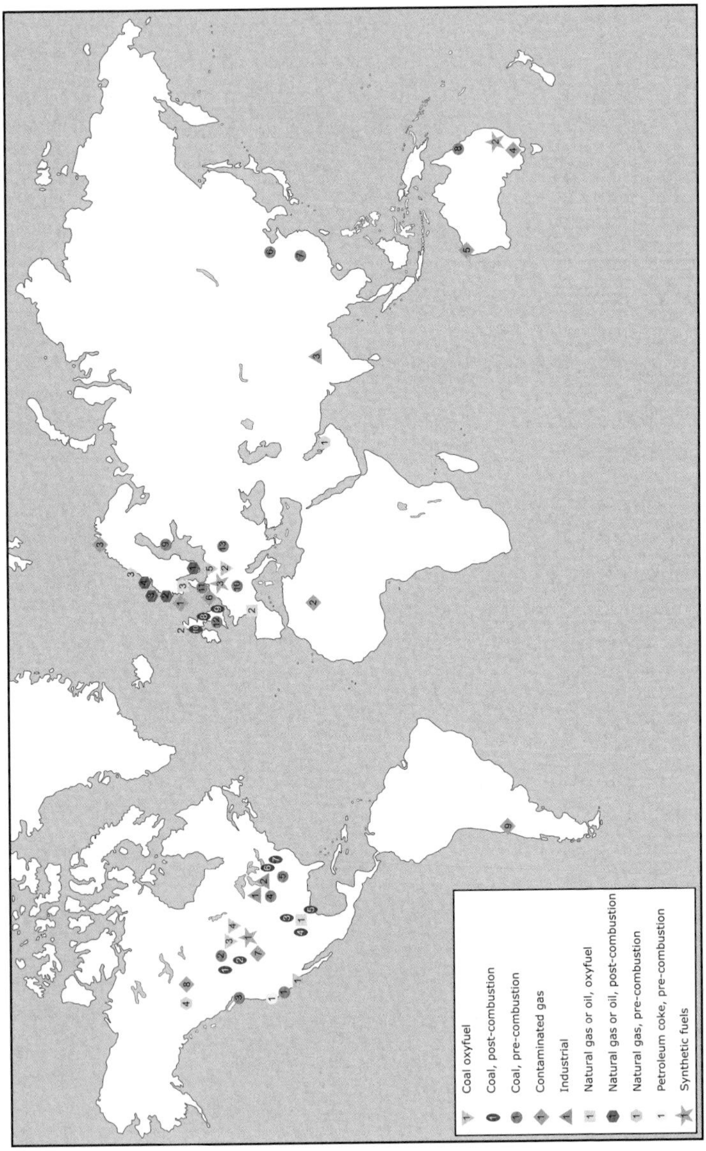

Figure 2.12. Selected carbon capture projects

Legend:
- Coal oxyfuel
- Coal, post-combustion
- Coal, pre-combustion
- Contaminated gas
- Industrial
- Natural gas or oil, oxyfuel
- Natural gas or oil, post-combustion
- Natural gas, pre-combustion
- Petroleum coke, pre-combustion
- Synthetic fuels

	Contaminated gas		Coal, post-combustion		Coal, pre-combustion
1	Sleipner (Norway)	1	Bow City (Alberta, Canada)	1	Carson (California, USA)
2	In Salah (Algeria)	2	Antelope Valley (North Dakota, USA)	2	PolyGen (Saskatchewan, Canada)
3	Snøhvit (Norway)	3	AEP Northeastern (Oklahoma, USA)	3	Wallula Energy (Washington, USA)
4	Otway (Victoria, Australia)	4	Tenaska (Texas, USA)	4	FutureGen (Illinois, USA)
5	Gorgon (Australia)	5	WA Parish (Texas, USA)	5	Appalachian Power (West Virginia, USA)
6	K12B (Netherlands)	6	Meigs County (Ohio, USA)	6	GreenGen (China)
7	La Barge (Wyoming, USA)	7	AEP Alstom Mountaineer (West Virginia, USA)	7	NZEP—technology to be decided (China)
8	Zama (Alberta, Canada)	8	Ferrybridge (UK)	8	ZeroGen (Queensland, Australia)
9	Cerro Fortunoso (Argentina)	9	Tilbury and Kingsnorth (UK)	9	Meri Pori (Finland)
		10	Longannet (UK)	10	Goldenbergwerk (Germany)
				11	Nuon Magnum (Netherlands)
				12	Teesside, Hatfield and Killingholme (UK)
				13	Kędzierzyn (Poland)

Coal, oxyfuel	Gas, pre-combustion	Gas or oil, oxyfuel
1 Kimberlina (California, USA)	1 Abu Dhabi (UAE)	1 ZENG Worsham-Steed (Texas, USA)
2 Schwarze Pumpe, (Germany)	2 Miller-Peterhead (UK—cancelled)	2 Lacq (France)
3 Boundary Dam (Saskatchewan, Canada)	3 Haltenbanken (Norway—cancelled)	3 ZENG Risavika (Norway)
4 SaskPower (Saskatchewan, Canada)	4 Fort Nelson (British Columbia, Canada)	
5 Janschwalde (Germany)		

Petroleum coke, pre-combustion	Synthetic fuels	Industrial
1 Hydrogen Energy (California)	1 Beulah (North Dakota, USA)	1 Decatur—ethanol (Illinois, USA)
	2 Monash Energy (Victoria, Australia)	2 TAME—ethanol (Ohio, USA)
	3 Ketzin (Germany)	3 Aonla and Phulpur—fertilisers (India)

Gas or oil, post-combustion

1 Karlshamn (Sweden)
2 Kårstø (Norway)
3 Sargas Husnes (Norway)
4 Mongstad (Norway)

Transport emissions are difficult to tackle with carbon capture, although large ships might be able to use CCS systems. Using CCS on plants which produce decarbonised fuel—such as electricity or hydrogen—is one possibility. Making carbon-based fuels from biomass or directly from atmospheric carbon dioxide is another. Employing direct air capture to offset transport is a third option.

3

TRANSPORT AND STORAGE

'We believe [underground storage] is safe; it is certainly technically feasible and really has very little environmental downside.'

Dr Andrew Chadwick, British Geological Survey[1]

'We're not going into a salt cavern, we're not going into an underground river. We're going into microscopic holes...Add it up and it's a large volume.'

Susan Hovorka, geologist at the University of Texas at Austin[2]

Having captured the carbon dioxide, we now have to decide what to do with it. This falls into two steps: transporting carbon dioxide from the site of capture to the storage point, and then storing (or sequestering) it—the S in CCS. As we have seen, carbon capture is the most technically challenging and expensive part of the chain. Yet it is the storage phase that has received most media attention and seems to attract the greatest concern. It is vital to know whether the carbon dioxide can be stored indefinitely, safely, at reasonable cost, in a way that society finds acceptable. If CCS is to play a major part in combating climate change, then we also need to know that there is enough storage space for many billions of tonnes of carbon dioxide.

Figure 3.1 shows the full chain. CO_2 is captured at a large emissions point. Here it is compressed and sent by pipeline to an injection site which may be onshore or offshore. Offshore injection may be from a fixed platform, or directly from a subsea installation. Alternatively

Figure 3.1. Schematic illustration of capture, transport and storage

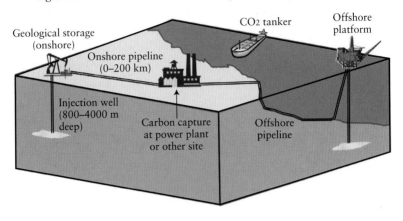

offshore transport may be done by ship. The injection site is a sub-surface structure chosen for its ability to hold the carbon dioxide safely over very long periods.

Transport

Transportation of carbon dioxide is a straightforward, mature technology. It should be very safe, and transport forms only a small part of the total cost of CCS.

The main method of carbon dioxide transport is by pipeline. This will probably be under pressure, so that it can be moved as a liquid, greatly reducing its volume. The carbon dioxide will emerge from the capture site under pressure, which will generally be sufficient to get it to the storage site. If not, booster stations can be spaced out along the line. Dry carbon dioxide is not corrosive, even if containing impurities such as sulphur dioxide or hydrogen sulphide, and ordinary steel pipelines can be used. Carbon dioxide pipelines have lower leaks than natural gas pipelines, and have caused no recorded injuries or fatalities. If required, then a chemical odorant could be added to the carbon dioxide so that leaks can be detected by smell (as is done for natural gas). Carbon dioxide is not explosive or flammable; indeed, it is used in fire extinguishers.

Currently about 50 Mt of CO_2 travel through 2,500 kilometres of pipelines each year,[3] nearly all in the USA, equivalent to about fifteen typical coal plants. In the International Energy Agency's view, the required carbon dioxide pipeline capacity might increase to 500 Mt per year in 2030, a large increase but one that seems eminently feasible. Probably emissions from several sources would be gathered into a single trunk pipeline. The largest existing carbon dioxide pipeline, running 800 kilometres from Cortez (Colorado) to West Texas and crossing the Rocky Mountains en route, has enough capacity to service three large (1,000 MW) coal power plants.[4] Distances of 300 kilometres between source and sink appear to be reasonable for commercial storage projects.[5]

There are no known offshore carbon dioxide pipelines today, but natural gas pipelines are common, in waters up to 2,200 metres deep, so offshore transport presents no special problems. Oil and natural gas pipelines have also been installed in virtually all challenging environments worldwide, including deserts, jungles, the Arctic and heavily populated areas. Large carbon dioxide trunk-lines would be of similar diameter to major natural pipelines.[67]

Carbon dioxide can also be transported by ship, in refrigerated, slightly pressurised tanks similar to liquefied petroleum gas (LPG) carriers, or in compressed natural gas (CNG) tankers. Typically, the carbon dioxide would be mildly refrigerated and under some pressure. This is likely to be a secondary transport method, but might be useful for collecting captured carbon dioxide from isolated sources (e.g. islands), or for delivery to remote offshore locations. It is only likely to be cheaper than offshore pipelines over long distances (>1000 km) and for small volumes,[8] and it will be rare for a coastal carbon dioxide source to be more than 1,000 kilometres from a suitable storage site. If dual-purpose vessels could be developed, capable of carrying both hydrocarbons and carbon dioxide, they could act in shuttle mode, to bring the carbon dioxide back from industrial sites for geological sequestration in or near the original fields where it was extracted. This could reduce storage costs and make shipping almost free.

As a comparison to the pipeline capacity mentioned above, a typical (500 MW) coal power station would require about ninety visits from a carbon dioxide carrier per year. Carbon dioxide would have to be stored while awaiting loading, requiring tanks which would be small[9] compared to oil storage tanks. Larger ships, similar to liquefied natural gas (LNG) tankers, could also be built, with about five times greater capacity and correspondingly fewer visits. A minor amount

(1–2% per 1,000 km) of the carbon dioxide would be lost due to boil-off on long journeys. As with pipelines, carbon dioxide ships would not be explosive. In the event of a rupture of the hull, carbon dioxide might escape and present a risk of asphyxiation in the immediate vicinity of the ship. The carbon dioxide would, however, then vaporise into air or dissolve in sea water, with no significant long-term environmental consequences (other than release of a relatively minor quantity of greenhouse gases).

If carbon capture were implemented on a distributed generation network (perhaps with fuel cells), an extensive network of small carbon dioxide pipes would be required, similar to the gas grid or the sewage system. The cost of such a collection network would be likely to add greatly to the cost of CCS for small sources. In principle, carbon dioxide could be moved by lorries or trains, but this appears expensive, logistically difficult, and likely to offset most of the carbon savings.

Similarly, a major 'air capture' plant would require a lot of small-diameter in-field flowlines. Offsetting the emissions of our prototypical coal plant might require some 60 kilometres of small gathering lines converging on a central point, or a long strip of systems following a main pipeline.

Practical Uses

The first option for carbon dioxide is to find a practical use for it. Carbon dioxide is employed in a number of industries: food and beverages (e.g. decaffeinated coffee, and carbonated drinks such as Coca-Cola), horticulture (enhancing plant growth in greenhouses), welding (providing an inert atmosphere), safety devices (such as fire extinguishers), and as a coolant in nuclear power plants and domestic heat pumps. As discussed, carbon dioxide can also be converted to synthetic fuels such as methanol.

Being a rather stable substance, carbon dioxide is not an ideal feedstock. Its uses, therefore, are relatively small, only about 100–200 Mt per year.[10] About two-thirds of this is urea production for fertilisers, which I have already addressed in Chapter 2. Most of the remainder only represent temporary storage, since the carbon in food, soft drinks, fuels and so on is rapidly returned to the atmosphere. Consequently, they do not represent any significant contribution to fighting climate change. However, today some of these processes use natural carbon dioxide, from underground sources, and it would be better to replace this with industrial carbon dioxide.

More permanent storage might be achieved in chemicals (if these are used to form long-lived substances such as plastics), fertilisers (adding carbon to the soil) and ceramics. The Kędzierzyn IGCC plant in Poland will make urea (for fertilisers), chemicals, and methanol from captured carbon dioxide. Even if all polycarbonate and polyurethane plastics worldwide were made from carbon dioxide, and any additional energy required were supplied from zero-carbon sources, total consumption[11] would be less than one medium-sized (500 MW) coal power plant.

Ceramics are perhaps more promising, since they are long-lived and used in large quantities. 'Supramics', an outgrowth of Los Alamos research into leach-proof cements for storing nuclear waste,[12] are a type of ceramic made from cement, which is then 're-carbonated' using supercritical carbon dioxide. This essentially re-creates the limestone originally used to make the cement, and could be considered a type of mineralisation (see below). Substituting 10% of ceramics with CO_2-derived materials would give potential annual demand of about 10 million Mt per year,[13] still small compared to emissions.

However, if CO_2-ceramics could make inroads into the approximately 4 billion tonne per year cement market,[14] then this could yield significant sequestration. One possibility has been developed by a company known as Novacem, a spin-off of Imperial College, London. It uses widely-available magnesium silicates to make a kind of cement that absorbs up to 0.6 tonnes of CO_2 per tonne of cement as it hardens, thus being carbon-negative.[15] Another alternative is Shell's 'C-Fix', a concrete substitute made from the heavy residues of oil refining. This not only avoids burning this carbon-rich material as low-grade fuel, but also saves on emissions from cement manufacturing. C-Fix, which I have held in my hand, is a rock-hard, apparently inert material, especially good for use in salty environments, such as sea defences, and in heavy-duty roads.[16]

Any new cement, though, is likely to take time to assess for safety and architectural requirements. Overall, it seems unlikely that industrial uses will take up more than a small fraction of captured carbon dioxide. We therefore require a more voluminous and permanent form of storage.

Storage Sites

There are four main options for storing carbon dioxide in the long term. Of these, carbon sequestration in the biosphere (in trees, soils, etc.) is dealt with in Chapter 4. The three other possibilities are: underground, where there is a variety of sites; in the ocean; and as stable minerals. A typical site would have to store several million tonnes of carbon dioxide annually, probably more if emissions from a number of plants are being gathered together. The minimum size of a useful storage site is probably about 2 Mt of CO_2 (a small point source, 0.1 Mt per year, operating for twenty years), equivalent to a small oil- or gasfield.

It is worth briefly describing the characteristics of underground reservoirs. Rocks can be divided into three main groups: igneous, meta-

Figure 3.2. Schematic view of storage sites

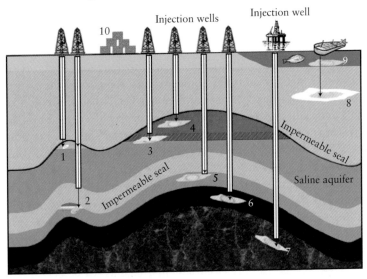

1. Use in enhanced gas recovery (pilot stage)
2. Storage in depleted oil- or gasfield (technically mature)
3. Use in enhanced oil recovery from residual oil zone (theoretical)
4. Use in enhanced oil recovery (commercially mature)
5. Storage in saline aquifer (commercially mature)
6. Use in enhanced coal-bed methane recovery (pilot stage)
7. Storage in basalt (small-scale pilot stage)
8. Use in methane hydrate recovery (speculative)
9. Ocean storage (small-scale pilot stage)
10. Conversion to solid minerals (research stage)

morphic and sedimentary. Igneous rocks, the most familiar of which are granite and basalt, were formed in volcanic processes from molten lava (above ground) or magma (below ground). Metamorphic rocks, such as marble and slate, were derived from an igneous or sedimentary predecessor by high pressures and temperatures. Igneous and metamorphic rocks generally have very low porosity—empty spaces in the rock that can be filled with fluids.

Sedimentary rocks, on the other hand, are formed at the Earth's surface, either 'clastic' rocks derived from the breakdown of pre-existing rocks, or by biological or chemical activity. They include sandstones, mudstones and shales, limestones, rock salt, coal and others. Evaporite minerals such as rock salt (halite), gypsum and anhydrite were formed by the evaporation of sea water or other salty water.

Many sedimentary rocks, sandstones and limestones in particular, tend to have high porosity and also permeability (a measure of how easily fluids can flow through a rock). A good hydrocarbon reservoir may have porosity in the range 15–30% (i.e. 15–30% of the rock is empty space and the remainder is mineral grains). Shales and evaporite minerals almost always have low permeability.

The pore spaces in sedimentary rocks are usually not empty but contain fluids. Typically this is water, but they can hold oil and gas,

Figure 3.3. Scanning electron microscope image of a Rotliegend sandstone from the Cleeton field, UK North Sea. (Photo: Professor Jon Gluyas)

collectively termed petroleum. Petroleum is formed from the action of heat and pressure on a 'source rock'—a sedimentary rock, usually a shale or limestone, which is rich in organic matter from the remains of prehistoric plants and animals.[17] Being less dense than water, petroleum will rise from the source rock until encountering an impermeable layer or 'seal', usually a shale, a low-permeability limestone, or one of the evaporite minerals (salt and anhydrite). Shales are by far the most common seals, but salt and anhydrite are more effective and less likely to be broken by faults. Some seals may trap oil but leak gas.

On reaching the seal, the petroleum then spreads out laterally. If the geological structure is suitable to trap buoyant fluids, such as a dome, then the petroleum will fill it and possibly form a commercially viable accumulation or 'field'. Petroleum may occupy about 70% of the pore space, the remaining 30% containing residual water. Otherwise, the oil and gas will continue moving until it enters a trap, or reaches the surface as a natural oil or gas seep (for instance, the famous 'eternal flames' of ancient Iraq and Persia). The field may contain oil with gas dissolved in it, oil with a 'gas cap' overlying it, or only gas. Commercial fields are found over a great range of depths below surface, from very shallow down to 6,000 metres or more, but 1–3,000 metres is typical. A large field, such as the major ones of the Middle East, may have an area of 1,000 km² (similar to New York City), and a hydrocarbon column of several hundred metres, comparable to the world's tallest buildings.

The major structure of oil- and gasfields can be determined by seismic waves—generating sound-waves (with explosives, air-guns or vibrating machines) at surface and measuring the reflections that return to surface, similar to the sonar systems used by submarines and to some medical imaging technologies. Seismic data can also give information about underground fluids such as oil and, particularly, gas. However, only drilling a well can conclusively demonstrate the presence or absence of petroleum. Success rates for exploration drilling have improved in recent years but still average only about 30%.

Petroleum wells are drilled to find oil and gas, and then to allow them to flow to the surface. Specialist electrical logging tools are lowered into the well to measure various properties of the underground rocks and fluids. Wells are 'cased' with a metal lining to prevent them collapsing and to stop unwanted fluids flowing into them. If a commercial field is discovered, usually further wells will be drilled to extract the petroleum. Several hundreds or even thousands of wells may be required to develop a large field, depending on its reservoir characteristics. Surface facilities are installed to remove impurities, such as water, sand and the contaminant gases carbon dioxide, hydrogen sulphide and nitrogen. Natural gas and oil are separated for transport. The petroleum is now in a suitable state to be moved to markets by pipeline or ship, refined into useful products such as petrol (gasoline), diesel, kerosene and fuel oil, and eventually sold to customers.

Of course, only a tiny fraction of sub-surface pore space contains oil and gas. Most is filled with water. Near the surface, this may be fresh, but in deeper rocks, it usually becomes increasingly salty, often much more saline than sea water. These 'saline aquifers' underlie oil- and gasfields.

Enhanced Oil Recovery

One of the most promising possibilities for early carbon storage is in enhanced oil recovery (EOR). Oil found in underground 'reservoirs' is extracted by drilling wells into this reservoir. It then flows out, either by its own natural pressure, or with pumping ('primary recovery'). In order to extract more oil, water or natural gas may be injected to maintain pressure and sweep oil to the producing wells, a process known as 'secondary recovery'. Typically, about 30–40% of the original oil in place underground is recovered by primary and secondary methods, although 60% or more may be extractable under favourable circumstances and with good technology, as in the North Sea. Therefore, large amounts of oil are left unrecovered, a concern at a time of high oil prices and limited access for the international oil industry to the largest fields in the Middle East and Russia.

One solution to this problem is 'enhanced oil recovery' or 'tertiary recovery'. This covers a wide range of techniques, including injecting steam, surfactants (effectively soap) and microbes. One of the most promising and mature EOR methods is carbon dioxide injection.

When carbon dioxide is injected underground at a depth greater than about 800 metres and a temperature exceeding 25–40°C, it forms a supercritical fluid, with properties of both a liquid and a gas. A volume of 420 m^3 of carbon dioxide at the surface occupies only 1 m^3 in the supercritical state underground. Supercritical carbon dioxide can be 'miscible' with oil, mixing with it to form a single phase of lower viscosity (stickiness), and causing it to swell, thus pushing it towards the production wells. Lighter (less dense) oils are more suitable for this technique, but miscibility has been achieved even for some heavy oils.[18] Even under conditions where carbon dioxide is not miscible with the oil, it can still improve recovery. Typically, water is injected concurrently, or in an alternating fashion in different parts of the field (Figure 3.4).

Figure 3.4. Schematic cross-section of carbon dioxide injection for EOR (based on Weyburn project)[19]

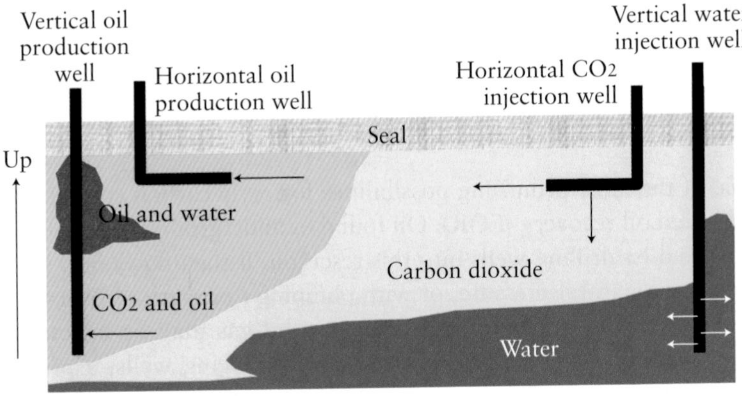

The typical incremental recovery from CO_2-EOR is of the order of 5–10% of the original oil-in-place, and careful management can increase this to 20%.[20] With some technological advances, additional oil in the USA alone could be some 265 billion barrels, about as much as Saudi Arabia's proved reserves. The value created can therefore help to pay for carbon capture. In the most favourable circumstances, CCS can have a negative cost—and therefore would be implemented even if there is no action against climate change.

CO_2-EOR was first trialled in the 1960s, and has been used extensively in the USA from the 1980s. The process has also been employed in Turkey since the early 1980s, injecting about 1 Mt per year of natural carbon dioxide, and Brazil has used it from 1981 onwards.[21] A very successful EOR pilot in Trinidad takes carbon dioxide from an ammonia plant. CO_2-EOR has been used in both carbonate rocks (limestones and dolomites in Canada's Weyburn and Texas's Permian Basin) and clastics (sandstones). A total of 106 projects are operating in the USA, mostly in Wyoming and the 'Permian Basin' of West Texas and New Mexico, amounting to some 5% of American oil output, and 18 are active outside the USA. More projects are being launched, encouraged by several years of high oil prices.[22]

Most of this CO_2 comes from natural sources, either from contaminated gas in Wyoming, or from underground reservoirs charged with carbon dioxide, typically from volcanism. There is no reason, though, why all these projects could not be converted to use anthropogenic carbon dioxide. Some is supplied from industrial sources, notably ammonia plants, and Saskatchewan's well-known Weyburn storage project receives its carbon dioxide from the Beulah synfuels plant in North Dakota. Of the four largest projects, one stores emissions equivalent to a large (1,000 MW) coal plant, while the other three are comparable to a 500 MW coal power station. This demonstrates the viability of CO_2-EOR on a large scale, big enough to make an impact on the climate challenge.

With the exception of Weyburn, these EOR projects are not designed for carbon storage. About 40% of the injected carbon dioxide eventually appears in the producing wells; the other 60% remains in the reservoir. The back-produced carbon dioxide is separated and re-injected. The amount of carbon dioxide remaining underground could be increased if operations are deliberately managed to maximise storage, while gels can be used to delay and reduce carbon dioxide recycling. The more oil produced, the more carbon dioxide can be accommodated in the field. In mature oilfields, large amounts of water are generally extracted with the oil (sometimes as much as 99 barrels of water for 1 barrel of oil), so space availability can also be increased by re-injecting produced water at a level other than the storage formation.

The amount of carbon dioxide required varies very much between fields, but typically some 3–6 tonnes injected liberate 10 barrels (about 1.5 m³) of oil. The EOR operators have to buy the carbon dioxide, usually for about $30 per tonne, so they recycle produced carbon dioxide rather than emitting it to the atmosphere. In the future, we can expect that oilfield operators will be able to demand payment for their service in disposing of carbon dioxide, rather than having to buy it.

When oil production drops too low to make continuing production economic, then the field will be decommissioned: wells will be plugged with cement, and surface facilities removed. If economic circumstances are right—when CO_2 emissions have a cost—then it will be possible to reconfigure a CO_2-EOR scheme as a pure carbon storage operation. The producing wells will be capped, and the separation kit can be removed for use elsewhere, but injection will continue, either into the oil reservoir, or into another rock formation (as I discuss below). Compressors and monitoring equipment will already be in place.

Major CO_2-EOR opportunities occur whenever oilfields are in proximity to large emission sources. This coincidence is found in several major oil provinces: the USA's Mid-West, Gulf of Mexico coast and California; western Canada; Mexico; the North Sea; the Persian Gulf, especially the United Arab Emirates; the Volga-Urals industrial region in Russia; Mumbai in India; Heilongjiang Province in north-east China; Java, southern Sumatra and peninsular Malaysia. Additional possibilities include Qatar and west Venezuela. Other good EOR locations like the rest of the Middle East, North and West Africa, West Siberia, Kazakhstan and South America are not near major carbon dioxide sources, though there are niche opportunities in each case.

There has been particular interest in EOR in some of the mature fields in the North Sea, notably with the Shell/Statoil plan to use carbon dioxide from a gas-fired power plant in mid-Norway in the Draugen and Heidrun fields (Haltenbanken), and BP's proposal for a hydrogen power plant at Peterhead in Scotland, whose carbon dioxide would be used in the Miller Field. Miller was particularly suitable, because its gas naturally has a high carbon dioxide content, and so the field's wells and facilities are made of stainless steel to resist corrosion. Unfortunately, the North Sea is a high-cost operating area, due to the rough offshore environment, and a lack of timely government support (in the case of Peterhead) caused the project to be cancelled. Lengthy offshore shutdowns for refits are also costly and problematic. Many of the big North Sea fields are approaching the end of their lives and will soon be decommissioned. If this happens, the opportunity for EOR using existing infrastructure will be lost, and these fields may well not be economic to reactivate later.

Another key project is found in the UAE. Here, Masdar, the Abu Dhabi-owned future energy company, hopes to work with the national oil company, ADNOC, to capture carbon dioxide from gas power plants and other sources. Abu Dhabi is currently short of natural gas due to rapid economic growth, and a large proportion of its production is re-injected into oilfields to maintain the pressure. Using carbon dioxide instead would substitute a waste product for this valuable gas.

The oilfields are large, onshore, low-cost and fairly near the emissions sources, so the situation is close to ideal. However, there are significant commercial issues with coordinating the three major players (Masdar, ADNOC and the power sector). For instance, ADNOC's new Shah 'sour gas' development will produce some 2 Mt of CO_2 per year, ideal for capture, yet this is as yet not part of the scheme. Relatively expensive capture from gas-fired power plants is also likely to be required, since there is no coal-fired power in Abu Dhabi. Nevertheless, success here could well be a template for the rest of the Middle East—freeing up valuable gas in Oman and Iran, rejuvenating mature oilfields in Qatar and Dubai. Qatar, with its two massive gas-to-liquids plants, has an ideal source of capture-ready carbon dioxide.

Some other EOR methods may emerge with growing availability of carbon dioxide. One longer-term possibility is to scrub out 'residual' oil, that is, remnants of oil that has migrated through rock on its way to a reservoir. This is often found beneath the oil-water contact,[23] with potential for 20 billion barrels recoverable in the USA alone (comparable to the country's current proved reserves). Another option is *in situ* recovery of oil sands. The 'THAI' method is being trialled in Alberta, and ignites some of the heavy oil below ground to heat it and drive it to producing wells. The associated carbon dioxide is mostly trapped underground. This is therefore a much lower carbon option than standard techniques using steam generated on surface, albeit very different from other CCS methods.

CO_2-EOR has so far not really broken out of its heartland in West Texas and Wyoming. This is partly due to the limited availability of cheap carbon dioxide elsewhere, a situation likely to change as carbon caps come into effect. The carbon dioxide has to be of high purity and contain little nitrogen. The projects (like most EOR schemes) are relatively expensive and were mostly uneconomic in the period of low oil prices, 1986–2000. Careful management of the carbon dioxide flood is essential. Where the reservoir rock contains strong contrasts in permeability, carbon dioxide tends to follow the path of least resistance, appearing in producing wells and leaving areas of bypassed oil. A lack of appreciation of reservoir architecture was, with a drop in oil prices, the main reason for failures in a number of early schemes.

A practical issue is that mature oil provinces, the onshore USA in particular, tend to be the domain of small oil companies who lack the capital and skills to implement such EOR projects. Even in large oil companies, EOR remains a specialist area, where many experts are coming close to retirement. Greater experience, and the establishment of a carbon dioxide pipeline network, will help in resolving some of these problems.

There is some debate whether CO_2-EOR is a solution or a problem, since it might be argued that it just liberates oil that would have remained unproduced, and so makes the greenhouse problem worse. Accounting for the permanently stored carbon dioxide versus the carbon dioxide released by burning the extracted oil, the carbon intensity of CO_2-EOR oil is between -0.17 and 0.13 tonnes CO_2 per barrel, compared to 0.43 tonnes per barrel[24] for conventional oil and as much as 0.55 tonnes per barrel for some high-carbon unconventional oil.

In any case, the idea of penalising CO_2-EOR projects for the oil they produce is a naïve view of the situation. It relies on the misconception that we will, in a rather short timescale, extract and burn all technically recoverable oil in the ground. Although possible, this does not sound very practical or likely, certainly not when compared to likely demand, and limits on carbon emissions, during the twenty-first century. Producing large quantities of CO_2-EOR oil will, no doubt, lower prices and so increase oil demand to a degree. But mostly, it will displace high-cost oil from other sources. If that oil is from mature fields with 99% water-cut,[25] or from Canadian oil sands, coal-to-liquids or oil shale retorting, then avoiding its use will greatly reduce the carbon footprint.

For a CO_2-EGR (enhanced gas recovery) or ECBM (enhanced coal-bed methane) project, the additional gas may well displace coal, hence giving a double climate benefit, and, by lowering gas prices generally, will improve the competitiveness of gas versus coal (and, of course, versus nuclear and renewables, which may lessen the emissions mitigation).

Anyway, if and when an effective carbon trading scheme or carbon tax is implemented, then the extra produced oil and gas will already have its climate impact included in its price.

Enhanced Gas Recovery

In the same way as for EOR, carbon dioxide injection can be used to extract more gas from mature gasfields—enhanced gas recovery (EGR). EGR, though, is as yet only an experimental technique. This is probably because gasfields are generally at less mature stages of exploitation than oilfields, and because gas recovery factors (the percentage of the gas originally underground that can be extracted) are much higher than for oil, 60–70% being typical. EGR has been used successfully on a small scale at the Budafa Szinfeletti field in Hungary, and a trial is continuing at the K12-B offshore gasfield in the Netherlands, at 0.48 Mt per year[26] (about half Sleipner).

Carbon dioxide has a higher density, viscosity and solubility in water than natural gas (which is mostly methane), so the injected carbon dioxide does not immediately show up in the producing wells. Instead, it tends to underlie the gas accumulation.[27] About 3–5 tonnes of natural gas are recovered per 100 tonnes of CO_2, so the carbon capture effect is high.[28] In any case, the extra gas produced will typically displace coal-fired power.

A somewhat different technique is to use carbon dioxide in underground gas storage (UGS) facilities. UGS is used to balance gas demand, typically between a high-demand period (winter in North America and Europe) when gas is extracted, and a low-demand period (summer) when surplus gas from fields is re-injected for storage. When such a site is commissioned, it needs a certain amount of 'cushion gas' to provide pressure. This cushion gas remains in the reservoir. Typically, natural gas has been used for this purpose. However, recent high gas prices have made this very expensive. It may be preferable to use carbon dioxide. This will slowly contaminate the stored gas, so some amount of separation will be needed on the surface.

EGR will be particularly attractive to Europe, given high gas prices and worries about over-reliance on Russian supplies. The supergiant Groningen field in the northern Netherlands, discovered in 1959, is a leading candidate.[29] Increasing the recovery factor here from 75% to 95% would be enough to cover all Dutch gas demand for more than fifteen years.[30]

Enhanced CBM

In recent years, there has been great success in extracting natural gas from coals, the technology known as 'coal-bed methane' (CBM). This started in the USA, spread to Australia, and is now being seriously examined in Canada, Europe, Indonesia, India, China and elsewhere. The gas is not trapped in pore spaces in the rock, as with a conventional reservoir, but is adsorbed on to the surface of the coal particles. This gives a very high storage capacity. To produce the gas, wells are drilled to remove the water from fractures in the coal. Once the pressure has fallen sufficiently, the gas begins to desorb from the coal and can be extracted. In the late stages, natural carbon dioxide also begins to be released, and this could be separated out and re-injected in a similar way to processing contaminated natural gas.

It turns out, fortunately, that carbon dioxide has a higher affinity for the coal particles than methane does. Carbon dioxide can therefore be injected into coal beds, where it becomes adsorbed and displaces the methane: this represents enhanced coal-bed methane recovery (ECBM). Between one and ten molecules of CO_2 are trapped for every methane molecule that is released, and about 20 tonnes of coal can hold 1 tonne of carbon dioxide. Due to this trapping, there is very little contamination of the natural gas with carbon dioxide.[31]

This is still a relatively experimental technique, although it has been trialled in the USA, Canada, China, Japan and Poland. For every 100 tonnes of carbon dioxide injected, about 6–18 tonnes of methane are released. This is therefore a more effective process than EGR, but the carbon dioxide cannot be too expensive; depending on future carbon policies, the ECBM operator may even be paid for disposal. Recovery factors can be increased from 77% to 95%, in one US example.[32] The coal should be fairly deep, more than 500 metres, but very deep coals (>1,000 metres) generally have lost their permeability. American coals in the San Juan Basin of New Mexico are particularly suitable, having high permeability, and commercial ECBM started here as far back as 1995. For European coals, the swelling that occurs when carbon dioxide is adsorbed may make the formation impermeable to further injection. That said, continuing advances in hydraulic fracturing may overcome permeability problems.

ECBM can only be applied to coal-beds that are not going to be mined (too thin, too deep or otherwise unsuitable). Some existing mines go as deep as 1,000 metres, suggesting that the depth window for ECBM may be rather limited. If mining is planned later, then it would be necessary to drain off the carbon dioxide to another storage site. With the current boom in unconventional gas in North America (mostly shale gas, but also non-enhanced coal-bed methane), gas prices there may be relatively depressed and there may not be much incentive to apply ECBM.

These caveats aside, ECBM appears promising. It could be particularly useful for areas such as Eastern Europe, eastern Australia, South

Africa and parts of China and India, which contain large polluting power plants, lack sizable oil- and gasfields, but have abundant coal-beds for storage. Ukraine and Russia, both big carbon dioxide emitters, Kazakhstan, Canada, the USA and north-west Europe also have sizable coal reserves. Liberating more natural gas could have a double climate benefit, by displacing coal used in power generation.

Depleted Oil- and Gasfields

Even when there is no prospect of enhancing recovery, old oil- and gasfields still make good storage sites. They are well understood, are proven to have a geological structure that can trap the carbon dioxide, and may have some remaining infrastructure that can be re-used, such as wells, pipelines, platforms, roads, accommodation and so on.

To avoid contamination of the remaining hydrocarbons, injection of carbon dioxide will generally start at or after the tail end of production (unless it occurs as the continuation of an EOR or EGR scheme). In a maturing basin, storage could start with abandoned fields and gradually move on to others as they are depleted. It would often make sense to begin with an enhanced recovery scheme, and to continue it even after hydrocarbon extraction is over, since the pipeline, injection wells and monitoring equipment will all be in place.

Some oil- and gasfields are produced by simple depletion—extracting hydrocarbons until the pressure has dropped too low to continue. In these cases, essentially all of the pore space that was originally filled with petroleum can be used for carbon dioxide storage. Some carbon dioxide will be stored by dissolving in unrecovered oil. The reservoir rock will have compacted as pressure was reduced, but carbon dioxide injection may partially reverse this. In many other fields, though, the petroleum is displaced by water, either flowing in naturally from a connected aquifer, or (for many oilfields) injected deliberately to maintain pressure. In these cases, a substantial amount of the original hydrocarbon-filled void space is not available for storage—that is, if the decision is taken not to increase the reservoir pressure beyond its original level before hydrocarbon production commenced. It is possible to inject carbon dioxide at pressures greater than the original, but this runs the risk of breaching the seal. Careful geo-mechanical modelling may partly mitigate this risk, but it is still preferable to avoid over-pressuring the reservoir. The loss of available porosity due to water invasion is probably around 30% for gasfields and 50% for oilfields.

The Lacq project in south-western France started carbon dioxide injection in June 2009 into the depleted Rousse gasfield,[33] while the old Altmark gasfield is the target for the Schwarze Pumpe project. The combination of enhanced oil and/or gas recovery and pure storage has the advantage that EOR/EGR schemes usually require a varying flow of carbon dioxide (typically starting high and then dropping off), while a power station provides a steady amount. It is better, therefore, if CCS schemes combine a variety of sites at different stages of the life cycle, some still undergoing hydrocarbon extraction, some purely designed for long-term disposal. Because of this, there is not a clear-cut division between EOR/EGR and pure storage in depleted fields. The overall capacity in depleted fields will, of course, increase as time goes by, and more and more fields cease production.

Saline Aquifers

For large-scale application of CCS, the storage space in coal-beds and oil and gas-fields is likely to be insufficient. Some industrial areas are also not near to suitable sites. The ideal solution is to inject the carbon dioxide into saline aquifers (sometimes referred to as deep saline formations, DSF). These water-bearing formations of porous, permeable rock are ubiquitous in sedimentary basins around the world.

Aquifer storage, and indeed other geologic storage, relies on suitably porous and permeable rocks, with sealing capacity to prevent the carbon dioxide leaking out. Old continental 'shield' areas, composed of igneous and metamorphic rocks with very low permeability, will generally not have good storage sites. Such areas include eastern Canada, Scandinavia, western Saudi Arabia, parts of Africa and Australia and much of Brazil. Active mountain-belts, such as the Alps, Andes and Himalayas, are also likely to be unsuitable because of extensive faulting and fracturing, the removal of seals by erosion, and in some cases volcanism. However, other tectonically active areas such as California, and the Zagros Mountains of Iran and Iraq, where earthquakes are common, are nevertheless world-class petroleum provinces. Sites, though, will have to be away from active faults which may slip during storage and create leakage paths.

In general, established hydrocarbon basins will be good places for carbon storage in aquifers, because they:

• are well understood;

- are generally tectonically quiescent;
- have proven capacity to trap fluids;
- may have opportunities for EOR, EGR and storage in depleted fields;
- and possess existing infrastructure that can be re-used.

One irony, that may enrage some environmentalists, is that new coal-, oil- and gasfields may be discovered while developing storage projects.

Saline aquifers contain high levels of salt and other minerals, often more than sea water, making their water undrinkable and useless for most purposes. They lie at considerable depths, well below potable aquifers, and isolated from them by impermeable rocks such as clays, shales, tightly cemented limestones and evaporite minerals.

Oil- and gasfields are often underlain (and overlain) by such aquifers, but they also occur in areas without hydrocarbons. There is no direct economic benefit from injecting carbon dioxide into aquifers, but there is also no dependence on petroleum infrastructure or the timing and needs of oil or gas production. Aquifer injection can, of course, be combined with EOR or EGR, with simultaneous injection at several levels, or, when oil extraction stops, converting an enhanced recovery scheme to aquifer storage. If carbon dioxide eventually finds its way into (possibly undiscovered) oil- and gasfields, then, as with ECBM, future production activities will have to drain it off and re-inject it.

At pressures above 73 atmospheres,[34] reached in aquifers deeper than about 800 metres, the carbon dioxide will go into its supercritical state, with a density 50–80% that of water. Shallower aquifers could be used for storage, but gaseous carbon dioxide occupies a much larger volume, and the risk of leakage would be greater. Shallower aquifers are also more likely to contain drinking-quality water. On the other hand, deeper aquifers will contain progressively more carbon dioxide and be more secure, but injection will be more expensive (as I discuss below).

There are several mechanisms for permanent trapping of carbon dioxide (which also apply to storage in oil- and gasfields). Initially, carbon dioxide will simply displace the water previously in the aquifer. The increase in pressure will compress the water slightly, creating space for the carbon dioxide. Some carbon dioxide will dissolve in the water, while the rest will stay as a supercritical fluid. In this state, it is less dense than water and oil (but denser than natural gas), so will tend to rise upwards until impeded by a barrier. Aquifers with multiple overlying barriers are preferable. It is a common experience in petroleum exploration to find two, three or more 'petroleum systems' stacked vertically and totally isolated from each other by sealing rocks. The same experience will probably hold for many aquifers. It is certainly true that deep aquifers are often cut off by 'aquicludes' or 'aquitards'[35] from communication with the shallower groundwater.

Injection sites will often be chosen at 'closed' structures, like domes or inverted saucers, in which carbon dioxide is trapped beneath sealing formations (the 'cap rock'). Such traps are well-known, since they are also the prime target of hydrocarbon exploration. With a historical success rate of hydrocarbon exploration of only 10–30%, there are very many drilled valid but water-bearing structures.[36] Traps encompass a wide variety of geometries, including those formed by folds, faults, lateral changes in rock type ('stratigraphic' traps) and various combinations.

If the carbon dioxide overfills its trap, it will slowly spread out laterally. Dissolved carbon dioxide will naturally sink out of the trap anyway. Over 1,000 years, a carbon dioxide plume will typically spread 5–12 kilometres from the injection well.[37] This is a relatively small distance: comparable to a medium-sized oil field, and well within the range of monitoring methods. As it spreads, some carbon dioxide will remain behind as residual bubbles in pore spaces, some will be stuck on the surface of clay particles, and some will be caught in local traps. Compare this to the small droplets left on a window in the rain, after the large drop has slid to the bottom. The movement of carbon dioxide will be affected by heterogeneities in the host rock, tending to follow the higher-permeability zones. Modelling and monitoring this uneven movement will be important; the techniques are well-known from the oil industry.

On a similar timescale, the carbon dioxide will gradually dissolve in aquifer waters. Within a few decades, as much as a third will dissolve. Injecting at several points in the structure, optimising the injection pattern using well-known reservoir engineering techniques, and injecting brine (salty water)[38] on top of the carbon dioxide can accelerate this dissolution. This greatly enhances the storage security at modest additional energy cost. Since waste-water from oilfield operations is already re-injected underground in huge volumes, it should be possible in suitable sites to combine carbon dioxide and brine injection at minimal extra cost. Storage space could also be increased by removing water and desalinating it for use, as is done (for non-CCS reasons) in desert areas of Jordan, although this tends to be an expensive process.

Aquifer water saturated with carbon dioxide is slightly heavier than the original water, so the CO_2-rich water will tend over time to sink out of the trap, being replaced with unsaturated water. Over several thousand years, this mechanism should dissolve all the injected carbon dioxide. The tendency to sink also enhances storage security by moving carbon dioxide further from the surface.

Carbon dioxide's solubility in water increases with pressure but decreases with temperature. Conveniently, this means that, for most practical situations, its solubility in underground waters does not vary much with depth,[39] other than reducing due to the generally higher salinity of deeper waters. We can therefore use aquifers down to great depths for storage, limited only by the expense of drilling deep wells, the higher pressures needed for injection, and the lower permeability and porosity of deeply buried rocks. Natural gas wells have been drilled beyond 9,000 metres depth, and production at 5,000 metres is fairly routine, so carbon dioxide injection at such depths is probably feasible, if expensive.

Aquifer waters themselves generally move very slowly, typically 0.5 metres-1 kilometre in 1,000 years,[40] so it will be millennia before the aquifers disperse the carbon dioxide far from the original injection point. Aquifers with slower water movement would be preferred. If the aquifer is not exposed within hundreds of kilometres of the injection site, then it will take up to several million years for fluids to reach the surface. Depending on pressure and density gradients within an aquifer, the waters may move down-dip (away from the surface), thus allowing hydrodynamic trapping of carbon dioxide.

In the very long term, the weakly-acidic solution of carbon dioxide combines with water to form bicarbonate ions (familiar from baking soda, sodium bicarbonate, when dissolved in water). In this state, it cannot leak as carbon dioxide. Ultimately, these ions will react with carbonates and silicates in the rock to yield solid mineral which immobilise carbon for millions of years. Even in a carbonate (limestone and dolomite) rock, such as that forming the Weyburn reservoir, there will typically be more than enough suitable minerals to react with typical injection volumes of carbon dioxide.[41] It has been shown that, in some natural carbon dioxide fields, the carbon dioxide has entered fractures in the cap rock, reacting there to form siderite (iron carbonate), so creating a 'self-sealing' trap.[42] Of course, such fortunate circumstances cannot always be relied upon, and in some situations carbon dioxide may dissolve carbonates and so permeate through a fault zone.[43] The formation of solid minerals is likely to be a slow process, and studies of known fields show it may be even slower than current theory predicts,[44] leaving dissolution in pore waters and the formation of ions as the main long-term trapping mechanisms.

Figure 3.5. Schematic picture of carbon dioxide trapping in an aquifer

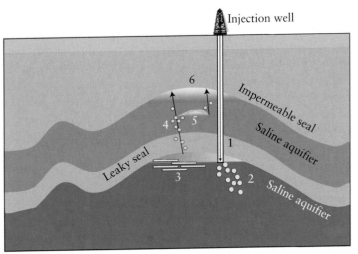

1. Injected CO_2 is trapped in a geological structure and begins to dissolve in pore waters.
2. Dissolved CO_2 is heavier than the native aquifer water and sinks out of the trap, gradually forming bicarbonate ions.
3. CO_2 reacts to form solid carbonate minerals.
4. Some CO_2 may leak from the primary trap (if it was not well chosen). Residual CO_2 is trapped as bubbles and sorbed on to clay minerals along the migration path.
5. CO_2 fills a small local trap; some remains, while the rest overflows.
6. Leaked CO_2 accumulates in a secondary trap.

Aquifers are the storage site for three of the world's four best-known CCS projects: Sleipner and Snøhvit in Norway and In Salah in Algeria (Figure 3.6). At Sleipner, the aquifer overlies the gas reservoir; at In Salah, it is the same formation but underlying the gas accumulation; while at Snøhvit, it is a deeper unit. Saline aquifers will also be used for storage at the Gorgon LNG project in Australia, a very large project planned to store some 5 million Mt per year (similar volumes to those from a typical coal power station). At Sleipner, original estimates suggested that only 2% of aquifer volume could be filled with carbon dioxide; but it now appears that this can reach 13–68%, greatly improving the prospects of having sufficient space.[45]

Figure 3.6. In Salah gas plant in Algeria (Photo: Statoil)

In all these cases, the carbon dioxide comes from the purification of the natural gas, which is re-injected at the same site. Some coal plant CCS projects, such as Callide A in Australia, will also use saline aquifers for disposal. Nagaoka, in Japan, has been the site of trials of aquifer injection, both onshore and offshore.

Other Underground Storage Options

There are numerous other, less tested possibilities for carbon dioxide storage: abandoned mines and salt caverns, oil and gas shales, gas-bearing aquifers, gas hydrates and basalts.

Abandoned mines are probably unsuitable. The capacity is small, particularly as they are at shallow depths where carbon dioxide would remain as a gas, and they would be very hard to seal adequately. **Salt**

caverns are used to store natural gas and various wastes, but are expensive to create, and their creation yields large quantities of brine that has to be disposed of somehow. However, they have the advantage that, because they are a cavity rather than pore spaces in a solid rock, they have a much smaller area than other geological storage options. They would be vulnerable to leakage as the salt slowly deforms and the cavern closes.[46]

Similarly to coal, **oil and gas shales** adsorb carbon dioxide on to the fine-grained organic matter they contain. Their capacity has not been quantified, but given recent breakthroughs in producing shale gas in the USA, it might be very significant. Low permeability makes injection difficult, but this problem has been overcome for gas production by fracturing the rock under high pressure. As with coals, 'enhanced shale gas recovery' may help pay for carbon storage. Processes have been proposed to extract oil from oil shales using supercritical carbon dioxide,[47] potentially unlocking this enormous resource, of which the US states of Colorado, Utah and Wyoming alone contain 2 trillion barrels (double current world conventional reserves).

Methane is often found dissolved in underground high-pressure **gas-bearing aquifers**. This resource is highly uncertain but potentially vast, perhaps 28,000 Tcf (trillion cubic feet) globally[48] (compared to total global gas reserves of 'only' some 6,300 Tcf). Aquifer gas has not yet been extracted commercially, partly due to the difficulties of processing huge volumes of salty water. Carbon dioxide is much more soluble in water than methane, so one possibility might be to inject carbon dioxide until the aquifer is saturated, simultaneously storing carbon dioxide and releasing methane.[49]

We have already mentioned **gas hydrates**: the curious, ice-like substance that forms at low temperatures and high pressures, in Arctic regions and under the seafloor. Gas molecules, typically methane, are trapped in a 'cage' of water molecules. Hydrates form a huge potential hydrocarbon resource, which could last for many centuries; and the use of such a gigantic quantity of carbon would certainly demand CCS. Hydrates carry also the threat of runaway climate change, since warming in oceans and northern latitudes can cause them to break down,[50] and, over a century, methane is twenty-five times more potent a greenhouse gas than carbon dioxide. Hydrate breakdown may have been implicated in some dramatic warming events in the geological past[51] and appears to be accelerating.[52]

Figure 3.7. Burning gas hydrate

Large resources were recently confirmed in the Gulf of Mexico, off-shore eastern India, and in the South China Sea.[53] There have been some experiments in extracting gas from hydrates in Alaska, by injecting hot water or solvents, or reducing the pressure. However, another promising route is to use carbon dioxide,[54] since carbon dioxide hydrates are more stable than methane hydrates, and 5 molecules of CO_2 can be trapped for every 1 methane molecule released. This is yet another route to swap carbon dioxide for natural gas. It could pre-empt the dissolution of hydrates[55]—even if the carbon dioxide is ultimately released, then this would be a much less serious problem than the methane. US oil company ConocoPhillips plans to start field trials in Alaska by 2011.[56]

Instead of injecting carbon dioxide into sedimentary rocks (sandstones, limestones, coal and so on), it could also be stored in volcanic rocks, notably **basalts** (Figure 3.8). Basalts are familiar as the hexagonal columns of the Giant's Causeway in Northern Ireland, and are globally widespread, notably in Iceland and many other volcanic islands, the Deccan Plateau of India, the Columbia River region of the north-western USA, the famous Palisades in New Jersey, the Paraná River area of Brazil, Ethiopia in East Africa, the Karoo of South Africa, and Russia's Siberian Traps.[57] These formations consist of enormous piles of lava, 1.8 km in Washington State and 3 km in the Deccan, inter-bedded with sedimentary rocks formed during temporary lulls in volcanic activity. The upper layers of oceanic crust are also made of basalt.

Figure 3.8. Basalt columns in Scotland (Photo: istockphoto.com)

Carbon dioxide can be injected into the thin sedimentary layers, which have good permeability. From here, it enters pores in the basalt formed when dissolved magmatic gases escaped, and cracks due to cooling, and reacts to form calcite (calcium carbonate) and other minerals. This happens on a rather short timescale, perhaps one to three years,[58] although the rate of mineralisation varies greatly between different basalts. It is anticipated that most of the carbon dioxide would be converted to solid minerals within a few hundred years. If carbon dioxide is injected into deep ocean basalts, the overlying sediment, mostly impermeable mudstones, and high pressures at the seabed would provide additional lines of defence in case some leaked.

Basalts have huge storage capacity for carbon dioxide, some one to five years of all global emissions in the Columbia River area alone, and ten years or more in the Deccan.[59] They provide a valuable additional option for regions that lack suitable sedimentary storage sites. However, there are two significant concerns about their use. Firstly, it is very hard to 'see' through basalt using seismic data,[60] which makes it difficult to identify suitable storage sites and monitor the movement of the carbon dioxide underground. Secondly, basalts are not prospective for hydrocarbons, and therefore we know much less about fluid flow in them than we do for sedimentary rocks. The key questions are how easily large volumes of carbon dioxide can be injected, and whether basalts can trap the gas for long periods. Fractures that provide permeability may also create leakage paths. Carbon dioxide storage trials currently underway in Washington State,[61] India and Iceland,[62] studies on- and off-shore in the north-eastern USA[63] and further research on mineral carbonation are all important in establishing the reliability of basalt storage.

Underground Coal Gasification

A final method is somewhat different from all those mentioned above. Underground Coal Gasification (UCG) is a process for converting coal to a syngas of carbon dioxide, hydrogen, carbon monoxide and minor methane and water, by injecting oxygen or air into the underground coal-beds. No mining is required. The carbon dioxide can be captured as extracted and re-injected, and the remaining gas used to run a combined-cycle gas turbine. UCG therefore achieves much higher efficiencies than pulverised coal plants, with lower pollution. CCS could be fitted to the gas power plant to reduce carbon dioxide emissions by a further ~85%, or the neat syngas could be fed to a pre-combustion capture plant, where its already high carbon dioxide content would make capture more efficient. Alternatively, the syngas can be converted to liquid fuels.

UCG can address coal resources that are too small, thinly-bedded, deep, sub-sea or otherwise inaccessible to mining. It avoids the surface disturbance of mining, the danger to miners, and the production of waste ash. The process is tricky to manage, and needs more experience, but recent advances in directional well drilling and hydraulic fracturing offer potential for significant improvements. UCG was widely used in the Soviet Union, with two sites (in Uzbekistan and Siberia) still operating, and is being trialled in the UK, Belgium, Spain,

South Africa, Canada, USA, Kazakhstan, India, China, Vietnam and Australia. Large-scale application could increase current world coal reserves[64] by about 60%,[65] making CCS essential to avoid further big greenhouse gas releases to the atmosphere.

Ocean Storage

Instead of storing carbon dioxide underground, it is also possible to store it in the oceans. Pipelines or ships can be used to deliver carbon dioxide to the deep oceans. Below about 3,000 metres, the carbon dioxide is dense enough to sink, dissolving in the water or forming pools of liquid carbon dioxide on the ocean floor. It will then slowly dissolve. About 40% of historical anthropogenic carbon dioxide has been taken up by the oceans, and most of that still resides in the surface waters. Over several centuries, these will reach equilibrium with the deep waters, and thus the entire ocean will be in balance with the atmosphere. This mixing could be accelerated by deep-ocean carbon dioxide injection, an idea with a long history, dating back to 1977.[66]

Some trials with small amounts of carbon dioxide have taken place, notably off California. Mixing the carbon dioxide with sea water prior to disposal, injecting at depths where it forms hydrates, or targeting depressions on the ocean floor would all limit the scope for leakage, which in any case is almost certain to be slow.[67] As discussed, leakage from carbon dioxide reservoirs beneath the seabed in deep-water sites would be delayed by pressure, hydrate formation and similar phenomena. That said, deep-water carbon dioxide geologic storage is likely to be very expensive, and shallow-water sites will be greatly preferred.

Ocean storage remains highly problematic. Modelling suggests that some of the carbon dioxide will return to the atmosphere as the oceans circulate, with 1% leakage after a century and 6% after two centuries. This would largely undo the carbon mitigation effect. As I discuss below, even leaky underground storage should do much better than this. Beyond several centuries, the ocean will be completely mixed, and there is therefore no reduction in atmospheric carbon dioxide levels in the long term. Ocean storage can mitigate the short-term peak in atmospheric concentrations but, as discussed in Chapter 1, this is less important for climate than the total carbon dioxide ultimately released.

There may be local mortality of marine life around the injection point, although dispersal as small droplets or from a moving ship could ameliorate this. Large-scale disposal would make the deep oceans increasingly acidic.[68] Given our limited understanding of marine biology, long-term consequences might be serious. Considerations like this will result in strong public opposition to ocean disposal, which has already arisen in response to the early experiments, for instance in Hawaii and Norway. Nor does it appear that ocean storage is a cheap option: estimated costs are at least as large as for underground disposal, even for sites near to shore.[69] Most sources of large carbon dioxide emissions are not near to suitable deep ocean disposal points,[70] one exception being Japan, which appears short of good geological storage options. A leaking geological storage site can be relatively easily cleaned up, while extracting dissolved carbon dioxide from the ocean, if its effects turn out to be unacceptable, sounds about as easy as taking the sugar out of sweet tea. Finally, ocean storage is a much less mature technique than geological storage, with which we have some three decades of experience.

An alternative approach would be to increase the oceans' ability to store carbon dioxide, increasing their alkalinity by adding carbonate minerals. On a timescale of about 6,000 years, the excess carbon dioxide could be mopped up by precipitating carbonate minerals to the seafloor, hence taking it permanently out of circulation. Unfortunately, easily soluble carbonate minerals, such as sodium carbonate, are not known in sufficient quantities to make much of a difference,[71] while the widely available calcium carbonate is rather insoluble and will not dissolve at all in surface waters. Alternatives include calcining limestone to calcium oxide (an energy- and CO_2-intensive process in itself), or mixing power station flue gases with sea water and crushed limestone to accelerate dissolution. It is yet to be demonstrated, though, that such a solution would not just re-precipitate calcium carbonate, hence releasing carbon dioxide, once added to the ocean. On a timescale of centuries, there is probably no net carbon dioxide removal from the atmosphere.[72]

Therefore, on this occasion, I agree with Greenpeace.[73] Direct ocean storage is probably not worth pursuing. The exception is that some basic research on increasing alkalinity may be worthwhile, particularly as it could also combat the problem of ocean acidification.

Mineralisation

On geological timescales, carbon dioxide is naturally removed from the atmosphere by reaction with silicate minerals, particularly calcium and magnesium silicates. This forms silica (the very common mineral quartz) and carbonate minerals, the main constituent of familiar rocks such as limestone and dolomite. Weathering is greatly accelerated, perhaps by forty times, through the activity of living creatures,[74] which break up the rocks with roots and burrowing, release carbon dioxide into soils via respiration, and create organic acids. The carbon dioxide is returned to the atmosphere when the ocean crust, carrying these carbonate rocks, descends 100 kilometres into the Earth's mantle at subduction zones (including Japan, New Zealand, the Andes and the area off Sumatra responsible for the devastating 2004 Asian tsunami). Carbonates decompose at high temperatures and release carbon dioxide from volcanoes. This forms the basis of the recycling of carbon through the atmosphere, biosphere and geosphere.

This weathering process can have an influence on climate. For instance, for long periods of its history, the Earth has not had large polar ice-caps. The climate of the Mesozoic era, when dinosaurs dominated, was much warmer than today and sea levels were hundreds of metres higher. The shift into the 'ice-house' climate of the last 30 million years, with its recurrent ice ages, may be at least partly due to the uplift of the Himalayas.[75] By exposing large areas of silicate rocks, this removed a substantial quantity of carbon dioxide from the air.

This, of course, leads to the idea that accelerating these natural weathering rates could be a way of reducing atmospheric carbon dioxide. If we are worried about having sufficient underground storage capacity (which is rather hard to assess), a straightforward mapping effort demonstrates that there are sufficient exposures of suitable minerals to react with many centuries of emissions. Locking up this carbon in rocks also has the clear advantages of being demonstrably safe and leakage-proof. The products are stable and non-poisonous, and can be used for bricks, road aggregate or, in a warming world, sea defences. Carbonation of toxic mine tailings can be used to clean up heavy metals and asbestos.

The reactions of carbon dioxide with silicates release energy and occur naturally, albeit slowly. The most favourable minerals, such as serpentine and olivine, are found in 'basic' and 'ultrabasic' rocks. Basic rocks include basalts, found in many locations as noted. Ultrabasic rocks are exposed in specific spots of the Earth's crust, such as Papua New Guinea, Oman,[76] Iran, Turkey, Puerto Rico, Cornwall (UK), Cyprus, the Ural Mountains in Russia and Kazakhstan, and the eastern

Adriatic, where the sea floor has been pushed up over the continents, and are also abundant in California, Washington State and the Appalachians.[77] The rocks in Oman alone could absorb all the carbon dioxide in the atmosphere some thirty times over.[78]

Ultrabasic rocks form thick ore bodies exposed at the surface, hence there is little overburden or unusable rock (very different from the situation in coal mining). Useful minerals, such as magnetite (iron ore), chromite, nickel, manganese and perhaps precious metals can also be extracted as part of the pulverising process. For instance, these rocks host a giant deposit of platinum and similar elements south of the Ural Mountains,[79] while the historic copper ores of Cyprus are also of this type. Consider, for instance, the Kempirsai ultrabasic rocks of Kazakhstan,[80] where the value of the by-product metals should be more than enough to cover the cost of processing the rocks for carbon sequestration.[81]

The main problem, though, is that these reactions happen very slowly, and the big challenge of mineralisation is somehow to accelerate the process. Simply spreading crushed olivine on fields and forests,[82] or dropping it in the ocean, requires centuries for a reaction.[83] It seems, therefore, that mineralisation would have to take place in some kind of industrial facility, where the rock could be pulverised. However, this process requires high pressure, likely to be expensive, and it is difficult to make productive use of the low-grade heat released.

Massive quantities of rock would have to be processed: to absorb a tonne of carbon dioxide requires at least a tonne and a half of rock. A typical 500 MW power station, burning some 3,000 tonnes of coal per day, would require about 30,000 tonnes of ultrabasic rocks to neutralise its emissions.[84] This immediately suggests that it would be necessary to pipe the carbon dioxide to the silicate mine, rather than transporting rock to the power plant. Ultrabasic rocks are present on all continents, but are very unevenly dispersed; this means that many or most emissions sources will be far from suitable silicates. It also implies a gigantic quarrying effort, with associated local environmental impact: clearing of vegetation, landscape degradation, dust and noise, and perhaps leaching of contaminants from tailings. Since the volume of mineral products is larger than that of the original silicates, they could not simply be used to fill in the quarry; an additional storage site would be needed, unless there is sufficient demand for them as construction materials.

One way to speed the weathering of silicates would be to use a different acid from the weak carbonic acid (carbon dioxide dissolved in water), such as hydrochloric acid, which reacts much more quickly.[85] This hydrochloric acid could be derived by the electrolysis of sea water or brine made from natural salt (halite) deposits. This electrolysis also produces sodium hydroxide, which reacts with carbon dioxide, removing it from the atmosphere. This could be done either by simply adding the sodium hydroxide to sea water, or by using it in air capture devices (as discussed in Chapter 2). Generating sodium hydroxide in this way eliminates the most energy-intensive part of the air capture process. If the sodium hydroxide is added to sea water, it will capture carbon dioxide in the short term, reversing ocean acidification, and precipitate carbonate minerals in the long term, hence permanently taking carbon dioxide out of the atmosphere. However, the environmental impact of sodium hydroxide on the ocean would have to be managed, possibly by dilution or by having numerous small outlet points.

An ideal location for this process might be volcanic islands, with plentiful zero-carbon geothermal energy which has no local market. These islands abound in rocks, such as basalt, containing the right minerals to react with the hydrochloric acid. To tackle 15% of global emissions (i.e. 1 Socolow wedge of 3.7 billion tonnes of CO_2 per year), would require a system equivalent to about 100 large sewage plants, which seems a feasible scale.

In order to avoid the quarrying of vast quantities of rock, it may also be possible to make the carbon dioxide react with minerals *in situ*. It has been shown in the well-studied Columbia River basalts of northwestern USA that injected carbon dioxide forms solid calcium carbonate.[86] This combines mineralisation with the underground storage in basalts mentioned earlier.

Oman's rugged, beautiful and baking hot mountains have long been a playground of geologists. I first visited in August 2000. Although my main focus was on increasing recovery from the Sultanate's oilfields, I spent some time walking over one of the most spectacular exposures of ultrabasic rocks in the world, including peridotite.[87] Now, years of fieldwork by Jürg Matter and Peter Kelemen have established that take-up of carbon dioxide by peridotite is surprisingly fast: the natural rate over the Oman exposures may be about 40,000 tonnes per year. One suggested method of speeding this reaction is to drill into the rock, fracture it under pressure, and inject CO_2-enriched water. The heat and increase in volume when the minerals react would cause further fracturing, and the reaction might need little further energy input from there on.

It could be possible to consume as much as 1 billion tonnes of CO_2 per year in Oman alone. However, this exceeds the likely capturable emissions in the neighbouring countries,[88] so might have to be supplemented by long-distance imports or local air-capture devices. On the other hand, the ultrabasic rocks in the Alps, Czech Republic, Greece and the Adriatic coast are reasonably close to major European industrial sites.

Storage Capacity

We can see, therefore, that the potential carbon dioxide storage options are varied and geographically dispersed. The wide range of possibilities means that we do not have to rely on a single trapping mechanism. It will probably make sense to begin with the best-understood storage capacity, that in oil- and gasfields and related aquifer traps, and then to move on to other saline aquifers, coal beds and, if required later or for certain sites, basalts, gas shales and perhaps hydrates, gas-bearing aquifers and mineralisation.

Now, do the various storage sites have sufficient capacity for the voluminous amount of carbon dioxide we may emit over the next century and more?

The estimation of storage capacity is still in its fairly early stages. The basic science is not particularly difficult, although it requires more practical verification. Standard techniques from the oil and water businesses for studying underground structures and their volumes are applicable. There is some useful experience from EOR projects, Sleipner and In Salah on the storage capacity of aquifers and oil reservoirs, but there is still a huge amount of work to be done to quantify capacity on a basin-by-basin or field-by-field basis. The oil industry, after all, has spent a century and a half cataloguing underground structures, and has not had to spend time on investigating water-bearing ones; the first carbon dioxide storage estimates date back little more than a decade.

Most existing global estimates are based on 'top-down', high-level methods, by mapping the area of sedimentary basins and applying average storage factors. This approach is vulnerable to over-simplification: narrow, deep basins may contain many promising reservoirs, while large, shallow ones may have no suitable cap rock.[89] Many regional assessments work from the bottom up, summing specific known structures, and generally working from existing petroleum industry data. For some areas, such as the North Sea and Canada, this data is easily available and of high quality. For others, such as the Middle East, Latin America and former Soviet Union, it may often be inaccessible or confidential, and little work has been done to date on bottom-up assessments. Even in well-known areas, key information such as the degree of water encroachment may not be readily available.

There is a risk here that non-hydrocarbon basins (more than half of known basins worldwide) are lacking in data, and hence their capacity is discounted. On the other hand, apparently promising basins may have to be rejected when more is known about them: they may, for instance, be shallow or without effective seals; be prospective for geothermal resources; be heavily faulted, or have active hydrodynamic flow that would return carbon dioxide to the surface too soon. Unlike most oil and gas, carbon dioxide trapping occurs not purely in defined structures, but is also effected via residual saturation, dissolution in pore waters and mineralisation, all mechanisms that are highly site-dependent and occur over different timescales.

Particularly when considering EOR/EGR and depleted oil/gasfield capacity, the assessor has to decide whether to include undiscovered volumes, which past experience indicates are likely to exist, but which are, of course, less certain. Conservative assumptions are normally applied: that all carbon dioxide dissolves over time, and that the formation water is highly saline. If some carbon dioxide remains as free fluid, and the aquifer water is less saline, capacity may be an order of magnitude greater, but this will have to be determined via basin-specific data and field experience.

Finally, only some (perhaps a small amount) of the technical potential for carbon dioxide capture is practical and economic. Very large basins in Siberia, the Arctic and Central Africa, for instance, are too remote from most emissions sources (unless used for air capture). Some sites may have low permeability, or be divided into many small compartments by faults, or be too small, located in deep water, sited under major cities or national parks, or suffer from other issues, so that they will not be economic. Local public opposition (as discussed in Chapter 6) may derail a number of projects.

The injection rate may also be a constraint, so that even if sufficient total space is available, it may be difficult to dispose of emissions from a large source fast enough. For instance, in some trapping situations, it may take time for carbon dioxide to disperse from the injection well, or numerous injection points may be required. The thick high-permeability aquifer at Sleipner requires only one injection well; In Salah, with a much thinner and lower-quality formation, requires three wells for similar injection rates. Some storage options have other constraints: for some of the larger estimates of ECBM potential, the associated methane production might amount to a sixth of current global gas output,[90] not implausible but requiring a major effort to develop the related infrastructure, and likely to depress natural gas prices.

As with hydrocarbon fields, storage will probably start with the largest, cheapest and most accessible sites first, and, as these fill up, gradually move on to tougher locales. The advance of technology and infrastructure will gradually widen the set of economic storage options. Carbon dioxide storage space can be considered as a resource, similar to any other sub-surface resource.

The required volumes of CCS are covered in more detail in Chapter 5, but we can observe here that the IPCC 'business-as-usual' scenarios for 1990–2100 range from 2,806 to 9,279 Gt of total CO_2 emissions. Compare this to maximum 'safe' levels of some 1,075 Gt from climate models. If we were to capture and store all emissions above the threshold, a very unlikely prospect, we would require around 1,700–8,200 Gt of storage space.

Estimates for storage capacity vary widely, but appear sufficient. They suggest minimum volumes of around 1,250 Gt and maximum around 11,000 Gt, with some estimates going up to almost 22,000 Gt. A more detailed breakdown is given in the Appendix note to this page. The vast bulk of this capacity is in saline aquifers, with perhaps around 10% in EOR and depleted oilfields, and smaller amounts in gas fields and coal. Only a few of these estimates include basalts, and none covers more exotic storage options.

As for oil and gas resources, estimates have tended to move up with time as more knowledge has been gathered. In particular, the very low estimates have disappeared, and doubt has been cast on the low figures (early estimates of only 2–6% of pore volume available for CO_2) taken for aquifer capacity by the encouraging experience at Sleipner. Careful study of a single formation in the USA, the Mount Simon Sandstone, indicates 160–800 Gt capacity, and a site-by-site estimate for Australia summed up to 740 Gt, strongly suggesting that global estimates of a few thousand Gt are pessimistic.

These estimates indicate that total storage capacity is in the range of the required amount. Considering some realistic estimates for the likely scope of CCS, the IEA's estimate of 80 Gt to be stored up to 2050 is only 6% of the lowest quoted figure for capacity; most of the estimates are sufficient to hold the IPCC's maximum requirement of 2,200 Gt over the century. Only in the case of low storage capacity, high baseline emissions and a large role for CCS in mitigation might we run short of space. By no means all business-as-usual emissions will be captured, since many other low-carbon options will be used, and we have not considered some of the more exotic storage options, so worldwide technically-available capacity should be ample.

A useful illustration of the size of the storage capacity required is given by the Netherlands. This small country, with a high density of energy use, storing all its emissions for the rest of this century would require a sub-surface space of about 160 x 160 km, a little over a quarter the country's area.[91]

It also seems clear that storage in oil- and gasfields and coal beds, although an important and potentially low-cost starting point, is a small part of the total capacity. A major storage effort will have to rely mostly on saline aquifers. This is also implied by geography, since most industrial sites are within a few hundred kilometres of aquifers, but suitable petroleum reservoirs are much rarer.

On a regional basis, the story is more complicated. One set of estimates is shown in Figure 3.9.[92] Here, the storage capacity in each region (in gigatonnes, Gt) is compared to the likely volumes available for capture up to the year 2100. Other capacity estimates, such as those quoted by the IEA, may vary significantly, either higher or lower. If major emissions locales have insufficient nearby storage, the requirement for shipping or long-distance pipelines will drive up cost and complexity.

North America appears to have plenty of storage space, enough for more than 500 years of emissions at today's rate.[93] American annual demand for EOR alone could be some 10% of its emissions. The USA has some 900–3,400 Gt of space in saline aquifers, about 170 Gt in unminable coal, and some 85 Gt in oil- and gasfields. Canada's capacity is particularly large compared to its emissions, suggesting that it could store American carbon dioxide, if there is no political opposition to being a 'dumping ground' for its larger neighbour. Canadian capacity estimates vary significantly, with some much higher figures than those shown in the figure, including larger potential in oil- and gas-

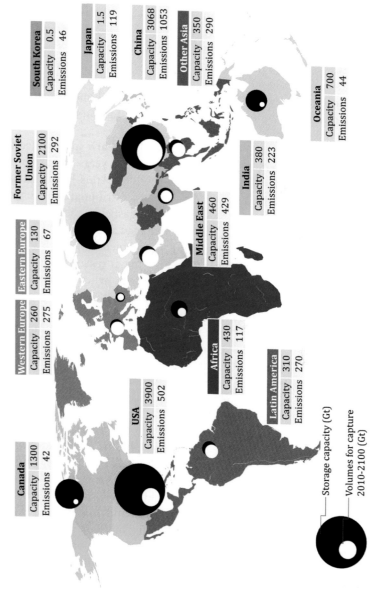

Figure 3.9. Regional carbon dioxide storage capacity

South Korea	
Capacity	0.5
Emissions	46

Japan	
Capacity	1.5
Emissions	119

China	
Capacity	3068
Emissions	1053

Other Asia	
Capacity	350
Emissions	290

Oceania	
Capacity	700
Emissions	44

Former Soviet Union	
Capacity	2100
Emissions	292

India	
Capacity	380
Emissions	223

Eastern Europe	
Capacity	130
Emissions	67

Middle East	
Capacity	460
Emissions	429

Western Europe	
Capacity	260
Emissions	275

Africa	
Capacity	430
Emissions	117

USA	
Capacity	3900
Emissions	502

Latin America	
Capacity	310
Emissions	270

Canada	
Capacity	1300
Emissions	42

Storage capacity (Gt)

Volumes for capture 2010-2100 (Gt)

fields (around 850 Gt), coal beds (up to 200 Gt) and aquifers (1,000–10,000 Gt).[94]

The much smaller storage space in **Latin America** versus the USA partly reflects geological conditions, but is probably partly due simply to less knowledge, since the IEA estimates Brazilian capacity alone as 2,000 Gt in aquifers, coal and oil fields. **Australia** is probably also amply provisioned, although much of its storage space lies in Western Australia, far from the big coal plants on the east coast. **New Zealand's** high level of volcanic and tectonic activity may limit storage, though there are some candidate oil- and gasfields. Emissions are small too, and could even be shipped to Australia.

Japan appears more limited, with barely a year's space noted in the figure, and a maximum storage of some sixty years of current emissions on the higher IEA figures. This partly explains Japanese interest in ocean storage. Japan's storage space is distributed between about 150 Gt in aquifers and 3.5 Gt in oil- and gasfields; there may also be some potential in coal beds. **South Korea** may also be short of capacity, although it could use offshore aquifers and potentially reach agreement with China, the large Chinese oil fields lying at reasonable distances.

India may be somewhat short of capacity, when we consider the significant growth in its emissions expected over the century. Most space is in saline aquifers, with around 350 Gt of capacity. The clusters of coal-fired power stations in the east of the country are rather distant from storage sites, although the possibility of using coalfields (5 Gt) and basalts (some 300 Gt), and recent major gas discoveries off the east coast, might change this picture. A modest amount of research on basalts could probably yield major advances in understanding, and so greatly clarify India's CCS potential.[95]

China has a large theoretical storage capacity, rather dispersed across the country and in nearby offshore basins, and mostly in saline aquifers. With Chinese emissions at about 6.1 Gt per year in 2006, storage capacity has been estimated as 3,000 Gt in saline aquifers, 5 Gt in depleted oilfields, 5 Gt in depleted gasfields and 12 Gt in unminable coal seams.[96] Obviously, the oil, gas and coal options can only hold a small portion of Chinese emissions, but they may be valuable, low-cost sites to start the storage effort. Many of the large-point sources are concentrated in coastal China and will have to use offshore storage sites. Opinion also varies as to how many large emissions points are close to known storage, with 40%-100% of Chinese sources within 300 km range.[97]

The capacity in **Europe** is uncertain, enough for only 8 years at the low end, but potentially many centuries at the high end, split between 260 Gt total in Western Europe and 130 Gt in Eastern Europe. For comparison, European emissions from all sources in 2004 totalled about 4.4 Gt of CO_2.

The UK has the advantage of having gained abundant data on off-shore aquifers during oil exploration, and aquifer capacity alone is estimated at more than 250 Gt, some 400 years of UK emissions,[98] or some sixty years of all EU CO_2, although it has been argued that these estimates may be optimistic.[99] The main aquifers considered were all offshore, in the North Sea, Irish Sea and Western Channel. Depleted oil- and gasfields add 7.5 Gt (more than a decade of all emissions), plus 2.3 Gt capacity in unminable coal-beds around Oxford, Cheshire (south of Manchester) and eastern England. Ireland has some small potential storage space in the Corrib gasfield off the west coast, and in other offshore gasfields, but its aquifers are generally too shallow to be suitable. Ireland might therefore need to be connected to a UK storage network.

To UK capacity should be accorded additional potential, probably at least as much again, in the North Sea sectors of Norway, Denmark and the Netherlands. Denmark's total capacity is estimated at 16–47 Gt, while the Netherlands has 10 Gt in gasfields (hosting Groningen, Europe's largest gasfield), 1 Gt in aquifers and 1 Gt in deep coal seams. Norway has 42 Gt in the Utsira aquifer alone (the host for Sleipner injection), with a total potential up to 473 Gt including 'open' (non-structurally trapped) aquifers, plus 13 Gt capacity in gasfields. Taking the higher estimates for Norwegian aquifers would double Western Europe's available space.

There are wide ranges of estimates for France, Germany, Italy and Spain. However, potential is fairly small compared to the UK and Norway. France, Germany and Italy each have in the order of 25 Gt of aquifer capacity, plus smaller amounts in coal-, oil- and gasfields. In Germany, the Altmark gasfield alone, the second largest onshore field in Europe, could hold 0.5 Gt. Earthquake and volcanic activity in some parts of Italy, with vents of natural carbon dioxide, suggest caution in siting storage. Spain appears to have the most storage space of these four countries, with about 45 Gt in seven basins near the main emissions sites, plus some coal potential.

Sweden has some capacity in the south and the south-west offshore, but Estonia and Finland are potentially short of space, and likely to have to reach agreement with neighbours. Latvia has limited capacity; Lithuania has rather more, in aquifers and some EOR potential.

In central and eastern Europe, Poland, a heavy coal-user, has several dozen Gt of capacity in depleted oil- and gasfields in the west, south-east and offshore in the Baltic, enhanced gas recovery potential at Kamien Pomorski and Borzęcin (site of a long-running acid gas injection scheme), the Jura and Kreda aquifers and potential ECBM in Krupinski and Silesia. Total Austrian capacity is estimated at 0.5 Gt in oil- and gasfields, Hungary has 1 Gt in poorly known deep aquifers, and about 0.65 Gt in oil-, gas- and coalfields, while Slovenia has small potential in aquifers. Romania, one of the world's oldest oil-producing nations, has 2.5 Gt storage in oil and gas and 3 Gt in aquifers. Greece, though, appears relatively lacking in capacity, and, being tectonically active, may suffer concerns about leakage.

Russia's vast territory encompasses some 2,000 Gt capacity, of which around 175 Gt resides in depleted oil- and gasfields in West Siberia, remote from the main emissions centres. Other sinks closer to sources include the Black Sea (Krasnodar), and EOR potential in the mature Volga-Urals oil province (around the cities of Ufa in Bashkortostan and Samara and Perm in Tatarstan). There is also ECBM potential in the famous Don coalfields of southern Russia. Basins in the Russian Far East might be usable by China, Korea and Japan, although distances are still significant, and nearby Sakhalin is tectonically active. The Caspian region has abundant capacity for its own emissions, but relative inaccessibility probably precludes other regions from exporting their carbon dioxide there.

The **Middle East** has particularly large capacity and high-quality evaporite seals. As with Latin America, the estimate in Figure 3.9 is probably conservative, and the IEA suggests room for 180–1,700 Gt, at least two centuries of current regional emissions. Enhanced oil and gas recovery are obvious starting points. Frequent tectonic activity in the Zagros Mountains of Iran and Iraq, despite excellent seals, may create concern for storage integrity. The plains of Khuzestan, in Iran, and most of the remaining Middle East, onshore and offshore, though, are geologically stable and rather simple, with numerous extensive reliable reservoir-seal pairs. It might ultimately make sense for the Middle East to be a carbon dioxide importer, probably via ship given its distance from big emitters, or a site for air capture operations driven by abundant gas or solar power.

The same, on a somewhat smaller scale, may be true of **North Africa,** which is close enough to southern Europe to be linked by carbon dioxide pipelines. Overall, Africa's storage potential greatly outweighs its likely capturable emissions. Storage potential is 6–220 Gt in aquifers, 30–280 Gt in oilfields, and 8–40 Gt in South African ECBM.

Of course, all these figures need to be taken with the caveat that we are not going to capture all emissions—probably rather less than 50%. Ultimately, if air capture can be implemented, then even areas without suitable storage capacity can pay to have their emissions mitigated elsewhere.

Alternatively, the successful development of mineralisation methods, as discussed above, and advocated by Klaus Lackner,[100] would relieve the pressure on underground storage, or at least provide a back-up. There is no practical limit to the amount of such minerals that could be stored. To give an absolute upper bound for the scale of the challenge, if we mineralised all global emissions above the 'safe' threshold, for the IPCC's highest scenario of fossil fuel use, and with the highest estimate for the volume of minerals produced per tonne of carbon dioxide, then over the next century we would cover a square 1,000 kilometres long on each side, the size of Egypt, with a 10-metre layer of minerals.[101] This is obviously a formidable amount, about fifty times the likely total cement production during the twenty-first century; it is, though, not inconceivable (it could, after all, be used to build artificial islands, defences against rising sea levels and so on).

Economic potential is, of course, much lower than technical potential, and this emphasises that CCS will compete with energy efficiency, nuclear, renewable energy and other techniques. In areas with good storage potential, CCS may be very competitive; in other areas, for instance Japan, the alternatives may predominate.

Probability of Leakage

One, perhaps the greatest, concern that is brought up about carbon storage is leakage. It is often thought, especially by the general public, that carbon dioxide leakage may occur, ruining the climate mitigation effect, and perhaps posing a danger to people or the local environment. In order to guard against leakage, storage facilities will have to have extensive monitoring efforts, covering the site itself, and the surrounding area. A related activity, verification, involves tracking the amount

Figure 3.10. Schematic picture of possible leakage paths from geological storage

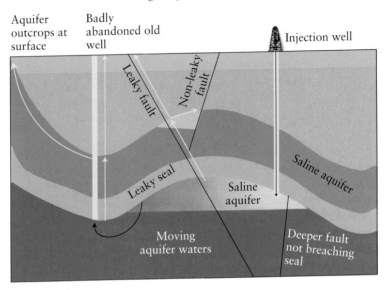

of injected and trapped carbon dioxide, so that emissions permits or other carbon accounting can be applied accurately.

Escape of stored carbon dioxide would certainly be problematic for the success of CCS. Firstly, large amounts of leakage might be dangerous or have serious environmental effects. Secondly, slow seepage would weaken or destroy the whole rationale for capturing carbon dioxide in the first place—to reduce climate change.

Leakage from a storage site can occur in a number of ways (Figure 3.10). If the primary seal at the initial injection point is weak, possibly fractured or faulted, or if injection pressures exceed its sealing capability, then carbon dioxide may move up through the cap-rock. (It is important to realise, though, that the mere presence of faults[102] does not mean a trap will leak. Indeed, many faults are actually barriers to fluid movement. Even leaky faults permit only slow upward seepage.) Moving aquifer waters may carry carbon dioxide out of the trap, to a point where a fracture or an improperly abandoned well allows leakage. The carbon dioxide may dissolve the cement in such wells, so finding a way through protective barriers. At some point, probably distant from the injection site, the aquifer formation may outcrop at the surface.

What is the evidence that geological structures can hold carbon dioxide safely over the long term? There are four main lines of argument to establish the security of storage:

- Natural analogues: underground reservoirs containing carbon dioxide and natural gas.
- Decades of experience in operating carbon dioxide pipelines and production facilities, and natural gas storage.
- Monitoring of existing CCS and other carbon dioxide injection projects.
- Modelling, computer simulations, geological research and study of seismic activity.

There are many natural analogues to carbon dioxide storage, which shows that underground reservoirs retain their charge over many millions of years. The vast amounts of natural gas should illustrate this clearly: methane is more buoyant than carbon dioxide, less viscous, less soluble in water, and does not generally react with minerals. It should therefore be much more leak-prone than carbon dioxide. Yet quadrillions[103] of cubic feet of gas have survived in underground reservoirs over geological time. Some natural gas fields in Oman and East Siberia may have retained gas for over a billion years. In a way, this should be obvious. Imagine a million-year-old gasfield, very young in geological terms. Even if it is now only 10% full from when it was originally charged, the implied leakage rate is minimal, just 0.0002% annually. Yet many, if not most, gasfields are full to the 'spill point',[104] showing that the leakage rate is even less than this.[105] Indeed, worldwide oil seepage is about 0.00002% per year, a tenth of our gasfield example.

Numerous **natural underground** carbon dioxide accumulations are known. High-CO_2 gasfields are common in South East Asia, notably Indonesia, Malaysia, Vietnam and Thailand, and also occur in many other hydrocarbon provinces including China, Australia, New Zealand, Colombia, Mexico, Canada, the North Sea, Libya, Algeria, Abu Dhabi and Pakistan. In the USA, Wyoming, Colorado and Mississippi are noted for carbon dioxide fields. Accumulations not associated with petroleum occur in Mexico, Nicaragua, Colombia and Kamchatka (Russian Far East).

Four significant fields in the USA, Big Piney-La Barge, St Johns Dome, McElmo Dome and Jackson Dome, which supply carbon dioxide for EOR activities, plus the giant Natuna D Alpha field in Indonesia (which is about 70% CO_2), hold between them almost 20 Gt of natural CO_2, equivalent to about 80% of annual global emissions.[106] Some of these fields also contain significant amounts of the very light and mobile gas helium, indicating that their seals must be extremely effective. The Pisgah Anticline, near Jackson Dome, has contained carbon dioxide for more than 65 million years, back to the age of the dinosaurs, with no sign of leakage. Similar occurrences in Utah, however, show a complicated pattern of carbon dioxide trapping and leakage, emphasising the need to build a good structural and hydrodynamic model of areas used for storage.

Near Weyburn in Saskatchewan, a key site due to its very well-monitored CO_2-EOR project, natural carbon dioxide entered carbonate rocks about 100 million years ago, but the overlying petroleum reservoirs contain only trace amounts of carbon dioxide, indicating that leakage over this vast span of time has been insignificant.[107] Similarly, in the Permian basin of West Texas, the main locus of CO_2-EOR activity, carbon dioxide has been trapped for up to 300 million years.[108] Another carbon dioxide storage trial, in the Otway Basin of Australia, is near to natural carbon dioxide accumulations. The existence of long-term traps for carbon dioxide is therefore not a fluke, but is a common and widespread phenomenon. So, rather than carbon dioxide storage creating a problem for our grandchildren, as some fear, we have been largely oblivious to vast quantities of underground carbon dioxide for all of human history.

Of course, we should not exaggerate the confidence given by such accumulations: there may well have been many natural carbon dioxide fields that have leaked and are now empty. These structures were probably filled much more slowly than artificial storage sites will be. They are also not penetrated by potential leaky wells (not until the last century or so, in any case). Not all sites are suitable: BP's plan for a coal gasification plant with CCS at Kwinana, south of Perth, Australia, had to be cancelled when seismic surveys revealed 'gas chimneys'.[109] These are signs of leakage of natural gas, making it impossible to be sure that the seal at the proposed storage location was effective. BP's decision to stop the project shows a responsible attitude towards ensuring storage integrity.

Long experience of transport, injection and production of carbon dioxide and natural gas shows a very good safety record. Leakage from pipelines and facilities is likely to be minimal, and it is improbable that carbon dioxide in air would reach dangerous levels. These installations, anyway, are carefully monitored. Leaks are quickly detected from direct measurements or from drops in pressure, and automatic shutdown occurs. Any escaping carbon dioxide will diffuse harmlessly into the atmosphere within minutes, as long as the area is well-ventilated.[110]

Similarly, underground gas storage, with repeated injection and retrieval of natural gas, was first carried out as far back as 1915, and is performed routinely at hundreds of sites throughout Europe, North America and the former Soviet Union (though there are nine documented cases of leakage, as discussed below). Natural gas is, in most respects, more vulnerable to leakage than carbon dioxide, and more dangerous if it does leak. Sour gas contains the highly toxic hydrogen sulphide, which is deadly at 0.1% concentration, a hundred times lower concentration than carbon dioxide, and is also corrosive. Sour gas is extracted in densely populated areas including Calgary in Canada, Emmen in the Netherlands and near Abu Dhabi in the UAE.

'Acid gas', a mixture of the waste products carbon dioxide and hydrogen sulphide from processing sour gas, is re-injected at more than thirty sites in Canada, and such disposal is being considered in Qatar and Abu Dhabi. Often, more than 90% of the acid gas is carbon dioxide, so these are effectively carbon dioxide storage projects. Most are small, but the largest, which started operations in 2002 in northeast British Columbia, is half the size of Sleipner.

Hazardous waste, about 34 million cubic metres annually, a similar magnitude to CO_2-EOR, is injected into saline aquifers in the USA alone. Safety procedures are strict and carefully enforced, and these activities lead to little concern. By these good practices, backed up by thorough communication, including engaging local people, explaining what is being done, and responding promptly to complaints, acid gas and toxic waste operators have managed to avoid serious opposition. It helps that many of the acid gas projects are in Alberta, a province with strong oil and gas employment, and familiarity with the industry.

Blowouts, uncontrolled flows of underground fluids through a well to surface, are also rare in the modern industry, and unlikely when drilling into well-known, low-pressure reservoirs. Similarly, even in California, a very mature oil province with many enhanced recovery operations and old wells, blowouts are exceptionally rare, and declining with time: only one blowout each year for every 100,000 idle or abandoned wells.[111] A carbon dioxide blowout would be much less hazardous than one involving flammable oil or gas, and procedures for capping a 'wild well' are widely known.

There is one example of a carbon dioxide blowout, at the Sheep Mountain carbon dioxide field in Colorado in 1982. The blowout was quickly detected, and there were no injuries or fatalities, but it took seventeen days to seal the well. About 0.2 Mt of CO_2 leaked,[112] about a month's output from a 500 MW coal power plant. If this had been a commercial storage operation, then, with current European carbon costs, this represents a $5 million penalty for the storage operator. Carbon dioxide blowouts, therefore, do not seem very likely or dangerous, and insurance could probably cover any leakage.

The most likely source of leaks, associated with depleted oil- and gasfields, and with enhanced hydrocarbon recovery, is from abandoned wells. When a well is no longer capable of commercial production, it is 'plugged and abandoned'. A better term may be 'decommissioned', since, particularly in modern practice, obsolete wells are carefully prepared to make them safe. The tubing, metal pipes for conveying oil or gas up the well, and 'Christmas tree', surface assemblage of valves and other equipment, are removed, the wellbore is filled with an inert fluid, and the casing (outer metal lining) is blocked with a number of cement plugs. The plugs are tested for integrity, the casing at the surface is capped, and the area is restored with soil and its location accurately recorded.

Standard practice is to seal all hydrocarbon reservoirs and freshwater aquifers, usually extending 15 metres above and below the zone. This means that saline aquifers may not themselves be automatically isolated, but there will be at least one cement barrier sealing shallow water-bearing formations. So carbon dioxide should not be able to reach the atmosphere, but it might be able to migrate up an abandoned well-bore into a higher saline aquifer. This might eventually provide a path, albeit a tortuous one, to potable water or to the surface.

Leakage may occur from a decommissioned well if it is imperfectly sealed, or, though this is unlikely, if carbon dioxide dissolves the cement—either the plugs within the well, or the cement used to fill the gap between the rock and the metal well casing. Older wells may have used lower-quality materials and procedures, and have had the most time to degrade. In the oldest fields, some well locations may not even be known, and they may be buried under later construction. Plugging requirements were introduced at different dates throughout the USA, starting as early as 1915 in California, and were standardised nation-wide in 1952. Before that, well abandonments were occasionally haphazard—including plugs made of tree stumps, logs, mud and even animal carcasses.[113] Between 1937 and 1950, well-cementing improved greatly, including additives to ensure that cement set properly at down-hole temperatures.

In intensively drilled basins, leakage through old boreholes is a significant long-term risk if carbon storage takes place on a regional scale. Western Canada contains some 350,000 wells. In Texas there are 135,000 'orphan wells', inactive wells, not properly abandoned but with no current owner, which a state government programme is slowly plugging. Assuming that carbon dioxide will spread about 10 kilometres from the injection point over several centuries, then for a given carbon dioxide storage project in Texas, there will be on average sixty orphan wells requiring plugging, costing the modest total amount of $270,000.

Such a high density of drilling is a rarity outside North America, and there are fewer old, poorly abandoned boreholes (with the possible exception of parts of the former Soviet Union and China). In the USA, any operator planning to inject fluids into the ground has to ensure that existing wells within 400 metres of the injection point are properly sealed. Even in highly drilled provinces, most older wells are relatively shallow, and deeper aquifers are therefore still fairly secure.[114] The main USA exception is a number of deeper wells that were not plugged due to bankruptcy of the operators following the 1986 oil price crash.

Leakage from old wells has been examined at the Weyburn Project in Canada, where it was concluded that, despite the presence of 1.000 abandoned wells, maximum leakage was only 0.14% over 5,000 years.[115] This was probably over-estimated, since it assumed that leakage would continue at a constant rate over all the 5,000 years, whereas in fact the rate would tend to drop off with time as available carbon dioxide is exhausted, even assuming that the leak is never detected and sealed. At the Rangely EOR project, hundreds of 1940s-era wells are still in use for carbon dioxide injection, with few problems.[116]

On the other hand, Crystal Geyser in Utah, USA, leaks significantly. Crystal Geyser is an unusual example of a cold-water carbon dioxide geyser, formed when an oil exploration well drilled in 1935 unexpectedly penetrated a shallow natural carbon dioxide accumulation at 215 metres depth. The geyser erupts several times each day. The emissions, a very visible sign of carbon dioxide escape, are about 11,000 tonnes per year, equal to about 0.1% annual leakage from a Sleipner-sized project. Of course, no attempt has been made in this case to seal the well, since the site is a tourist attraction.

The amount of seepage through a poorly sealed well is limited more by the reservoir conditions around it than by the condition of the well; in other words, there is not much difference in escape rates between a moderately badly sealed well and a very badly sealed one. For the early phase of leakage, say the first fifty years, most or all leaking carbon dioxide dissolves in intermediate aquifers and so does not reach the surface.

Thus a crucial part of storage in depleted fields and in aquifers in old hydrocarbon provinces will be to locate all the existing wells, make sure they are plugged properly, plan for any sub-surface migration that might take place via the borehole, and monitor the wells. The first few projects in a given basin will have to bear most of the burden. The task will become easier once the most problematic wells have been identified and sealed. As for ensuring the longevity of wells, CO_2-resistant cements exist,[117] and perhaps the 'supramics' mentioned previously could also be used. It would useful to have a better understanding of the processes of casing and cement corrosion by carbon dioxide, and how the sealing properties of cement degrade with age. It is worth noting, though, that existing CO_2-EOR operations, after some thirty years, have not had significant problems with cement corrosion.[118] Experience suggests that the key is doing the cementing effectively, not whether CO_2-resistant rather than normal cement is used.[119]

There is a wide variety of tools for detecting leaks and weak points in well cements and casings. Insecure wells, once identified, can be easily re-sealed. Of nine cases of escape from underground natural gas storage, five were due to leakage from wells. In the 1970s, at the Leroy natural gas storage facility in Wyoming, gas began leaking via a corroded well casing into an adjacent old well and from there to the surface. The well had eventually to be sealed permanently. A serious accident occurred at the Hutchinson natural gas storage facility in Kansas in 2001, where gas escaped through a damaged well and exploded,[120] killing two people. However, although carbon dioxide might leak in this way, it cannot explode.

Storage in aquifers or basalts would involve a small number (perhaps ten or so[121]) of injection wells, which would be modern and carefully constructed and monitored. The main risks in this case are from leaks via undetected faults and fractures; the accidental fracturing of the cap-rock due to over-pressuring the reservoir or to the corrosive effect of carbon dioxide; and from eventual migration of the carbon dioxide to areas where the aquifer formation is not sealed, reaches the surface or is connected to a drinking-water reservoir. In this case, though, leaks would be likely to be small and highly dilute, unless focussed by a fault or similar feature. Leakage would also be unlikely to be constant over time: the percentage escaping would drop with time, as pressure is bled off, and the amount of mobile carbon dioxide decreases, leaving a sub-stantial residue (probably the majority of injected gas) trapped in vari-ous ways. Interestingly, at the University of Texas's test site at Dayton, the researchers deliberately attempted to use normal natural gas pro-duction methods to pump out the carbon dioxide that they had previ-ously injected, but were unable to extract it again.[122]

At the previously mentioned Leroy facility, gas was detected leaking through the top-seal of the reservoir.[123] This occurred at a depth of 900 metres, shallower than most likely carbon dioxide storage sites, in 1978, when monitoring methods were not as advanced as now. Nev-ertheless, the seep was quickly detected. Annual leakage was about 0.7%, above acceptable levels (see below). It appears, though, that Leroy was simply a badly chosen site. Better understanding today should avoid such a choice; if leakage is detected within the first few years, then remedial action will have to be taken, or the injection shut down and a new sink located.

In hydrocarbon-bearing areas, fractures leading to the surface can be detected by searching for oil and gas seepages, which was, indeed, the main method of early hydrocarbon exploration. Trace amounts of gases can be detected with modern tools. Major faults can be detected on seismic data, as can sub-surface natural gas leakage, seen as 'gas chimneys' where seismic imaging is disturbed. Areas with such faults and fractures cutting through to the surface would preferably be avoided. Pressure and salinity data can indicate whether aquifers are connected. For instance at Weyburn, numerous lines of evidence show that the formation for carbon dioxide injection is isolated from shal-low drinking-water aquifers.[124]

Hydro-fracturing is a technique used increasingly in the oil industry to enhance production from low-permeability reservoirs by injecting under high pressure a fluid made of water and various speciality chemicals. This fractures the rock, creating paths which allow oil or gas to flow easily to wells. Hydro-fracturing has been crucial, for instance, in establishing shale gas as a commercial energy source. If a hydro-fracced reservoir is later to be used for carbon dioxide storage, or if fracturing is used to make it easier to inject carbon dioxide into a low-permeability formation, very careful modelling and monitoring will be required to make sure there are no leak paths through caprocks.[125] However, used appropriately, hydro-fraccing can allow carbon dioxide injection at reasonable rates into low-permeability rocks.

Modelling of natural fractures suggests that leakage would be low. For instance, studies examining the effect of a fracture 8 kilometres from an injection well suggest that it would start to leak after 250 years, and total leakage might be 10–20% over the next 2,000 years,[126] i.e. less than 0.01% per year. This is acceptable, particularly when considering that most sites, if well-planned, would not be near such fractures. Since some 30% of carbon dioxide dissolves within decades, and the heavier CO_2-saturated water will tend to sink, it seems implausible that, even in a site with no seals, total leakage could ever exceed 70%. To give three examples:

- For the Forties oilfield in the North Sea, even over 1,000 years, fluid dynamic simulations suggest that only 0.2% of injected carbon dioxide would enter overlying layers, and even in the worst case, it would reach only halfway to the seabed.
- Similarly, at Sleipner, it is calculated that there will be no leakage into the North Sea for 100,000 years, and even after a million years, annual escape would be only 0.001%.
- At Weyburn, calculations show 95% confidence that leakage over 5,000 years will be no more than 0.00026% annually, with an average total release over this period of 0.2% of injected volumes.

Current storage projects are all rather new, and so give only limited evidence about possible leakage. Since no escape has been detected at Sleipner or Weyburn after ten years of injection, these sites appear to meet a limit of 0.01% escape, meeting the standard described below. If there is still no detectable escape by 2023, this will indicate that leakage is less than 0.001%. Furthermore, movement of carbon dioxide out of the original trap should not be confused with seepage at the surface. Most basins will contain multiple layers of shales (which constitute about half the volume of continental sedimentary rocks[127]), salt, anhydrite and other sealing rocks, particularly when injection is into a deep formation. Studies at Weyburn have concluded that, of an insignificant amount[128] of injected carbon dioxide that will diffuse into the cap-rock over 5,000 years, none would reach a secondary seal which exists above the primary cap-rock.

As Figure 3.11 shows, assuming a (rather high) 0.1% leakage annually from each seal—as if each one leaks as much as the spectacular Crystal Geyser—about 50% of injected volumes are retained for 1,000 years by a single seal. If two seals are present, so that injected carbon dioxide has to pass through both of them before reaching the surface, retention is almost 85%. And if there are three seals, nearly 97% is trapped.

Figure 3.11. Retention of carbon dioxide in storage with 1, 2 or 3 seals[129]

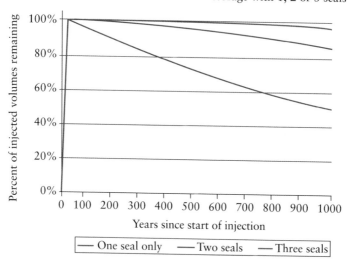

At the Rangely oilfield in Colorado, an EOR operation not specifically designed for storage, small amounts of leakage have been detected, less than 0.0008% annually of the total stored to date. It is not clear whether this is occurring through old wells, which could be sealed, or the cap-rock, which is more problematic.[130] Still, at this rate, less than 8% will be lost over 10,000 years (approximately the time since the last Ice Age).

A large earthquake at the Nagaoka test site in Japan did not cause any leakage,[131] although obviously it would be preferable to use storage sites away from seismic and volcanic activity. In the USA, with the exception of western California and local areas along the west coast, the New Madrid fault zone in Missouri and coastal New Carolina, there is negligible to low seismic risk.[132] The same minimal level of seismic danger applies globally, to most of Canada east of the Rockies, most of South America east of the Andes, northern Europe and Russia, most of sub-Saharan Africa away from the Great Rift Valley, the Middle East (excluding the Zagros and Oman Mountains), most of India south of the Himalayas (excepting a belt running east-west across the country at the latitude of Mumbai), most of peninsular South East Asia (excluding Myanmar) and Borneo, South Korea, and eastern Australia.[133]

China and Japan are the two clear CCS candidates where seismic activity might be problematic: in Japan's case, across most of the country; for China, in the west (which has sparse population and industry), and around Beijing, in the south-west, and in some coastal areas south of Shanghai. In these areas, major faults will have to be carefully mapped and the risk of disruption to storage sites assessed.

Injection of fluids (including oilfield water, natural gas, carbon dioxide and other waste-water) into the sub-surface has been associated with micro-seismic events, detectable only by instruments, and might have induced moderate local seismicity in Denver and Ohio. The larger events were not due to CO_2-EOR, and in general, despite the James Bond film *A View to a Kill*, significant earthquakes linked to fluid injection are extremely rare.

Hazards of Leakage

If carbon dioxide did escape, what are the hazards? The main issues are: danger to people through asphyxiation; contamination of drinking water; local environmental impacts; mobilisation of toxic metals; and greenhouse gas emissions.

Higher concentrations of carbon dioxide, above 7–10%, are toxic to people via asphyxiation. Low concentrations of carbon dioxide, 1% or less, have virtually no known adverse affect on humans, animals or plants, even over the long term. In fact, most international classification systems, including European, American and Canadian standards, regard carbon dioxide as non-toxic and non-corrosive. Carbon dioxide is, of course, not explosive or flammable, a point which needs to be made clearly to people living near CCS sites. Offshore injection would involve little or no risk to human beings and would be less vulnerable to NIMBY-style protests. It is, however, more expensive than onshore storage.

It should be recognised, though, that sometimes the carbon dioxide streams to be disposed of will also contain other pollutants, notably hydrogen sulphide. Co-disposal can significantly reduce capture costs, but H_2S is highly toxic, and so this activity would raise much greater safety concerns.[134]

The well-known 1986 Lake Nyos disaster in Cameroon, in which 1,746 inhabitants died, occurred when natural carbon dioxide was suddenly released from the bottom of a volcanic lake, where it had been accumulating for many years. Numerous natural volcanic vents in southern Italy release carbon dioxide, which occasionally causes fatalities, usually to animals, when it builds up in caves, depressions or cellars. Almost 0.5 Mt per year of CO_2 emerges at the Mammoth Mountain volcano in California, killing trees around the crater, and it has led to at least one near-lethal incident with a ranger in a ski-hut where the gas had accumulated.[135]

However, a CCS-linked recurrence of the Lake Nyos tragedy is highly unlikely. This disaster occurred under very specific circum-stances: a small, steep-sided tropical lake, with volcanic carbon dioxide seeping in at the bottom and with only one outlet, which concentrated the carbon dioxide cloud.[136] Lake waters outside the tropics overturn seasonally, hence releasing any accumulated gases. It would be straight-forward to outlaw CCS from taking place in tropical regions near similar lakes. Lake Nyos itself now has a simple system installed to remove carbon dioxide continually, and monitoring devices have been installed to watch for signs of danger.

Similarly, in Italy and at Mammoth Mountain, monitoring devices and air pumps have been installed, and daily life continues near major carbon dioxide vents, releasing up to 150 tonnes per day without sig-nificant problems.[137] Crystal Geyser, as noted, is a tourist attraction. Even next to the well, carbon dioxide concentrations during eruptions average only 0.6–0.7%, far below dangerous levels, and visitors stand nearby without concerns.

Seeps could **contaminate drinking water or affect sensitive ecosys-tems**, although natural carbon dioxide vents show only minor and localised effects on fauna and flora.[138] Biological communities, mostly bacteria, do exist in the deep underground, and are not well understood, partly due to the difficulty of getting at them. The likely impact of carbon dioxide on these organisms is unknown, but they are not likely to play a major role in global ecosystems, and carbon dioxide injection will affect only a small part of the vast sub-surface volume.[139] Even if carbon dioxide did accumulate in potable water, it is not, in itself, poisonous—as we know from drinking Coca-Cola. It would just have to be removed as part of the purification process, and re-injected in a more secure site.

Water saturated in carbon dioxide can **leach dangerous levels of toxic metals**, including arsenic and lead, and this might be a problem if saline aquifer waters are seeping into a potable aquifer. In the 2004 Frio pilot in the Texas Gulf Coast, substantial amounts of metals were dissolved from mineral grains in the reservoir.[140] Simulations suggest that injection of carbon dioxide directly into shallow lead-bearing minerals might cause contamination a few hundred metres around the site; this is an extreme case, since storage sites will be chosen to avoid such toxic elements. Nevertheless, baseline surveys of potential storage sites should establish the amount of toxic metals in groundwater before injection starts.

The more insidious long-term threat to CCS is that **slow leakage can dent the mitigation effect**. Perfect storage is emphatically not required, nor can it be assured. Our knowledge of the underground is never complete—mere uncertainty is not an excuse to rule out carbon storage. The oil business has developed sophisticated approaches over many years for dealing with sub-surface uncertainty, and in other fields—climate modelling, economic forecasting, predicting technological change—we have to deal with imperfect knowledge of even greater magnitude.

Very slow leakage will not affect the climate significantly, because there will be time for the excess carbon to be mopped up by the natural cycle—converted into plant matter and minerals. Even in badly chosen sites, the majority of injected carbon dioxide will be retained for geological time, millions of years, in pore spaces and dissolved in formation water. Anyway, we cannot make any sensible predictions about the energy system, economy, society or climate 10,000 years from now. The sun's output may have risen or dipped, humanity may be so technologically advanced that carbon is a forgotten problem only recalled by archaeologists, or we may have become extinct or gone back to living in caves.

However, at higher leakage rates, we would have only delayed the problem, not solved it. Imagine, for instance, storing all emissions, then suffering a leakage rate of 1% of cumulative storage. After 100 years the leaks would be equivalent to the global output of carbon dioxide, and so we would be back where we started. A 1% level of leakage is clearly unacceptably high; but what is a reasonable upper limit?

For CCS to be an attractive way of tackling greenhouse gases, it has been calculated that 0.1% leakage or less per year is probably tolerable.[141] So 0.01% should be sufficient even for scenarios of very high fossil fuel use and a strict target for the stabilisation of atmospheric carbon dioxide. Overall, the IPCC has estimated that underground storage is 'likely' (better than 66% chance) to leak less than the 0.1% annual limit.

Since natural carbon dioxide additions from volcanoes are on the order of 1% of anthropogenic levels, then, even if we were to capture all emissions for a century, 0.01% leakage would be equal to natural levels, with which the climate system coped perfectly well in the past. Removal of carbon dioxide from the atmosphere and ocean by weathering occurs on a timescale of 3–7,000 years.[142] With 0.01% leakage, 50% of stored carbon dioxide would still be trapped after 7,000 years, suggesting that natural weathering would be fast enough to mop up the excess.

More stringent limits of 0.001% per year (93% retention over 7,000 years) are required by Texan regulations and have been suggested by the World Wildlife Fund.[143] This standard seems excessive.[144] It makes little sense to set excessively tough limits to save the climate in seven millennia, if we thereby fail to save it in seven years. There is also an obvious inconsistency in prescribing very strict targets for geological storage, but then promoting forestation (as described in Chapter 4), which takes a significant time to reach maximum rates of carbon take-up, and leaks 100% over a century or two unless renewed.

However, as discussed above, storage sites probably can achieve even these strict targets. There is also a reasonable argument that we should plan for a sufficient margin of safety, and one tough enough to support public confidence.

When considering the advantages of CCS versus biological carbon sequestration such as in forests, we should bear in mind that the retention time for underground carbon dioxide is at least 1,000 times longer than for that in the biosphere.[145] Mineralisation, of course, is entirely secure, since carbonate minerals do not release carbon dioxide until recycled into the Earth by geological processes over millions of years.

If it becomes apparent that leakage is occurring after some decades of CCS, then air capture forms a useful 'last line of defence'. Air capture machines could be deployed in enough quantity to offset the leakage. Alternatively, biomass CCS plants or biological sequestration could be used. So the occurrence of small amounts of leakage is not fatal to the success of underground storage. It would, though, be extremely useful to demonstrate at least one of these forms of air capture on a large scale and set an upper limit on its cost. That would greatly help in determining the maximum permissible leakage from geological storage.

Any sign of significant leakage from an early CCS project would obviously pose a big threat to the future success of the method. It would cast doubt on the long-term integrity of storage. Even if it were clearly demonstrated that the leakage could be fixed, or that the reasons for escape were understood and could be avoided in future sites, such an event might worry the public, especially those living near future CCS sites, and would undoubtedly be seized upon as ammunition by those who are already opposed to carbon dioxide storage. As discussed in Chapter 6, good management of stakeholder relations, including concerned NGOs and local communities, will be very important, as it is for other energy developments, including renewables and nuclear plants, and megaprojects such as mines, roads and bridges.

It is thus essential that the good record of Sleipner, In Salah and Weyburn continues, and that the next generation of CCS projects continue these high standards of site characterisation and monitoring. The Hutchinson natural gas tragedy was ascribed to poor management, bad engineering and lax regulation; although carbon dioxide storage is inherently much safer, such errors must be avoided. As we gain a better understanding of the key drivers of site integrity, we will refine the suite of monitoring tools, and will no doubt become better and quicker at identifying good and bad locations.

Monitoring and Remediation

In order to avoid leakage, remedy it if it does occur, and accurately account for stored carbon dioxide, storage sites will be monitored in any variety of ways. These methods are well developed from the oil industry and other uses, and some modest adjustments make them suitable for monitoring carbon dioxide.

Numerous monitoring techniques are available, including acquiring seismic data, as mentioned (using sound waves to detect changes in underground fluids), tracking reservoir pressures, measuring the electrical resistivity of groundwater and soils, and detecting elevated levels of carbon dioxide in the air. An advancing carbon dioxide front is signalled by geochemistry, including changes in alkalinity, electrical resistivity and carbon isotopes, before it even arrives at wells.[146] Carbon dioxide that has leaked to shallow levels and returned to gaseous form (from a supercritical fluid) stands out very strongly on seismic data. Satellite imaging can detect bubbles of carbon dioxide rising through sea water, and areas where vegetation is affected by leaks. Recent advances in electromagnetic surveys can highlight sub-surface oil and gas, and may be effective for carbon dioxide too, although they can currently only be used offshore. Gravity data can detect shallow carbon dioxide accumulations, while tilt-meters and satellite altimeter surveys pinpoint how the surface is rising[147] (by a matter of millimetres) in response to increases in underground pressures.[148]

Geo-mechanical techniques for estimating the strength of cap-rocks have been long-used in the oil business, and are already employed in the various pilot projects to ensure that pressures in the storage reservoir do not approach dangerous levels where they might fracture the seal. Micro-earthquakes can also be tracked, for any evidence that the seal is cracking. The combination of a variety of these techniques, most of them well-tested in the petroleum business over many years, can give a clear picture of where injected carbon dioxide is in the sub-surface, and where any leaks may be occurring.

Mathematical and computer modelling of fluid flows in the sub-surface has been used in the petroleum and water industries for many years, and, with the advent of high-speed computing and advances in 3D visualisation, has become highly sophisticated. Dynamic reservoir models can be used to predict the movement of injected carbon dioxide, and this can be compared to ongoing measurements and monitoring to detect any mismatches.

Current simulation packages are designed for studying oil and gas flows; they can be adapted reasonably to use for carbon dioxide injection, but there are still some required refinements, such as adding mineralisation and ionisation reactions, capturing the changes in density of aquifer water as carbon dioxide dissolves, modelling slow leakage through abandoned wells, and including potentially very large attached aquifers, which would be ignored in oilfield simulations but into which carbon dioxide may slowly move over centuries. As opposed to a typical oilfield simulation, which might extend 50 years into the future, carbon dioxide injection studies will have to make predictions up to 1,000 years or more away. Modelling will also become more complicated when several injection projects, possibly widely spaced, are using the same aquifer, or nearby aquifers which may leak into each other. Neighbouring operators will have to share data to ensure robust modelling results and avoid competing for the same storage space.

A key part of monitoring will be to detect when a storage site is becoming full. Locations will probably be chosen so that a conservative estimate of their capacity is more than enough for the anticipated need. As injection proceeds, it may become apparent that capacity is insufficient (in which case, a new location, preferably nearby, would have to be found). More often though (assuming the basic science is reasonably well understood), the initial cautious estimate of storage space will prove to be too low. There may then be the possibility of connecting another capture source.

Baseline surveys should be conducted to establish the natural flux of carbon dioxide at a location, over the yearly cycle; excess carbon dioxide in the future can then be detected. Fossil fuel carbon dioxide is isotopically different from carbon dioxide in the biosphere and atmosphere, so leakages from storage can be distinguished from natural systems.[149] Small rates of escape, of the order of 0.01% annually for 100 Mt total storage, would still be well in excess of natural fluxes, and should be easily detectable and quantifiable.[150]

Nevertheless, demonstration of methods for removing carbon dioxide from soils and groundwater would be useful to assure the public that any leaks could be cleaned up. The approach would presumably be to extract the water, and either clean it for reinjection or use, with the carbon dioxide being shipped to a more secure storage site, or to inject all the carbonated water back into a saline aquifer. Alternatively, slow seepage could be tolerated, as long as it did not threaten populated areas or ecosystems, and could be offset by more carbon capture elsewhere. The injection point responsible for leakage could be closed down, or reconfigured.

To enhance the security of chosen sites, proven barrier materials can even be injected sub-surface to form a 'wall' around the injection site.[151] This method is used, for example, in underground tunnelling operations. It could be done to ensure that carbon dioxide does not migrate towards an area of leaky faults or old wells.

Results of monitoring of existing projects to date are promising. Even small amounts of injected carbon dioxide change the density and the speed of sound waves through subsurface formations, and so can be picked up by seismic surveys. At Sleipner, layers only 1 metre thick can be 'seen' clearly, giving confidence in monitoring methods. Typically, it is hard to calculate the exact amount of carbon dioxide present, once its concentration rises above 5–10%, using only seismic data. However, it is very easy to detect the arrival of even small amounts of carbon dioxide where there had previously been none. The minimum amount of concentrated carbon dioxide detectable with seismic data is less than a day's injection at a typical storage project,[152] or even less at low concentrations.

The carbon dioxide at Sleipner has risen rapidly through some mudstone layers within the aquifer, leaving remnant accumulations behind

Figure 3.12. Seismic monitoring of carbon dioxide injection at In Salah (Photo: BP)

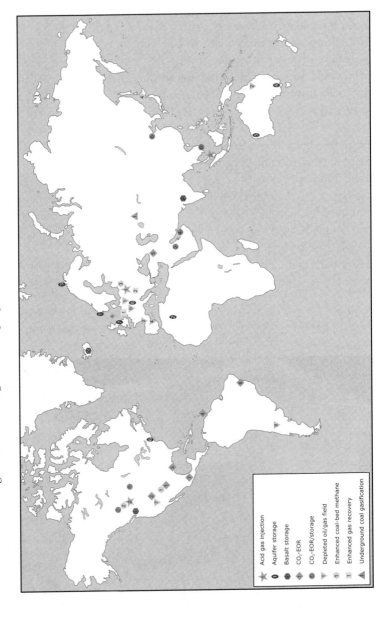

Figure 3.13. Existing and some proposed carbon dioxide storage projects

Aquifer storage

1. Sleipner (Norway)
2. In Salah (Algeria)
3. Snøhvit (Norway)
4. Otway (Australia)
5. Gorgon (Australia)
6. Mountaineer
7. Ferrybridge (UK)
8. Ketzin (Germany)

Acid Gas Injection

1. Borzęcin (Poland)
2. Canada (Zama and others)
3. Malaysia

CO₂-EOR/Storage

1. Weyburn (Saskatchewan)
2. Spectra (Alberta)
3. Ghawar (Saudi Arabia)
4. Abu Dhabi (UAE)
5. GreenGen (China)
6. White Tiger (Vietnam)

Basalt

1. Columbia River
2. Iceland
3. Deccan (India)

CO₂-EOR

1. Texas/New Mexico (multiple)
2. Rockies (multiple)
3. Eastern Gulf Coast (multiple)
4. Trinidad
5. Bati Raman (Turkey)
6. Recôncavo (Brazil)
7. Mexico (multiple)

Enhanced CBM

1. San Juan (New Mexico)
2. Fenn Big Valley (Alberta)
3. RECOPOL (Poland)
4. Hokkaido (Japan)

Depleted oil/gas field

1. Schwarze Pumpe/Altmark (Germany)
2. Callide A (Queensland, Australia)
3. Lacq (France)
4. Tarragona (Spain)
5. Germany (multiple)
6. North Sea (multiple)
7. Riley Ridge

Underground Coal Gasification

1. Angren (Uzbekistan)

Enhanced Gas Recovery

1. K12-B (Netherlands)
2. Budafa Szinfeletti (Hungary)

153

in each case, the main part being trapped against the overlying seal. The changes in repeat surveys ('time-lapse' or '4D' surveys, where time is the fourth dimension), taken at intervals of a year or more, show how the carbon dioxide is moving and highlight areas where even a small amount has accumulated. Seismic monitoring is critical for managing the effectiveness of EOR schemes, minimising the amount of carbon dioxide that has to be separated and re-injected, and potentially warning if carbon dioxide is approaching old wells or faults that may form leakage paths.

Conclusion

A variety of potential storage sites exist for captured carbon dioxide. Oil- and gasfields are the best-known, and can help pay for CCS via increased hydrocarbon production. Coal-beds offer a similar possibility. For a large CCS industry, though, these storage sites are probably insufficient in volume, and may not be located near the main points of emission. Therefore saline aquifers will also be required, potentially along with less-studied sites such as basalt formations. Ocean storage appears costly, short-term, environmentally problematic and probably unacceptable to the public. Converting carbon dioxide to solid minerals is feasible, and offers unlimited safe storage, but is slow and likely to be expensive, barring technological advances. Existing and some proposed projects are shown in Figure 3.13.

Transport of carbon dioxide is a safe and mature technology. Several lines of evidence and existing projects indicate that carbon dioxide can be stored safely below ground. Any leakage is unlikely to pose a major safety threat. Seepage could imperil climate targets, but likely leakage rates are acceptably low. Numerous monitoring and mitigation techniques exist to discover and tackle leakage. Indirect capture of carbon dioxide from air is a potential backstop in the case of some escape from storage.

4

BIO-SEQUESTRATION

'The world's forests need to be seen for what they are—giant global utilities, providing essential public services to humanity on a vast scale. They store carbon, which is lost to the atmosphere when they burn, increasing global warming.'

Prince Charles[1]

'It's very hard to know who killed someone twenty years ago to get a piece of land and who just arrived recently.'

Denis Minev, planning secretary for Amazonas State, Brazil[2]

There is not room here for an exhaustive discussion of all the various biological sequestration opportunities—that would require a separate book. But it is important to compare bio-sequestration[3] with techno-logical carbon capture, since the two are often presented as if they were alternatives. In fact, as we will see, they are complementary, and each has its own opportunities and problems.

Since the human addition of carbon dioxide to the atmosphere is a small fraction of the natural fluxes,[4] it follows that we could balance out our industrial activities by slightly increasing the rate at which the oceans, soils and biosphere absorb carbon dioxide. This is shown in Figure 4.1. The boxes indicate total stocks of carbon dioxide equivalent[5] and the arrows indicate annual flows, all in billion tonnes (Gt).

Figure 4.1. The carbon cycle[6]

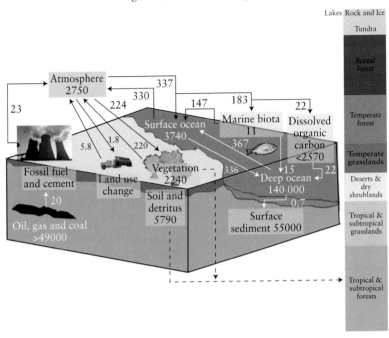

We can see that most biosphere carbon is held in dissolved form in the deep ocean. Additional massive quantities exist in fossil fuel reserves, which are being steadily added to the atmosphere. Vegetation takes up about 50% more carbon dioxide than it releases, with tropical, subtropical and boreal (northerly) forests being the main stores. However, land-use change, particularly deforestation, is releasing additional carbon dioxide to the atmosphere. Excess atmospheric carbon accumulates in the surface waters of the ocean, from where it gradually finds its way to the depths. This is partly offset by deep water emerging at the end of many thousands of years' journey in ocean currents, where it re-releases stored carbon dioxide to the air.

During the 1990s, deforestation contributed about 5.8 Gt of CO_2 annually to the atmosphere, 20% of global emissions. This greenhouse effect was supplemented by 3.3 Gt CO_2e[7] of methane and 2.8 Gt CO_2e of nitrous oxide from agriculture.[8] Over the industrial era, about 1,000 Gt of CO_2 has been emitted from fossil fuel burning and cement, about

500 Gt has been released by land-use change, mostly deforestation, and about 840 Gt has been absorbed, in roughly equal proportions, by the oceans and by land vegetation. So during this period, plant life has been a modest net source of carbon dioxide to the atmosphere.

Given that deforestation is such a major source of carbon dioxide, far larger than air travel, for instance, it is remarkable that it has received so little attention from business, government and many leading environmental groups.[9] We hear often about carbon dioxide emissions from industry in China and the USA, or from flying. Yet there is much less comparison made with the massive emissions from forest clearance. Renewable energy, nuclear power, CCS and efficiency do not address this major source of emissions at all. Therefore bio-sequestration is an essential part of creating 'carbon space', allowing us to allocate our limited ration of carbon dioxide emissions to wherever it is genuinely indispensable.

The way in which all life affects climate is very complicated. Vegetation both emits and absorbs carbon dioxide. Living creatures, including microbes and tree-roots, are crucial in speeding the weathering of rocks, the major long-term sink for atmospheric carbon dioxide. Biological activity affects climate on local, regional and global scales: by altering albedo (reflectivity), since dark trees absorb more heat than desert or white snow; changing surface roughness (and hence wind patterns); influencing precipitation and cloud formation; shading the ground surface; anchoring soil and holding back floods; releasing methane through bacterial activity and decay. Marine organisms mediate cloudiness over the oceans.[10]

Similarly, climate and the atmosphere affect life: plants initially grow more effectively with higher levels of carbon dioxide and warmer temperatures, but at some point they either lack water or nutrients, or become stressed by heat and reduced or unpredictable rainfall. Rising sea levels turn coastal areas into wetlands, and changing temperatures spread pests and diseases. For instance, warmer winters have allowed the mountain pine beetle to spread into British Columbia, where it has killed a billion trees and turned the province's forests from absorbers of carbon into net emitters.[11]

Humans, of course, have a major influence on the biosphere, and hence affect climate both directly and indirectly. There is chilling evidence for this even in the pre-industrial period. When Europeans arrived in the Americas during 1500–1750, the consequent massacres and pandemic diseases caused the native population to collapse. Farmland was abandoned, and this reforestation may have sucked carbon dioxide out of the atmosphere and so contributed to the 'Little Ice Age' of sharply lower global temperatures.[12] In the sphere of agriculture and land-use, it is worth mentioning:

- Deforestation (releasing carbon dioxide both directly, when trees are burned or rot, and indirectly as soils dry out).
- Re-forestation (replanting trees on previously deforested areas).
- draining of wetlands, which releases carbon dioxide but may reduce methane emissions.
- Use of fertilisers which enhance plant-growth (sometimes in unintended places, including the ocean) but emit the powerful greenhouse gas nitrous oxide (N_2O). We recall that use of captured fossil-fuel carbon dioxide to make the fertiliser urea can lead to carbon sequestration in soils.
- Forest fires (set deliberately for land clearance, or triggered accidentally at times of high temperatures and drought), which release carbon dioxide, nitrous oxide and black carbon, but whose sulphate aerosols probably help to cool the planet, unhealthy though they are for those in the vicinity.

At the moment, land-use, primarily agriculture, takes little or no account of its carbon impact, even in environmentally aware, developed countries.

Using biology to tackle carbon dioxide has five main components:

- Replacing fossil fuels with biomass/biofuels.
- Preserving and restoring forests.
- Enhancing carbon storage in soils.
- Burying or storing carbon from biomass.
- Enhancing ocean take-up of carbon dioxide.

Since both bio-sequestration and CCS involve managing atmospheric carbon dioxide, it is important to be able to compare their potential, costs, policies and impacts. As we have seen, the two are often seen as alternatives. In fact, biological carbon sequestration is in most ways very different from carbon capture and storage (Table 4.1). The two are therefore much better seen as complements, both key parts of climate strategy.

Table 4.1. Differences between biological sequestration and technological carbon capture

Biological carbon sequestration	Technological carbon capture and storage
More in developing countries that have large forests and/or agricultural sectors	More in developed countries, and some large industrialising countries
Relatively low-cost at first, rising over time	Relatively high-cost; costs probably fall in the medium term
Low capital costs, high operating costs and high opportunity costs (competing land uses)	High capital costs, sizable but lower operating costs
Technologically simple	Technologically quite complicated
Many small projects	A few large projects
Relatively easy to reverse	Investments, once made, are largely sunk
Removes carbon dioxide from the air	Prevents carbon dioxide entering the air (except for air capture)
Side benefits can be substantial, including improved soil fertility; reduced erosion and flooding; lower fertiliser run-off; preservation of biodiversity; ecotourism	Side benefits relatively few, but can include enhanced hydrocarbon recovery, and decreased emissions of other pollutants. Power plant research might also yield breakthroughs in efficiency
Large land area required	Small land area required
Total capacity limited by alternative land uses	Total capacity limited by underground storage space (probably very large)
Long-term sequestration insecure	Long-term storage expected to be highly secure
Somewhat difficult to verify exact quantities sequestered	Straightforward to verify exact quantities stored
Popular with environmental groups, but challenging social issues with implementation	Environmental groups have mixed opinions, and implementation may face public opposition
Key players: NGOs, agriculture/forestry companies, farmers, local communities	Key players: oil & gas companies, coal mining, power utilities

Biomass

Biological material can be used to substitute for fossil fuels. Waste plant matter (wood chips, rice husks, corn stalks and so on) can be burned for heat and power. Crops and trees can be grown specifically for energy: wood, burned in highly efficient stoves, is popular for domestic heating in Scandinavia, while fast-growing energy crops such as switch-grass and miscanthus, growing on marginal land, can be used in power stations. Finally, transport can be powered by biofuels. Sugars from crops such as sugarcane and corn can be fermented to give ethanol; oil from a variety of plants including oil palms, oilseed rape and jatropha can be processed into biodiesel; and 'second generation' biofuels break down the hard, woody material cellulose into usable fuels. These biofuels are now typically blended with fossil fuels, but could be used neat.

Biomass and biofuels are not themselves directly linked to carbon capture, except when biomass is used in CCS power stations or if algae become a viable means of carbon capture (Chapter 2). However, they are complementary to CCS in a carbon-constrained future. Decarbonising large-scale power, heat and industry is a huge task, but not technically so difficult. We have a good idea of the plausible solutions: renewables, carbon capture, nuclear and efficiency. Transport is much more problematic: we might have a breakthrough in electric cars in the next few years, but there is no obvious alternative to liquid hydrocarbons for fuelling aeroplanes. Shipping is also problematic, while the use of hydrogen as a fuel still seems a long way off. Views on the future availability of biomass vary widely, but agriculture will have to contend with climate change while feeding a global population of perhaps some 10 billion.

Industrial biofuels can equally be a contributor to climate change. Old-growth forests are cut down to establish monoculture plantations, and fossil fuels are used to grow corn and convert it, at low efficiency, to ethanol. There is a concept that biofuels can be grown on 'marginal' or 'degraded' land to avoid competing with forestry and food, but, in practice, this makes an already expensive fuel even more costly. In reality, most biofuels are grown in fertile areas, and displace forests more than farmland. The conversion of rainforests, peat or grasslands to biofuel plantations can release between seventeen and 420 times the carbon dioxide that these biofuels save annually, thus creating a 'carbon debt' that takes decades or centuries to pay off.[13] Rapidly growing energy crops such as hemp and sugarcane[14] tend to be hungry for nutrients and water, and fertilisers themselves emit greenhouse gases, both in manufacture and use.

The development of advanced biofuels would benefit greatly from genetic modification (GM), and if GM were used to boost crop yields, that would ease the competition between food and fuel.[15] However, GM is strenuously opposed by many people and environmental organisations.[16] Improvements in conventional farming techniques can also increase yields substantially, up to half for Indonesian and Malaysian palm oil, for instance.[17]

Due to the low conversion efficiency of biomass into biofuels, it therefore seems preferable to use biomass in electricity generation with CCS. This can either drive transport with electricity (or hydrogen), or be used for 'carbon negative' power which offsets transport's carbon dioxide emissions indirectly. Since biomass has to be gathered from a wide area, the practical size of biomass-fired power stations is much smaller than for fossil-fuelled plants, and so carbon capture is more costly, not benefitting from economies of scale. One solution is to co-fire coal plants with some biomass, as discussed in Chapter 2. For large parts of Africa and South America, though, where coal is currently little used, this may require a lot of infrastructure development and be very unpopular with environmentalists. In poorer countries with underdeveloped transport systems (roads, railways, inland waterways and ports), transporting biomass is even more costly, and so the effective area each power plant can draw on is smaller. A visit to rural Pakistan during harvest time vividly illustrated to me the chaos of moving the crop to markets on inadequate roads.

Another option, which avoids competition for fresh water and land, is to use algae. They grow quickly, can tolerate pollutants, and use saline or waste water. They can be harvested, and burned for power (with CCS, or carbon dioxide recycling to grow more algae) or used to produce biofuels or biochar. Sealed tubes of algae, 'photobioreactors' (PBR), could be bolted on to buildings,[18] with the added benefit of improving insulation or (in hot climates) shading the interior, and fed with carbon dioxide and waste heat from the building's generators and boilers. At the moment, though, PBRs are expensive, and would need more development and study of integrating such a system into an urban environment.

Forestation

Forests, and vegetation in general, contain almost as much carbon as the atmosphere. To avoid the ugly acronym ARD,[19] I group here under the heading 'forestation' the same concepts of avoiding deforestation, reforestation (replacing forests that were recently removed) and afforestation (growing forests in areas that were not previously forested). Obviously the dividing line between reforestation and afforestation is somewhat arbitrary, and depends on the length of time we define since the land was forested (typically twenty to fifty years is used). Forestation also covers changes in logging and forestry management to increase carbon sequestration, and niches such as urban forestry and agro-forestry (combining growing trees and crops).

Forests have been cut down throughout human civilisation, and approximately 10% of all land is now cropland, 22% pastureland and 1.5% urban areas.[20] Tropical forests are the richest store of above-ground carbon, and tropical deforestation alone currently accounts for about a quarter of greenhouse emissions, with about 130,000 square kilometres being lost annually,[21] equivalent to some 2.5% of the Amazon rainforest. A square kilometre of tropical forest releases about 25–35 thousand tonnes of CO_2, if burned,[22] implying that the Amazon holds some 130 Gt of CO_2-equivalent. Of the total carbon stocks in tropical forests, 50–65% is held in Latin America, 25–40% in Africa and 10–15% in South East Asia—the wide variation in these figures being indicative of some of the uncertainties in definitions and measurement. Moist temperate forests, which grow fairly quickly and rot more slowly than tropical forests, are also important,[23] while tundra (the cold, sparsely vegetated areas of northern Russia, Canada, etc.) are rich stores of underground carbon due to the low decay rates.

Deforestation is currently concentrated in some tropical countries (Figure 4.3). As Figure 4.2 shows, in a similar way to coal use, the leading emitters are responsible for four-fifths of emissions: Indonesia and Brazil in particular. This is reasonably encouraging for the prospects of success of a concentrated effort. However, despite two decades of campaigns, only a small fraction[24] of tropical timber comes from sustainable sources.[25]

Figure 4.2. Carbon dioxide from deforestation, by country[26]

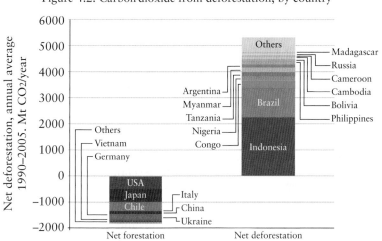

Figure 4.3. Annual rate of deforestation, 2000–2005. Green (negative numbers) indicates net reforestation; yellow, orange and red (positive numbers) net deforestation[27]

	>3%		1% to 1,5%		− 0.5% to 0%
	2% to 3%		0.5% to 1%		− 3% to − 0.5%
	1.5% to 2%		0% to 0.5%		− 7% to − 3%

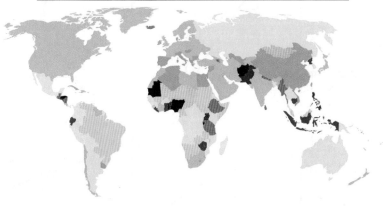

In the absence of further action, deforestation will continue, albeit at diminishing levels, through to 2100.[28] By that year, worldwide projections suggest deforestation will constitute only a sliver[29] of anthropogenic carbon dioxide, since (in the business-as-usual case) other emissions will have grown. The rate of deforestation gradually diminishes, as accessible old-growth forests are removed, a slowing in population growth and increasing agricultural productivity reduce pressure on farmland, and wealthier societies place higher value on forests, for recreation, climate and ecological services: an example of the 'environmental Kuznets curve', discussed in Chapter 6. In the USA, land-use change was a large source of carbon dioxide in the nineteenth and early twentieth centuries, as forests were cleared for agriculture and fuel, but by the second half of the twentieth century, land became a net sink of carbon. However, the rate of absorption peaked around 1960 and has fallen subsequently.[30] In Europe, regrowing forests take up about a tenth of the continent's carbon emissions.[31]

In a warmer, CO_2-rich world, forests overall may be more luxuriant and hold more carbon, but in some areas, drought will cause forests to

Figure 4.4. Deforestation in Brazil (Photo: istockphoto.com)

die back to grasslands, or, at least, sparser forests thinned out by dry seasons and frequent fires. This is particularly a threat for the Amazon.[32] Some recent studies suggests that tree-lines in general are advancing at higher altitudes and latitudes, as winters are warmer and trees are more able to survive.[33] Changes in forest cover feedback on climate, either negatively (cooling) if growth wins out, or positively (enhancing warming) if drying and desertification dominate. If forests are growing more strongly, then forestation becomes a more attractive method of carbon mitigation. These coupled effects of climate and economics are highly complex, and modelling is only just beginning to tackle them.[34]

Deforestation can be reduced or halted by initiatives including:

- Stopping illegal logging.
- Giving the poor land rights.
- Encouraging the sustainable use of forest products.
- Setting aside forestry reserves.
- Finding non-destructive uses of forests, such as tourism.
- Paying for forestation and/or taxing deforestation.
- Reducing firewood use with more efficient stoves and alternative fuels.
- Introducing agricultural practices that make more productive use of available farmland.
- Encouraging agro-forestry, a useful compromise between food and Carbon sequestration.
- Employing reduced-impact logging to replace the current, highly wasteful practices. This can cut carbon losses by a third.[35]

Similar incentives can be used to encourage reforestation and afforestation.

Young trees take some time to reach their maximum rate of carbon uptake: twenty years for loblolly pine in the southern USA, but seventy-five years for Ponderosa pine in the Rocky Mountains.[36] The rate of absorption for US examples is 800–4,000 tonnes of CO_2 per square kilometre per year. Trees also do not absorb carbon forever; new forests continue gaining carbon for typically twenty to fifty years after planting,[37] but they contain much less carbon per hectare than old forests.[38]

As the forest matures, carbon storage in soils, roots, litter (forest floor detritus) and undergrowth may continue to increase, although at a slower pace. At some point, older trees cease to grow, those that die are replaced by younger trees, and the forest reaches steady-state. The capacity of forest sequestration is therefore limited by the amount of land that can realistically be covered with trees. For these reasons, although reforestation measures can be implemented immediately, they take a long time to give their full benefits. This complicates comparing them to other carbon mitigation methods (CCS, shifting away from fossil fuels, and avoiding deforestation) that take effect immediately.

With the gradual elimination of old-growth forests, wood is now supplied increasingly from renewable plantations, which are replanted after felling. Plantations can be established on marginal land and degraded soils, in which case they sequester carbon and have positive impacts including erosion prevention. In some areas of the world, notably Australia and the Sahel in Africa, large-scale tree planting might be sufficient to flip the local climate into a wetter state and hence permanently reverse desertification.[39]

Rising carbon prices would encourage a slower rotation of renewable forests (i.e. they would be allowed to grow older before being harvested) and would also make hardwood more attractive relative to pines.[40] This could enhance carbon sequestration at modest additional cost. Carbon dioxide fixation can also be increased by genetic engineering; Taiwanese and American scientists have developed a strain of the tropical eucalyptus that takes up to three times more carbon dioxide than normal varieties. It has more cellulose (the hard, woody material) and less lignin (the 'glue' that binds the cellulose together), and is therefore more suitable for making paper and biofuels.[41]

However, monoculture plantations are not really comparable to the natural forests they replace, a point forcefully made by 'Gaia' guru James Lovelock.[42] They may appear 'sustainable' and do have significant value for sequestering carbon and providing useful products. However, plantations of fast-growing pine contain less carbon than slow-growing hardwood or natural pine forests. They are not real ecosystems, and so are not resilient in the face of climate change, disease or fire. They lack biodiversity and have very little attraction for ecotourism. Therefore the key role of plantations is to prevent further logging of old-growth forests; they should not be used to replace natural woodlands. These issues can be partly addressed by planting mixed stands rather than monocultures, which improves the plantation's biodiversity and pest resistance. As Indian agro-forestry pioneer Shonil Bhagwat suggests, mixed plantations can be interspersed with natural forests to create a continuous habitat for wildlife.[43]

In Africa, in particular, fuel-wood is an important driver of deforestation. This can be tackled by substituting commercial fuels (in rough order of increasing income: kerosene, LPG, natural gas/electricity) or renewable cooking methods such as solar ovens, introducing more efficient wood stoves, and possibly via biochar (see below). In Cameroon, I commissioned an example of the Kenya ceramic jinko.[44] It is one example of a simple cooking stove that can be made by local artisans, so creating employment and recycling materials. By saving wood, it not only reduces deforestation, and saves back-breaking labour in wood-gathering, but also reduces smoke inhalation. Such low-tech approaches require a lot of grassroots work (literally) to encourage people to use them. They are, though, within reach of the budgets and technological capability of millions of householders and small workshops.

In South America, 'slash and burn' agriculture may be more of a problem, where farmers clear an area to grow crops for one to three years, until the fertility is exhausted, and then leave the land fallow for two to twenty-five years. They move on to somewhere else, leaving the forest to re-grow (or permanent settlers to move in). The low productivity of this kind of farming suggests that it could be tackled by encouraging longer-lived occupation of the land with investment in modern agricultural practices. This, though, requires giving the poor secure tenure over their property, and protecting them from powerful landowners.[45]

In many parts of the world, commercial logging has not succeeded in achieving either environmental protection or poverty reduction. For example, timber has been a source of revenue for rebel groups in West Africa.[46] In many tropical countries, most logging is illegal—some four-fifths of total deforestation in Indonesia.[47] Reversing deforestation is therefore connected to issues of international development, strengthening governance and local democracy and tackling corruption and instability.[48] Given that four-fifths of Brazilian logging was illegal in 1997, the improvement to no more than three-fifths by 2006 does encourage us to believe that such policies can have some success.

Verification of forest carbon sinks is challenging. Forestation can be monitored from space. But definitions can be somewhat problematic, since forest has generally tended to be measured by a simple parameter of canopy cover. This can disguise considerable differences between sparse woodland, a degraded forest and a dense forest in good condition. Harvesting of smaller trees or removal of small pieces of wood can leave the canopy untouched but considerably reduce the carbon stock. Carbon in soil and roots requires on-site monitoring to check, and may vary considerably across a region.

Forests can also be destroyed by natural processes (fires, floods, pests) or re-grow on abandoned lands; this has an effect on the atmospheric carbon balance, but should a country be able to count this towards its international climate targets? In some cases, it may be very hard to disentangle natural and human-caused changes. Of course, if a country or company has received carbon credits for establishing a forest, then it should also be liable for the subsequent destruction of that forest, whether naturally or artificially. Large year-by-year variations in forest cover, with vegetation being destroyed by fire or drought one year and re-growing the next, would lead to volatile carbon payments to and from the country, probably hard for a poor nation to manage.

Rising carbon prices will inevitably encourage forestation, reducing the land available for agriculture and hence increasing food prices. Forestry products and food are traded worldwide, so simple mandates to convert certain lands to forests will be mostly or entirely offset by deforestation to arable land elsewhere, unless measures are implemented on a near-global scale. The success in stopping deforestation in China and Vietnam has been matched by a corresponding increase in logging, often illegal, in the neighbouring countries of Laos, Cambodia and Indonesia.[49]

Balancing the need for food, fuel, water and carbon sequestration is likely to be a major twenty-first-century challenge.[50] Those commentators and members of the public who, when confronted with climate change and carbon capture, blithely reply, 'Plant more trees', are clearly unaware of the complex issues of forestation in the real world.

Land-Use and Soils

Options for reducing carbon emissions from land-use include enhancing soil carbon via conservation tillage and other methods; preserving peat; and extinguishing underground coal fires.

Soils contain organic carbon in the form of microbes, decaying plant matter, aggregated particles and chemical complexes. The amount of soil carbon varies greatly between different land types, and has tended to decrease with intensive agricultural use. It used to be thought that organic waste from farming should be ploughed back into the fields, but it now appears that this disrupts the structure and microscopic life within soil, and so causes loss of carbon.

Conservation tillage leaves the crop residue on the surface, and can increase soil carbon content by up to half.[51] Low- and zero-tillage increased significantly in the USA since 1985, driven by government incentives, advances in planting techniques and weed control, and drives for improved soil conservation and water quality. One square kilometre of farmland can sequester about 1,100 tonnes of CO_2 using no-till agriculture.[52] There are additional savings in greenhouse gases due to lower use of farm machinery, and less requirement for artificial fertilisers. Yields can also be higher, yet only a third of agricultural lands have adopted conservation tillage.[53]

Conservation-based tillage has to be applied over several years to build up carbon stocks. But a single season of intensive tillage (due to weather or crop change) can release all of the previously stored carbon.[54] Conservation tillage may also be ineffective in the tropics, where high heat and humidity destroy the organic matter, and tends to decrease yields on well-drained soils. Yields are also lower in the early years, and more variable with time, while farmers perceive higher risks from minimum tillage as against conventional methods. This, with deficits of knowledge and capital, explains the low level of adoption so far. Conservation tillage requires a much larger acreage for the same carbon storage as forestation, but, of course, allows agriculture to continue.

Peat bogs, water-logged areas filled with partly decayed plant material, cover about 3% of the world's land surface. These are rich stores of carbon, some 2,000 Gt CO_2, the bulk in Russia and the Americas, and some in South East Asia.[55] Peat and permafrost in northern latitudes are also decomposing, mainly due to rising temperatures, releasing not only carbon dioxide but also more potent methane. Peat soils store 30 times more carbon than the same area above-ground in tropical forests. Peat areas are being cleared, particularly in South East Asia, for logging, transporting logs in drainage canals, and establishing palm oil and other plantations. As the peat dries out, it decomposes and burns, producing smoke clouds that blanket neighbouring countries. Almost half of Indonesia's peat-lands are being degraded in this way. As well as massive carbon dioxide emissions, this leads to pollution and respiratory diseases across a broad swathe of South East Asia, and the destruction of biodiversity,[56] including the orang-utan in Borneo and the Sumatran tiger.

Destruction of peat can be reduced by similar methods as for fighting deforestation: eliminating incentives for logging and biofuel cultivation, and repairing wetlands sustainably by establishing fish-ponds, and closing drainage channels with dams. The key is to restore the water table and keep the peat wet, since then it is very unlikely to burn.

Underground coal fires present a somewhat different problem. These occur where natural deposits of coal catch fire, during mining, when old mines were abandoned, from combusting waste tips on top of coal seams, or perhaps from lightning strikes or spontaneous combustion. Coal fires in Indonesia are often ignited by forest or peat fires above them, so combining two problems. They can burn for decades or even centuries: the oldest known, Burning Mountain in New South Wales, Australia, has been going for 6,000 years.[57] They are known from the USA (especially Pennsylvania), but are most widespread in China and India. As well as the carbon dioxide emissions and the loss of a useful natural resource, these coal fires produce other pollution, contaminate groundwater and cause severe subsidence.

Extinguishing these fires is dangerous and expensive. It is done by a combination of methods: for surface mines, removing coal around the burning site; and for underground fires, sealing the air vents so that no oxygen enters and the fire's own carbon dioxide smothers it, introducing an inert gas such as nitrogen, and pumping in water or mud to cool the coal.

Carbon Sequestration and Biochar

Carbon sequestration can be applied directly to biomass in three ways:

- Biomass can be used in power stations, possibly co-fired with coal, and the resulting emissions captured to give 'carbon-negative' power. This option has been discussed in Chapter 2.
- Unburned biomass can be stored or buried, removing its carbon from the biological cycle.
- Biomass can be partly burned for energy, yielding a carbon-rich product, biochar (essentially charcoal). Biochar can be buried, or added to soils.

A large amount of harvested wood is used for products: timber for buildings, furniture and so on. About a third of wood is made into **long-lived products** (greater than twenty-five years) and slightly less into shorter-lived products.[58] Carbon sequestration could be considerably increased by using more wood in buildings, and by recycling wood from demolished structures. Increasing timber construction has the additional advantage of saving on energy-intensive steel, aluminium, glass, brick and concrete. However, action to prevent deforestation will tend to drive up the price of wood products and so discourage their use.

It has been proposed that **unburned biomass**—crop residues and so on—could be simply buried rather than burned. This is somewhat similar to the original geological processes that formed coal. Three methods have been proposed: grinding the waste to a slurry which could be pumped underground; burying it in river deltas, such as the Mississippi, where it would be quickly covered by sediment; and sinking it in the deep ocean with weights attached.[59] The first method seems likely to be expensive and it would be difficult to transport the waste. The second has the problem that decomposition to carbon dioxide and methane is likely to be relatively fast, perhaps within a few years. The third is therefore the most plausible. The deep seas are cold and contain minimal oxygen, so breakdown would be very slow.

However, the overall concept appears flawed from the viewpoint of energy balance. If we burn 1 tonne of carbon from biomass, we gain enough energy to run a 100 watt bulb for about a year.[60] If we bury this carbon, and burn an equivalent quantity of natural gas, we could run the bulb for two years, while also remaining carbon-neutral. But if we do both—burn the biomass and capture and store its carbon dioxide—and then burn the natural gas, we can light the bulb for almost three years,[61] Indeed, it makes little sense to be digging up fossil fuels to burn, while simultaneously burying un-burned biomass without extracting its useful energy.

'**Biochar**' is an approach that has received much attention recently. Wood, dedicated fast-growing energy crops, waste matter and other biomass can be burned in simple stoves to give charcoal. Once mass-produced, these stoves should be comparable to or cheaper than alternatives that burn kerosene or LPG.[62] Suitable waste biomass includes rice husks, nutshells (groundnuts, coconuts, peanuts, walnuts, etc.), sawdust, residues from pulp mills, bagasse from sugarcane and so on. Pyrolysis, heating in the absence of oxygen, can also yield usable liquid fuels and gases. More advanced modern pyrolysers can use the waste gases to provide the necessary heat. About a quarter of the carbon in the biomass is retained in the biochar, and a further quarter goes into the useful bio-oils and bio-gases—this can be tuned by varying the temperature. Other feedstocks can also be used—for instance, plastics and sewage sludge—although more research is needed on possible heavy metals and other contaminants. As mentioned in Chapter 2, it may even be possible to use biochar and biogases in a scrubbing process for carbon capture in power plants.

Figure 4.5. Biochar. (Photo: Christoph Steiner, Biochar.org)

When buried in soils, this biochar not only sequesters carbon, but is also claimed to have other benefits, including reducing leaching of nitrogen into groundwater, possibly reducing emissions of the greenhouse gas nitrous oxide from fertilisers, retaining nutrients and hence enhancing fertility, moderating soil acidity, increasing water retention and preserving microbes and fungi that are essential for root systems. Even forest fires will only burn the top few centimetres of soil, preserving the deeper biochar.

Biochar is potentially attractive for carbon sequestration because it is simple and relatively cheap, requires no breakthrough technologies, can be applied in developing countries (hence improving the situation of the poor, and potentially providing supplementary income), appears to store carbon for long periods, and has other non-climate benefits. The very fertile *terra preta*[63] soils of the Amazon Basin were apparently created from 5000 BCE to 1500 CE by farmers who accumulated charcoal, manure and household wastes.[64] The Japanese 'charcoal soils' are similar. A biochar pilot is taking place near Port Orford in southern Oregon, USA, while an industrial tree plantation in Indonesia sequesters 0.2 Mt CO_2 per year of biochar,[65] and in the Maldives, on the front-line of sea-level rise, a plan has been launched to make biochar from coconut shells.[66]

Consider that 1 km^2 of land can sequester about 1,100 tonnes of CO_2-equivalent if cropped with zero-till agriculture, around 30,000 tonnes if covered with tropical forest, but as much as 92,000 tonnes if treated with biochar. The land can then still be used for agriculture or forestry.

However, used in humus-rich moist soils (Sweden, in the example studied), biochar can cause loss of soil carbon;[67] it is, therefore, more appropriate for nutrient-depleted tropical soils. Some studies showed that crop yields were initially higher but fell away after three or four years. It has still not proved possible to replicate the *terra preta* soils, and it may take a century or more for soils treated with biochar to convert to *terra preta*. Furthermore, there is still a lot of conflicting evidence about how much black carbon accumulates in soils, and how and at what speed it breaks down.[68] Up to a fifth of carbon in biochar is rather unstable and decomposes rapidly. A large-scale application of biochar, which has a high energy content, might lead to the expansion or evolution of microbes capable of eating it.

Pyrolysis also creates a wide variety of chemicals, some of them carcinogenic, which might accumulate in soils and food. Some of the biochar will be eroded and carried into rivers and the ocean, where it may either degrade or be buried permanently in sediments. The process of actually burying the biochar might break up the soil and encourage erosion, while if production is done badly, or if its dust blows around, then biochar might contribute to global warming, since black particles absorb the sun's heat effectively. Almost a third of biochar can be lost while it is being applied, as field trials in Canada showed.[69] Therefore considerably more research is required into the lifetime of biochar, its effects on fertility and soil carbon, and the climates where it is most effective.

Accounting for these long-lived pools of sequestered carbon is rather tricky. It is analogous to the case of leaky geological storage discussed in Chapter 3. Sequestration in wood products is short-lived, and net removal of carbon dioxide ceases once wood use stops growing.

The half-life of stable biochar (the time taken for half to break down to carbon dioxide) is thought to be about 5,000 years in cold wet environments,[70] equivalent to about 0.015% leakage annually. This is of a similar timescale to that for removal of carbon dioxide by natural weathering, and in line with some suggested limits for geological storage, so appears acceptable. The lifespan may be shorter, though, in other environments, perhaps as little as a century for less stable biochar exposed at surface.[71] A biochar programme could always include a small allowance for the ongoing degradation of a small part of the stock to carbon dioxide. However, since about half of the carbon in biomass is released as carbon dioxide when biochar is manufactured, then the short-term climate situation only gains if we use feedstock that was going to decay rapidly or be burned anyway. It would make no sense, for instance, to pyrolyse long-lived carbon stocks like trees or plastics, since the 'carbon payback' time is too long.

Ocean Fertilisation and Other Geo-Engineering

A completely different approach to bio-sequestration is to look at the sea, rather than the land. As Figure 4.1 shows, the oceans are a vastly greater reservoir of carbon than terrestrial ecosystems, and have historically taken up about half the excess carbon dioxide pumped out by human beings. Yet marine biological productivity over most of the oceans is low, averaging only as much as a desert, partly due to lack of necessary nutrients.

The choice of fertiliser depends on what the limiting nutrient for growth is. One plan, put forward by Australian scientist Ian Jones, suggests using the nitrogen-rich chemical urea[72] in coastal waters. James Lovelock has proposed using thousands of small pipes to bring up nutrient-rich water from the deep ocean. Other schemes involve phosphates; already fertiliser run-off from land may be accidentally contributing to sequestering as much as 1 Gt CO_2 per year.

However, most attention so far has been given to iron. Iron is attractive because it is only required as a trace nutrient: 1 tonne of iron can theoretically fix 83,000 tonnes of CO_2.[73] The iron is taken up by algae (microscopic plants) which use it to grow, taking up carbon dioxide via photosynthesis. When they die, they sink. Part reaches the bottom, becoming incorporated into the sediments, and so the carbon is permanently removed from circulation. Another part dissolves in the deep ocean where, as we have seen in Chapter 3, it is isolated from the atmosphere for several centuries. In parts of the Southern Ocean and equatorial Pacific distant from land, iron is the limiting nutrient, being supplied only in small quantities by wind-blown dust and fall-out from volcanic eruptions.

Ocean fertilisation can have some additional benefits. Plankton emit dimethyl sulphide, a volatile liquid that promotes cloud formation over the seas. This can assist cooling. Increased algal growth can also provide food for fish and hence potentially for humanity.

Nevertheless, experiments on ocean fertilisation so far have not been very promising. Fertilisation efforts at some sites were limited by the lack of other nutrients, particularly silica, and the blooms created were quickly devoured by fish and so had no carbon sequestration effect.[74] Fertilising in short pulses may allow the algal material to sink before the slow-growing animal life can multiply enough to eat it. It appears, though, that only a small fraction of algal particles actually sink out of surface waters.[75] Phosphorus may also be a limiting nutrient, but if mixed together phosphorus makes the iron inactive, so they have to be added separately in low concentrations. Phosphate would be needed in large quantities, and is relatively costly.

A major algal bloom may deplete the oxygen in deeper waters, affecting marine life, and consume other nutrients which would normally be carried by currents to different parts of the ocean. Hence, we may enhance fertility in one place, at the cost of diminishing it in another. Algal blooms may feed creatures such as jellyfish and disrupt marine food chains, including whales, while the decomposing organic matter can cause 'dead zones' in the deep sea, depleted of oxygen.

The idea of bringing up ocean water via pipes seems in practice counter-productive, because the deeper water contains more carbon dioxide which is liberated when it reaches the surface. The required mixing rate to have any real impact on climate is inconceivably huge,[76] and on such a scale would probably disrupt marine ecosystems.

Verification of the total amount of carbon dioxide sequestered is also highly problematic,[77] as is potential leakage some centuries hence when bottom waters return to the surface. Ocean fertilisation has attracted significant opposition from environmental groups—although it is, arguably, a less drastic modification of marine ecosystems than that which we routinely inflict on the terrestrial environment. For these reasons, continuing activity is low and it seems unlikely that ocean fertilisation will emerge as a major mitigation technique in the near future.

Ocean fertilisation is one of a group of techniques that can be called geo-engineering: the purposeful modification of the Earth on a large scale for human benefit.[78] Most geo-engineering techniques proposed to date have been concerned with preventing or slowing global warming. Air capture (Chapter 2) is another. Both air capture and ocean fertilisation are Carbon Dioxide Removal (CDR) techniques.

Somewhat similar to air capture would be methods for removing non-CO_2 greenhouse gases from the atmosphere. I have not seen this idea proposed elsewhere, and it is not really a carbon capture technique. However, in the case that hydrates begin to decompose due to rising temperatures, releasing vast amounts of methane, it would be essential to find a way of eliminating the gas from the atmosphere. Methane is a much smaller component of air than is carbon dioxide, but it is also much more reactive and easier to destroy. Speculative possibilities for removal might include some kind of biological method, since some existing bacteria metabolise methane; adaptation of carbon dioxide air capture to destroy methane as well; or release into the atmosphere of small quantities of a suitable catalyst.

The other geo-engineering methods proposed to date fall under Solar Radiation Management (SRM), by reflecting or blocking a small amount of the sun's rays. SRM methods do not involve storing or sequestering carbon, and so I do not consider them in detail in this book. They include putting giant 'sunshades' in space to block out a portion of the sun's rays; injecting sulphate particles into the upper atmosphere to reflect solar radiation; spraying sea water drops into the turbulent air under marine clouds to increase their reflectance;[79] and using more reflective materials for urban areas (which has only minor potential). Geo-engineering techniques have the advantage of potentially being much cheaper than other large-scale mitigation methods, and SRM can be significantly faster than CDR.

SRM approaches, though:

- Do not address the underlying problem, of rising atmospheric carbon dioxide (and hence ocean acidification).
- In the absence of any large artificial sink for carbon dioxide, would have to be maintained and gradually wound down over centuries.
- Lead to abrupt warming if stopped.
- May have substantial unintended consequences, since we are tampering with the whole Earth system of which we have a very limited understanding.
- Solve the temperature problem, but can lead to other climate change by affecting rainfall and cloud cover. Some nations may therefore suffer and others gain.
- Are likely to meet major public opposition.

Geo-engineering approaches are, then, potentially problematic, but it is probably worthwhile giving them some research attention now, which could be done very cheaply. This would provide some insurance against the most frightening and disastrous climate scenarios. As environmental economist Martin Weitzman says, 'With the unfortunately limited information we currently possess, geo-engineering...looks at first glance like an incredibly cheap and effective...emergency response—albeit with largely unknown and conceivably nasty unintended consequences... there is an acute—even desperate—need for a more pragmatic, more open-minded approach to the prospect of climate engineering.'[80]

5

SCALE, COSTS AND ECONOMICS

'*Energy efficiency is the lowest-cost solution, but CCS is not far behind.*'
Steven Chu, United States Energy Secretary[1]

'*CCS is likely to be cheaper and quicker than building onshore and offshore wind.*'
Stuart Haszeldine, Professor of Geology, University of Edinburgh[2]

We have established that carbon capture should be technically feasible, safe, and capable of playing a major role in fighting climate change. But how big a role, and, crucially, at what cost? If fossil-fuelled power stations with carbon capture can deliver electricity at a lower price than renewables, it is almost inevitable that they will be a major part of twenty-first-century energy. And do forestation and agricultural practices offer large and low-cost abatement? Should we be focussing our attention on bio-sequestration rather than technological fixes?

CCS's Role in Emission Reductions

It is sometimes argued that CCS cannot be built fast enough to make a difference in the short term, or, alternatively, that it can never make up a large part of emissions reductions, or that the scale of a complete CCS system would be unfeasibly large.

Firstly, Greenpeace and others often repeat the claim that carbon capture cannot be ready before 2030, while large emissions reductions are needed immediately. But it takes about six years to build a new coal plant in Europe or the USA; allowing for some additional time for permits suggests a total of between six and ten years. Installation in industry and in many developing countries could be faster. So it is eminently feasible to have a substantial demonstration programme in place by 2020.

Several reports indicate that CCS could already be a major source of carbon abatement by 2030, up to 15% of current emissions.[3] To get the same amount of carbon savings with wind power would require more than 1 million large wind turbines. Figure 5.1 shows McKinsey's view, where red bars indicate emissions, and green show potential cuts. This chart indicates that carbon capture could be the second-largest contributor to emissions reductions in the power sector by 2030. CCS's contribution here is somewhat less than that of energy efficiency, but way ahead of renewable power.

On the second point, several comprehensive studies show that CCS has the potential to be a large contributor to emissions reductions, at

Figure 5.1. Potential emissions reduction in the power sector, 2002–30[4]

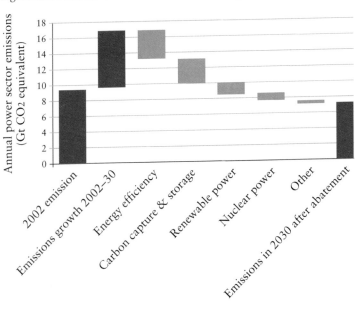

least on the scale of other approaches such as wind power. Climate change is such a big issue that any meaningful approach to tackle it has to be on a huge scale. All the leading contenders are, in magnitude, similarly plausible or implausible, depending on your point of view.

Large sources provide most emissions. For instance, four-fifths of Chinese emissions from major stationary sources come from just 600 or so facilities.[5] Globally, 8,100 large fixed sources produce some two-thirds of emissions. This makes the task of fitting CCS seem more manageable—certainly, several thousand carbon capture projects do not seem more daunting than the several million wind turbines that would be required for comparable carbon reductions.

Total current technically capturable emissions from large stationary sources amount to some 19 Gt of CO_2 annually (Figure 5.2). Of this, somewhat more than half comes from coal-fired power stations, by far the largest opportunity; about a quarter from industrial sources, and the remainder from power stations burning gas, oil, petroleum coke, biomass and waste. Of the industrial sources, about a third is represented by iron and steel, and another third by cement, for both of which it is rather challenging to implement capture. The remaining third is a mix of smaller opportunities, including some (such as ammonia, hydrogen, contaminated gas and DRI) which are ideal for capture, and others (carbonation of alumina, fly ash, waste concrete and steel slag) which use by-products to implement low-cost, albeit rather small-scale, capture via mineralisation.

There are, of course, numerous scenarios for emissions growth in a 'business-as-usual' case, for the possible levels of reductions, and how they could be achieved. It is useful to see these figures in terms of Socolow wedges, each wedge achieving around 3.7 Mt per year of reductions by the year 2050. The IPCC suggests that, in a low fossil fuel world, CCS might represent a significant contribution, 4.7 Gt, about one-and-a-half wedges. In a high fossil fuel world, CCS is the majority of the solution, as many as ten wedges (37.5 Gt). Other studies suggest a reasonable maximum around 11–16 Gt,[6] between three and four-and-a-half wedges.

To meet, for example, the IPCC's lower target, CCS would have to be fitted to fewer than 1,000 typical coal-fired power stations.[7] This figure sounds substantial, but the equivalent of about 200 such stations are now being built each year. If this were to continue to the year 2050, equipping a little more than 10% of new plants with CCS would be enough to reach the target. To reach the IPCC's high case for emis-

Figure 5.2. Total current CCS potential from power and industrial sources[8]

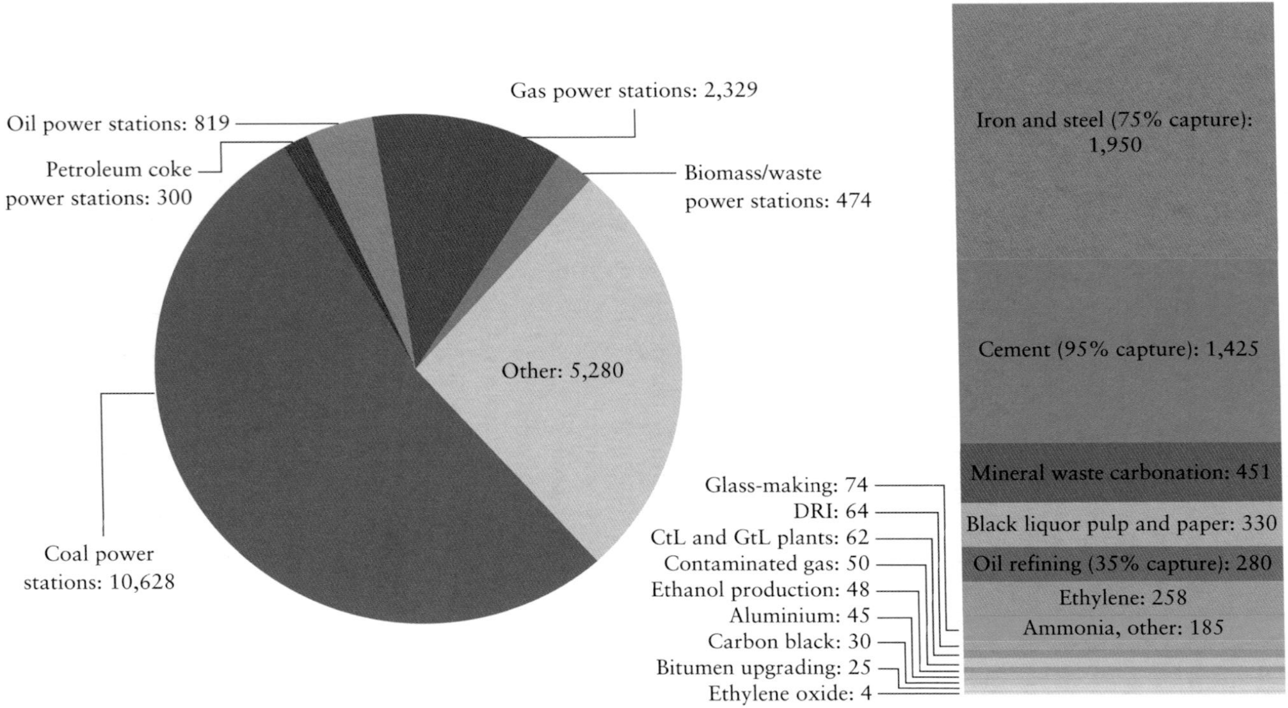

Iron and steel (75% capture): 1,950

Cement (95% capture): 1,425

Mineral waste carbonation: 451
Black liquor pulp and paper: 330
Oil refining (35% capture): 280
Ethylene: 258
Ammonia, other: 185

Glass-making: 74
DRI: 64
CtL and GtL plants: 62
Contaminated gas: 50
Ethanol production: 48
Aluminium: 45
Carbon black: 30
Bitumen upgrading: 25
Ethylene oxide: 4

Biomass/waste power stations: 474

Gas power stations: 2,329

Other: 5,280

Oil power stations: 819

Petroleum coke power stations: 300

Coal power stations: 10,628

sions, then all these new coal plants would have to include CCS. This is obviously challenging, but again conceivable, certainly if combined with retro-fits of existing plants, and with CCS implementation on gas power stations and industrial sites as well. For instance, the IEA's scenario for strong climate change mitigation would fit CCS to half of iron, steel, cement, pulp and paper and ammonia plants by 2050.

Figure 5.3 shows a scenario for CCS implementation to 2030, culminating in almost 800 active projects. This compares to about 100 currently in existence or under development, which, even conceding that many current projects are small-scale pilots, still gives reasonable encouragement that the task is manageable. Approximately one-third of the projects are industrial rather than power plants.

Figure 5.3. CCS project build-up to 2030[9]

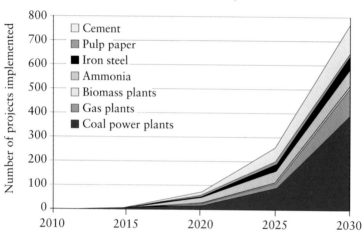

The IEA envisaged that about a third of the CCS capacity in 2050 would be from retro-fitting existing plants, with the remainder being new-build facilities, mostly IGCC. Plants built before 2005 are generally too inefficient to make retro-fitting worthwhile. In general, the higher the cost of carbon, the more worthwhile it will be to retire old plants and build new ones equipped with CCS, rather than retro-fitting.

Existing projects at Sleipner, In Salah, Weyburn and Snøhvit plus various pilots and EOR schemes currently capture about 15 Mt per year. If all current plans go ahead, then this will increase by an order of magnitude, to 150 Mt per year by 2014. The slow progress so far means that this date will almost certainly slip back, and inevitably some projects will be delayed further or cancelled. New projects, though, will be announced, and could probably come on-line from 2016 onwards if started today. Some existing operations, particularly those involving natural gas processing, have scope for rapid and straightforward expansions.

The plausible rate of project build-up over the next decade (Figure 5.4) is just about fast enough to match one Socolow wedge. Existing projects are in green, planned projects in red and proposed ones in grey. For comparison, I have shown the carbon mitigation effect of solar power up to 2008, assuming that it replaces inefficient coal plants.[10] It might be surprising to some that CCS, even at its current early stage of deployment, already plays a larger part in preventing climate change than does solar power, despite solar's rapid growth in the past three years.

Figure 5.4. Build-up of existing, planned and proposed CCS projects

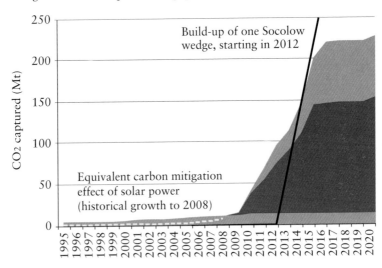

Scale of a Global Carbon Capture Industry

The scale of CCS is sometimes advanced as implausibly large. But there are some analogues that indicate the task is manageable: fuel and cooling water handling at power stations; sewage systems; global oil production; gas pipelines; and oilfield water reinjection.

Large **power plants already handle big volumes of material.** Our typical 500 MW coal plant burns about 5,000 tonnes of coal daily, releasing more than 18,000 tonnes of carbon dioxide. Coal, being a solid, is considerably harder to transport and process than fluid carbon dioxide. The power station also requires some 4 million tonnes of cooling water daily. Granted that the cooling water is usually surface water or sea water, and does not require extensive processing or transport, it is still some 200 times more in volume than the waste product carbon dioxide.

For an analogue on the scale of a whole economy, we can compare to another big system for handling liquid wastes: the **sewage system.** The average person in the UK produces 81 m³ of waste water per year,[11] so 30 British people could fill an Olympic swimming pool, if the exercise were not too revolting to contemplate. Average carbon dioxide emissions are only 9.4 tonnes[12] (equivalent to 12.5 m³ in supercritical form).

So British waste water treatment is more than seven times greater than CCS will be—particularly since not all carbon dioxide will be captured, more than a third of emissions being small scale or transport-related. For a US example, waste water injection underground in Florida alone is 500 Mt per year,[13] twice as much as the state's carbon dioxide emissions.[14] The sewage system has been built over a century or so in the UK and about half a century in the USA, almost unnoticed, without any need for dramatic lifestyle changes, an 'Apollo Program', a 'War on Excrement' or anything similar.

Another useful comparison is that one wedge of carbon capture (3.7 Gt per year) represents a flow of carbon dioxide into the group about equal in volume to **global oil production.**

At first sight, the re-creation, over fifty years, of an equivalent to the world's oil infrastructure is a colossal task. However, consider that essentially 80% of global oil capacity was built in the period 1950–2000,[15] and that this includes fifteen years (1979–93) when production was below peak due to oil shocks and recessions. The required rate for one wedge of CCS is therefore somewhat less than the rate of building oil infrastructure. When the build-up of modern oil infrastructure gained pace, in 1950, world GDP was less than a quarter the size of today's,[16] and technology far less advanced. And this should be seen in the context of challenges faced by the oil business in delivering this system.

Transportation capacity is sometimes advanced as a limitation on CCS, but in fact carbon dioxide pipelines required might increase to

Table 5.1. Comparison of challenges for building up oil
and CCS infrastructure

Challenges to delivering oil industry infrastructure	Comparable issues for carbon capture and storage
Geological risk: only some 10–20% of exploration wells discover a field	Suitable storage sites much more common; many already known from oil and gas exploration
Engineering risk: extracting oil (in particular) requires careful reservoir engineering, and both oil and gas need sophisticated facilities including de-gassing, de-watering, sand control, gas sweetening, etc.	Injection requires less sophisticated reservoir management, and surface facilities are simple (just compression)
Political risk: many oilfields are found in remote and/or unstable regions, and are vulnerable to wars, nationalisations, etc.	Most storage will be in stable, developed countries near industrial sites
Accident risk: oil and gas are explosive/flammable and oil causes environmental damage if released	CO_2 is neither explosive nor flammable, causes little environmental damage (other than greenhouse effects) and is only toxic in high concentrations
Logistical risk: oil and gas have to be transported around the world, sometimes from land-locked countries or through unstable regions or tough terrain	Most storage will be within the same country, and probably within 300 km of the emission point
Decline: after a few years, production from oilfields begins to decline, typically at 2–10% per year, and has to be replaced by new fields	CO_2 emissions from a plant remain at a consistent level as long as the plant operates. Storage sites, however, may become full at some point, requiring a new location
Decommissioning: oil and gas fields are expensive and complicated to decommission, especially offshore, but do not require monitoring once properly abandoned	CO_2 storage sites are cheap and simple to decommission once full, but will require long-term monitoring, albeit this is low cost and straightforward

about 25,000 km worldwide by 2030. This is barely half the amount of new **natural gas lines** to be built in North America alone over the same period.[17]

One final analogue for the magnitude of the job is the **re-injection of oilfield water** produced together with oil. Mature oilfields tend to yield increasing volumes of saline water, ultimately up to 100 times the oil. This water is usually re-injected, partly to maintain reservoir pressure, partly because it is usually highly saline and contains oil droplets and other contaminants which would cause pollution if discharged to surface. Globally, 158 million barrels of oilfield water were re-injected in 1999,[18] compared to 72 million bbl of oil extracted. This is the same volume as 6.4 Gt of CO_2 per year, about a quarter of emissions. It would appear, therefore, that the size of the task is eminently manageable.[19]

We should also recognise that decarbonising the whole energy system is a vast undertaking, however we do it. That is another argument for deploying all feasible solutions.

The Cost of CCS

So, volumetrically, CCS appears capable of making a sizable contribution to fighting climate change. It could even be the largest single approach. But this will only be feasible if it can save emissions more cheaply than other leading techniques, and still deliver competitively priced electricity and other products.

The cost of CCS can be expressed in several ways. The most obvious is how much it costs to avoid emitting 1 tonne of carbon dioxide to the atmosphere. This can be divided into capture, transport and storage costs. Of these, the capture costs are the most uncertain, and the most sensitive to future advances in technology. Transport costs are well-known from long experience. Storage costs may be at least partly offset by enhanced oil and gas recovery.

The cost of CO_2 avoided (in \$/tonne) can be used in three ways:

- It can be compared to CO_2 permit prices or carbon taxes to determine whether an investment in CCS is worthwhile.
- It can be set alongside the CO_2 abatement costs for other options—such as biological sequestration (forestry, etc.), energy efficiency, nuclear and renewable power—to see whether CCS is a better abatement option than some other approach.

- Different types of carbon capture can be compared: is it more effective to target cement plants, or oil refineries? This comparison can also be made by looking at the price of the product (for power plants, electricity), with and without CCS, when non-CCS plants are appropriately penalised for their carbon emissions.

It is a difficult task to estimate the cost of CCS systems, especially the capture component. There is substantial experience with the components of CCS systems but, as yet, no full-scale power plant using carbon capture. Reported values are based on many different assumptions:

- What is the technology employed?
- What is the size of the plant, given that CCS has strong economies of scale?
- What is the assumed plant lifetime?
- What is the target capture rate? Capturing less than 90% of emissions may reduce costs.
- What is the reference (non-CCS) plant? For post-combustion capture, the answer is simple: the same plant without the capture system. But what about an IGCC which is not the standard type of coal plant: should it be compared to a conventional pulverised coal plant, or to an IGCC without capture? The best comparison (though it is not always made) is with the standard plant that would be built in the absence of carbon concerns; or, in the case of a retro-fit, the original power station. However, public pressure and government regulations may cloud this comparison, for instance if the standard non-capture pulverised coal plant is simply unacceptable. Furthermore, if the CCS plant is replacing an older facility which is to be shut down, it can take some credit for increased efficiency, even before considering the effects of capture. This issue is thornier when considering new concepts such as fuel cells and chemical looping combustion, particularly as they are more risky than conventional designs, even though they may hold the promise of being cheaper. The answer may also vary by region: for example, a CCS-equipped coal plant in China may be displacing a non-CCS coal plant (CO_2 savings around 85%), but one in the Middle East, where coal is rarely used, will replace a non-CCS gas-fired plant (savings about 70%).
- Are credits given for any reduction of non-greenhouse pollutants?
- Where is the facility to be built? Construction costs in India and China may be 30% lower than in the USA.[20] Similarly, costs may be

affected by location factors such as terrain, altitude (which changes plant performance) and the availability of water,

- What are the technical specifications of the plant? For instance, is an oxyfuel plant also designed to be capable of air-firing (which may increase uptime but adds heavily to costs)?

- What uptime is used (i.e. what percentage of the time is the plant available for operation)? Advanced systems tend to have lower uptime than mature technologies due to technical hitches. Within this uptime, how often is the plant actually running? Due to market demand, power stations, especially gas-fired ones, do not run all the time; at 3 a.m. in the summer in Edinburgh or Seattle, demand tends to be low.

- Is carbon dioxide compression at the plant included, or lumped with transport costs?

- What fuel type is assumed? Different grades of coal, in particular, may yield very different results, due to variation in the content of moisture, energy and sulphur. High-quality coals (such as bituminous) will give higher efficiencies and lower capture costs than low-quality fuels such as lignite (brown coal) or biomass.

- Is the reported number calculated per tonne of CO_2 captured or per tonne mitigated (as discussed below)?

- Is the cost estimate only for the plant itself, or including impacts on other plants, upstream fuel supply and the electricity grid, or concerning a whole country's energy system? For example, carbon avoidance at a national level might include the effect of replacing some wind power with CCS. Increased grid stability from this option might be taken as a credit against capture costs. On the other hand, increasing coal use due to the lower efficiency of CCS plants might drive up coal prices.

- Particularly relevant for 'breakthrough' technologies, is this cost estimate for a plant built today, or a hypothetical one to be constructed some time in the future when we have gained experience? If for a future plant, what learning factor is used? Learning factors are highly uncertain but, for CCS-type systems, costs typically drop by 12–20% for each doubling of installed capacity.[21] On the other hand, the first one or two systems tend to be much more expensive (up to twice) than predicted, due to teething troubles and unforeseen problems.[22]

- Are recent rises and falls in the costs of materials and engineering included? The cost of power plants jumped sharply in recent years, taking off since 2006, due to steep increases in raw materials costs (energy, steel, copper, cement, etc.) and engineering services. This was driven by the global economic boom, and competition for similar inputs from other industries such as oil, petrochemicals, infrastructure and real estate. Power plant construction costs rose by some 130% from 2001 to late 2007.[23] But costs plunged again in 2009 as the global recession took hold. The CCS costs of the IEA (and others) were updated for increases, which now appear transient, so recent estimates of CCS costs may be unduly pessimistic. On the other hand, older estimates may be too low unless the right inflation factor is used.

- Particularly embarrassing for the US Government's FutureGen programme:[24] what is the year of the cost estimate, and hence what correction has to be made for inflation?
- What fuel prices (coal, oil, biomass and gas) are used? Volatile energy prices mean that fuel costs vary greatly with the date of the estimate, as do projections of credits from enhanced oil and gas recovery. If some electricity has to be bought by the facility, as for a hydrogen plant, what price is used? Similarly, for poly-generation plants, what are the assumed sales prices of products such as synthetic fuels, heat, electricity, hydrogen etc.?
- What interest rate or discount rate is used to bring future costs back to the present?

This uncertainty in cost estimates is, though, common to most future energy systems, whether solar photovoltaics, solar thermal, offshore wind, tidal barrages,[25] biofuels or next generation nuclear plants. Cost estimates can vary widely even for mature technologies. Plausible costs for most leading future energy contenders overlap widely. This emphasises even more strongly that, in our current state of knowledge, it is close to impossible to 'pick winners'.

Capture Costs: Principles[26]

Two key measures are applicable for looking at the costs of CCS. The first is the cost of the product supplied—typically electricity, but it can also be cement, steel, hydrogen, synfuels and many other industrial outputs—with and without CCS. Hence, what is the increase in production costs due to adding CCS? This cost can be further divided into capital costs (the money required to design, purchase and install the system) and operating costs, including the increased fuel required to cover the efficiency penalty, and other expenses such as maintenance, staff and insurance.

The second metric is the cost per tonne of CO_2. When comparing carbon dioxide costs, we have to distinguish the 'cost per tonne captured' and the 'cost per tonne avoided' (or mitigated). The cost of capture is simply the total cost of the capture system (discounted appropriately), divided by the total amount of carbon dioxide to be captured and stored over plant lifetime.

The cost of avoidance is the total cost of the capture system, divided by the reduction in total carbon dioxide emissions that the plant achieves with capture, versus a reference plant producing the same useful output but without capture. Carbon capture has an efficiency penalty. If we apply CCS to a facility, we increase its carbon dioxide output, and so the amount of carbon dioxide we have avoided emitting is less than the total amount captured. The cost per tonne of CO_2 avoided is therefore always higher than the cost of carbon dioxide captured. However, transport and storage systems work on actual carbon dioxide captured, not on hypothetical amounts avoided, and so we cannot simply add transport and storage cost estimates to separate capture costs; the results have to be calculated consistently, even if they are later reported separately.

To illustrate this, consider a power plant (Figure 5.5). The facility emits 10 Mt per year CO_2 (the first bar). Now we fit carbon capture with a 40% energy penalty. Emissions increase by 4 Mt per year (the second bar), but we now capture 90% of the total, 12.6 Mt (the third bar). 1.4 Mt per year still escapes. However, the emissions avoided are only the difference between the plant without capture and with capture, i.e. 8.6 Mt per year (the fourth bar).

Now let us say that this capture system has an annualised cost[27] of $500 million. The cost per tonne of CO_2 captured is 500/12.6 = $40. But the cost per tonne avoided is 500/8.6 = $58/tonne, almost 50% higher.

Now we have to calculate the cost for transport in a pipeline. The pipeline costs, say, $2 per tonne for transport, but it must, of course, carry the captured CO_2, a larger quantity than the avoided CO_2 (which is really a theoretical figure). The annual cost of transport is therefore 12.6 */2 = $25 million.

Figure 5.5. Illustration of carbon dioxide captured and avoided

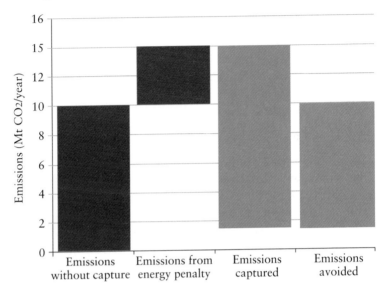

This makes it clear how important it is to apply CCS to highly efficient plants, and to continue to upgrade the efficiency of the capture process. By contrast, increasing the percentage of carbon dioxide captured is a secondary goal, only worthwhile if it can be done at low cost and without further compromising plant efficiency.

As noted, there are strong economies of scale. The capture cost per tonne halves with each ten times increase in plant size.[28] It therefore makes sense to co-locate facilities, to share carbon dioxide scrubbers, oxygen separators, waste heat, pipelines, and so on. A group of CO_2-emitting facilities, which might individually be too small for economic capture, can be combined into a cluster. A variant on this is to co-fire a CCS coal-fired power station with 10–20% of biomass, to overcome the small size of most pure biomass plants.

Cost calculations generally assume that the reference plant is downrated by the parasitic power required by the capture system. However, unless we are considering retro-fits, this probably overstates costs. If we were building a new plant today, with CCS, we would size it for demand in its locality. Larger plants are relatively cheaper, so this saves on costs.[29]

Capture Costs—Current Systems

The first step is to establish which of today's technologies is most attractive, and what the likely capture costs are.

Figure 5.6 indicates carbon avoidance costs in a range of $13–51 per tonne for coal and $37–74 per tonne for natural gas. As a piece of supporting evidence, we know that coal post-combustion capture costs should not be more than $100 per tonne, since the Shady Point coal plant in Oklahoma captures a small portion of its emissions commercially, and sells to industrial customers for about this price.[30]

Figure 5.6. Cost of carbon dioxide mitigation from coal- and gas-fired power stations, with today's technology

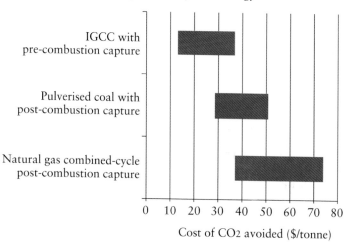

Cost of CO2 avoided ($/tonne)

It also shows that the cost of capturing a tonne of CO_2 is lower for a coal plant than for a gas-fired one, because of the higher emission intensity from coal. It therefore makes sense to apply carbon capture preferentially to coal-fired plants, but specific circumstances may encourage its early use in gas-fired plants. For example, Abu Dhabi possesses good geological conditions for carbon dioxide use in EOR, and its power generation is almost entirely gas-fired (with some oil); it has no coal plants.

Of the two more mature capture technologies, post-combustion (using solvents) appears generally more expensive than pre-combustion in an IGCC. However, in the near term, improvements in capture costs may come mainly in post-combustion, particularly in using superior solvents and better heat recovery systems.

Figure 5.7 shows the cost of electricity produced by coal and gas power plants, with and without carbon capture. Plants without capture are shown in blue and those with capture in green. We can see that, for non-capture plants, natural gas provides the cheapest electricity, although this is very dependent on circumstances. The high gas prices of 2005–8 led to something of a resurgence of coal in the USA, and some countries, especially India and China, have access to low-cost domestic coal but are short of affordable gas. Pulverised coal plants are generally cheaper than IGCCs, explaining their dominance.

Figure 5.7. Cost of electricity from gas and coal plants with today's technology, with and without capture

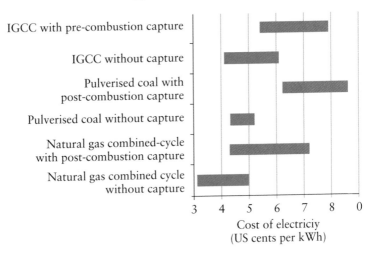

With carbon capture, the picture is different. Gas is still generally the cheapest option, but coal IGCCs gain considerably in competitiveness. IGCCs become clearly preferable to pulverised coal plants on the grounds of carbon mitigation cost, cost of electricity and energy efficiency. IGCCs have the advantage that at zero or low carbon costs they are not much more expensive than pulverised coal, and at higher carbon costs they are clearly cheaper. So if an electricity generator has any expectation that the cost of emitting carbon is likely to rise, then

IGCCs are preferable, particularly given the relatively lower cost of capture-readiness (as discussed below).

This suggests that, with more familiarity and some improvements in reliability, IGCCs could become the dominant choice for new coal power plants. However, this partly depends on fuel type. IGCCs may be preferable for high-grade (bituminous) coals, while for cheaper, low-grade brown coals (lignite), post-combustion capture may be superior.[31]

One option that has been discussed is for 'partial capture', trapping perhaps only 50% of emissions. This would make coal-fired plants as clean as gas power stations, and possibly significantly reduce the capture cost. It might also have benefits for reliability, and, by bringing down up-front capital requirements, reduce the risk of deployment and allow for more demonstration projects. However, different studies disagree on whether this idea really would reduce per-tonne costs.[32] It might also lead to PR issues, since partial capture plants would still be significant polluters.

The utilisation factor of the power plant is also important. Gas-fired power stations have relatively low capital cost and can be fired up quickly. They are therefore preferred for providing power at peak times, while coal and nuclear plants are used for steady baseload power. It is costly, though, to fit a peaking plant with expensive carbon capture equipment which will only be used infrequently.[33] This may present utility companies with a difficult decision, since the balance of plant utilisation depends on the relative levels of coal and gas prices, CCS operating costs, residual carbon dioxide emissions and the level of carbon taxation or permit costs. We would expect that, assuming the cost of carbon is high enough to encourage the utility to fit CCS in the first place, they would then prefer to run a capture-equipped gas plant rather than a non-capture plant. Older coal-fired plants without CCS, now running as baseload, may come to run as 'intermediate load' (between peak and trough periods).[34]

Specific industrial sources have carbon dioxide mitigation costs which can be below those of power stations.[35] We can divide industrial capture into three groups based on cost, comparing it to the long-term $30–40 per tonne cost of carbon capture on new coal and gas power stations.

- Cheaper than $30–40 per tonne: contaminated natural gas; synthetic fuels, including the production of hydrogen, methanol and DME;

direct reduced iron manufacture; and possibly black liquor (pulp and paper). These low-cost sources can be early adopters of CCS, particularly when combined with value-added storage opportunities.

- Around $30–40 per tonne: hydrogen from natural gas; steel mills. These sites could be fitted with CCS at the same time as capture is taking off for most new power stations.
- Above $30–40 per tonne: oil refining, biomass synfuels, ethanol and cement. Given that corn-derived ethanol for transport fuel is only economic with heavy subsidies, the addition of carbon capture would make it even less competitive. CCS on cement plants is particularly costly ($75–100 per tonne). Widespread installations on these sources may require improved technology, clustering capture facilities to save costs, and the start of large-scale retro-fit programmes on older coal plants, which are comparably expensive.

Capture Costs: Future Systems

Moving on to compare today's technologies with some future options, we come to Figure 5.8, showing the cost of carbon mitigation.[36] Present-day coal plants are shown in blue, future coal and pet-coke plants in red, and natural gas plants in green.

Five key points emerge:

The cost of electricity generation rises significantly with CCS, but retail prices are cushioned. A cost of electricity of, say, 4 ¢ per kilowatt hour (kWh)[37] from an IGCC without capture rises to 5.5 ¢ per kWh when capture is included. This can be compared to wholesale US electricity prices around 3–8 ¢ per kWh during 2008–9. With carbon capture using current technologies, the cost of electricity generation might increase by about 70% for new pulverised coal plants and 40% for IGCCs. However, the increase in consumer prices will be much less than this. Firstly, about half of a consumer's bill pays for transmission and distribution, the cost of which will not be changed by CCS.[38] Secondly, CCS will only be fitted gradually, and will target the lowest-cost plants first, so that the marginal cost of electricity, generated at inefficient (and probably non-CCS) plants, will not change.

Capture on new, efficient plants is much cheaper than retro-fits to older models. The cost of carbon capture drops very significantly when, instead of the older sub-critical coal plants, we fit capture to the

Figure 5.8. Cost of carbon dioxide mitigation from coal- and gas-fired power stations, comparing today's technology with advanced options

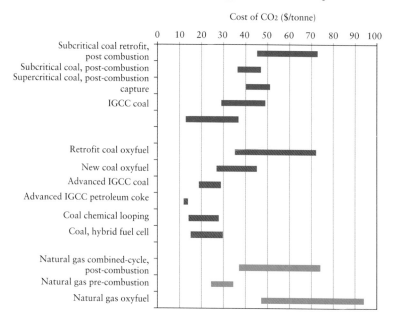

Cost of CO₂ ($/tonne)

more efficient and modern coal plants, such as supercritical and ultra-supercritical, and especially IGCC. Present-day coal plant designs without capture all have a similar cost of electricity, with IGCC perhaps somewhat better. Existing plants, with their capital costs already amortised, have of course the lowest generation cost. When carbon capture is fitted, IGCCs are clearly the cheapest generation option. Inefficient old coal plants perform very poorly on both capture and generation costs.

Retro-fitting existing subcritical coal plants is particularly costly, about 35% more than building a new plant, whether with post-combustion capture or with oxyfuel. 'Brown-field' modifications in general tend to be expensive, due to the need for design compromises and 'working around' existing kit. Current plants also have a limited remaining lifespan, which may prematurely cut short the use of the carbon capture equipment. And the plant will have to be shut down while the retro-fit is going on, hence losing electricity sales.

Indeed, it seems that retro-fitting non-capture-ready plants will be so expensive that it will often be cheaper to replace them entirely, although retro-fitting the most efficient existing plants will be important if CCS is to be implemented rapidly. This also implies that new non-capture coal plants risk rapidly becoming 'stranded assets'—uneconomic to operate—as carbon costs rise.

Because of these factors, it would be much more effective to ensure that all new coal power stations have CCS, then probably to move on to new gas power stations, before beginning retro-fits. More likely, the current fleet of old coal plants could be progressively rebuilt and repowered, fitting CCS at the same time. This would be a relatively efficient approach, since existing brown-field sites can be re-used, with the legacy transmission systems, infrastructure, operating personnel, permits, local familiarity and so on, and perhaps salvaging some equipment.

It will become increasingly attractive to ensure all new plants are 'capture-ready'. A plant can be called capture-ready if specific preparation has been made for adding capture, and if it would be cheaper to retro-fit capture rather than building an entirely new power station. Building a capture-ready IGCC plant is slightly more expensive[39] than non-capture-ready, but will almost always be preferable[40] since carbon costs are expected to rise rapidly. Adding capture to a capture-ready plant is overall about 17% cheaper than trying to retrofit a non-capture-ready plant.[41]

New capture technologies can be more cost-effective, though the range of uncertainty is wide. On current figures, new coal oxyfuel appears a more expensive method of carbon dioxide mitigation and power generation than IGCC. However, the uncertainties and room for improvement are great enough that it cannot be written off yet. These figures, based on the IPCC's report, are relatively more pessimistic on oxyfuel than the IEA[42] and Statoil.[43]

IGCCs burning petroleum coke stand out as an especially low-cost approach, with an estimated cost of CO_2 avoided of only \$13/tonne, and technologically quite mature, since already several are in operation. Pet-coke is a by-product of processing heavy oil, and its output is growing worldwide much faster[44] than oil or coal use, due to increased refinery complexity to produce the more valuable light products such as gasoline and diesel.

Chemical looping combustion (CLC) and hybrid fuel cells also offer low carbon avoidance costs, as good as or better than advanced IGCCs, but are still at the research stage. The cost of electricity from CLC appears competitive with that from other CCS plants; that for fuel cells, although potentially cheap, has a wide range of uncertainty and may equally be extremely expensive.

Carbon capture on biomass- and gas-fired plants is more costly than for coal. Biomass-fired plants produce expensive electricity even without capture; with CCS, they are, by a considerable margin, more expensive than any other option, even fuel cells. This is mainly due to the small size and hence lack of scale economies of biomass plants, and supports the case for instead using biomass as co-firing in capture-equipped coal stations. Much larger biomass plants, though, may have capture costs closer to those of coal.

Capture on gas-fired power stations is more costly than for a coal IGCC or some of the advanced coal systems. Pre-combustion capture appears the clear winner, superior to post-combustion capture from all but the most efficient coal plants. It is, though, more expensive on carbon capture than the leading coal CCS options. Post-combustion and oxyfuel capture are much more expensive mitigation options for natural gas.

Overall it would appear, then, in a world of strict carbon constraints, that coal IGCCs and pre-combustion gas-fired plants will fight it out for the generation technology of choice. Of course a mix of both will be used, the balance depending on local fuel prices and availability, demand patterns and many other factors. However, given the relative technological immaturity of the systems, and the potential for breakthrough advances, we cannot yet rule out that oxyfuel, greatly improved post-combustion capture, or one of the advanced systems such as fuel cells, may be the ultimate winner.

Capture Costs: Potential for Cost Reductions

Capture costs are expected to fall with time, due to experience, technological advances and perhaps economies of scale in manufacturing some components. The first generation of CCS plants may well be very expensive, due to 'teething troubles', engineering conservatism and redundancy, the smaller scale of the early installations, non-optimisation and so on. Similar high-tech plants have had problems. Cost overruns, safety worries and long delays for nuclear power plants are famous, of course, though the French programme was rolled out successfully during the 1980s. The first generation of coal IGCCs in the USA suffered initially from low reliability, though European ones burning pet-coke were more successful.

Gas-to-liquids plants, sharing some components with pre-combustion capture and oxyfuel, have a chequered record. The first, at Bintulu in Malaysia, exploded and had to be rebuilt, though this was the fault of soot from forest fires clogging air vents, not of any problem with the technology itself. The Oryx plant, in Qatar, operated very little during its first year, because of problems with its catalysts. And Shell's Pearl project ran spectacularly over budget, the initial cost of $6 billion ballooning to $20 billion,[45] albeit this was partly explained by soaring materials prices.

Indeed, there is a risk that heavy cost over-runs on early facilities could dampen or destroy enthusiasm for CCS.[46] The US's FutureGen was cancelled due to cost inflation, though it may now be resurrected.[47] If early projects do turn out to be much more expensive than forecast, the industry and government will have to show great resolution to persist. CCS opponents are likely to seize on any early hiccups to brand the technology as a failure.

For every doubling in capacity, costs are estimated to fall by 10%. This 'learning factor' is consistent with, indeed somewhat lower than that achieved by liquefied natural gas, and sulphur dioxide and nitrogen oxides scrubbers.[48] By 2030, compared to the very early generation of demonstration plants, the cost of electricity generated and carbon captured by CCS plants may drop by as much as 40%,[49] as shown in Figure 5.9. The cost of conventional pulverised coal power, a mature

Figure 5.9. Learning curves for capture- and non-capture power plants[50]

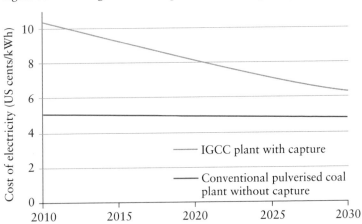

technology with a large existing base, falls only slightly over the period, so that CCS-equipped IGCC makes up most of its initial cost disadvantage.

With a large CCS programme, the cost savings may be even greater than this.[51] However, the cost of electricity from non-capture coal plants could fall more, with IGCCs being marginally the most cost-effective option by 2030. If this does occur, then it would present not only CCS but other forms of power generation (including renewables) with a serious challenge. Since carbon capture consumes significant energy, scenarios assuming rising fuel prices will be relatively unfavourable for it, versus both non-capture plants and alternative low-carbon generation.

Carbon capture involves increases in both capital and operating costs (opex). In this, it is distinct from most renewable technologies (such as solar and wind) and nuclear power, which have very high capital costs but minimal to low opex. Typically 40–70% of the cost of a CCS project is capital and the remainder operating and storage costs.[52] It is therefore possible to argue that carbon capture represents less of an irreversible, up-front commitment to an unproven technology than some other low-carbon generation options.

Transport Costs

Transport costs are a small part of the total. For a 100 km pipeline onshore, carrying 2 Mt per year (about half our typical coal plant), the cost is $2–6 per tonne of CO_2, while a larger, 10 Mt line would cost half as much.[53] These considerable economies of scale argue for centralised gathering and storage centres. Offshore pipelines are 25–100% more expensive than onshore. As a comparison, the large oilfields in the Central North Sea, potential storage sites, are up to 300 km from major coal-fired power plants in northern England.

For smaller quantities and over longer distances, ships are superior, with a cost of $15 per tonne over 1,000 km. However, most capture operations will have a suitable site closer than this. For ocean storage, ships will probably be cheaper than pipelines if the deep-water disposal site is 500 km or more from the source, and indicated transport costs are around $6–7.5 per tonne.

Once pipeline infrastructure of a few million tonnes per year is established, future transport and storage costs per tonne are roughly constant. Capture costs for pipeline transport will probably not fall much with technological advances; shipping may be more amenable to future improvements.

Storage Costs

The cost of storage is more than that of transport, but is still much less than typical capture costs. Storage costs have three main components:

- Construction and operations.
- Site selection and monitoring.
- Any credits (negative costs) for enhanced oil or gas recovery.

There is a wide range of costs for **constructing and operating** a storage facility. Key factors for cost include:

- Location, particularly offshore or onshore, and also the remoteness and terrain of the site. Populated areas will tend to be more expensive due to higher land costs, additional safety precautions, and so on. In general, onshore storage is cheaper, largely because onshore wells costs much less (typically, 5–10 times less) than offshore wells. To the extent that offshore facilities are required (for example, compression, and carbon dioxide separation for EOR projects), these are also much more expensive offshore. Offshore projects do have the advantage, though, of greater public acceptability (out of sight, out of mind) and higher safety, not being near populated areas. Seismic monitoring is also cheaper and of higher quality offshore.
- Reservoir quality and configuration, which controls the number of injection wells required.
- Reservoir depth, deeper wells being much more expensive than shallower ones and requiring higher pressures for injection. Deeper reservoirs also tend to be of poorer quality.
- Any contaminants in the injected carbon dioxide, particularly hydrogen sulphide.
- Size of the project, larger projects being cheaper per tonne due to economies of scale in surface facilities.

It appears that there is a large amount of storage capacity available for between $0.5 and $8 per tonne CO_2,[54] with most of it towards the low end of this range. Relatively low transport costs imply that it is worthwhile going longer distances to find a good site: for example, a large onshore USA project might be prepared to pipe carbon dioxide 200 km to find a good site (storage cost $0.4 per tonne) rather than use a poorer nearby location (storage at $4 per tonne). This conclusion is supported by the rather long distances that existing pipelines in the USA cover to reach EOR fields. These carbon storage costs, although greater than transportation, still represent only a modest part of the total CCS system.

Site selection and monitoring is a minor part of total cost. Selecting a suitable onshore site has been estimated to cost about $1.7 million; appraising it to a state where it is ready for storage costs $15 million or more.[55] This is a trivial amount when averaged over the ~100 Mt of a full-scale storage project. The very thorough site selection exercise at Weyburn, together with the first few years of monitoring and verification, cost about $26 million[56] (spread over an ultimate ~30 million tonnes CO_2 injected), but much of this work remains relevant to similar storage projects in the vicinity, such as the new Midale EOR project. Offshore sites are significantly more expensive, mainly due to the much higher cost of offshore wells ($15 million or more each in the North Sea). If the area has not been previously explored for oil and gas, then seismic data will be required, costing $4 million or more per site.

Monitoring and verification at In Salah cost $2 per tonne, but that was largely because it was a pathfinder project and the monitoring effort was exceptionally comprehensive. Larger projects will also benefit from economies of scale. A range of $0.1–0.3 per tonne will probably be more typical in the future,[57] especially once we have determined the most effective monitoring methods from the wide set of possibilities available today. Surprisingly, increasing the time span of monitoring results in only a minor increase in cost. The modest cost for monitoring suggests that existing EOR projects should install such systems to ensure that their injected carbon dioxide is retained, thus potentially qualifying for carbon credits.

Storage costs can be partly or completely offset by **enhanced oil and gas recovery**. However, there are some additional costs associated with EOR and EGR. Firstly, the extra barrels are not free: they incur incremental operating costs (and, depending on the jurisdiction, royalty and tax payments to the government or mineral rights owner). Secondly, the additional oil does not emerge immediately: there is a delay of some years before the full reservoir response is seen in the producing wells. Thirdly, carbon dioxide appears in the producing wells and is recycled, which incurs a cost, and means that not all the carbon dioxide injected is 'new'.

In the late 1990s, when oil prices were very low, CO_2-EOR was viable at a cost of $19–38/tonne CO_2. This can be interpreted as the maximum capture and transport cost that could be sustained, plus any cost for emitting carbon dioxide. With higher oil prices today, partly offset by higher capital and operating costs, this price can be expected to increase.

For instance, in one field example I have calculated,[58] the project is economic at a CO_2 price of $35 per tonne and an oil price of $50 per bbl. This is also consistent with recent prices of carbon dioxide delivered to a well.[59] The EOR credit that could be available to help pay for the carbon capture system—or, looked at another way, the negative cost (benefit) of such a storage project—is $35 per tonne. This amount would probably be sufficient to pay for capture from new coal plants. If we assume a higher oil price in line with OPEC aspirations and average 2009 levels around $70 per barrel, the possible CO_2 price increases to over $100 per tonne. This is enough to pay for even the more costly and experimental capture options.

EOR on offshore fields is considerably more expensive. Offshore wells can easily cost $10–20 million each (as opposed to the $3 million assumed here for onshore wells), and engineering, especially modifying existing platforms, is expensive. Lengthy shutdowns for modifications are also very costly offshore. These factors, with some disappointments in calculations of incremental oil benefits, and delays in agreeing government incentives, largely explain the cancellation of offshore European projects such as Miller (BP) and Haltenbanken (Shell/Statoil). The North Sea is, though, due to its harsh weather, a particularly high-cost area, and the fields there are increasingly mature. More favourable areas for offshore EOR might be found in the Middle East, shallow-water Gulf of Mexico, and perhaps South East Asia.

To continue operating an old EOR facility in pure storage mode (or storage with some continuing by-product oil) can be done at low costs, probably $6–11 per tonne of CO_2 stored,[60] and can potentially double the ultimate storage.

There are still few enhanced coal-bed methane projects. One example, in the San Juan Basin of New Mexico, was uneconomic at the low gas prices of 2002, but would have been profitable at recent higher prices.[61] The implied credit was about $14 per tonne of CO_2. Even allowing for some transport costs, this does suggest that ECBM could be profitable without carbon credits, if it could source cheap carbon dioxide from capture-ready industrial sources.

Figure 5.10 indicates a plausible range of storage costs for different sites, with aquifers in blue, depleted oilfields in pink, depleted gasfields in pale green, a range of Chinese sites in orange, EOR in red, enhanced gas recovery in green, enhanced coal-bed methane in green and ocean storage in pale blue.

The large credit for EOR, especially in onshore USA, stands out. The potential benefit of enhanced gas recovery is also notable, though with only two pilot projects worldwide we might question why it has not been taken up more widely, especially during the recent period of high gas prices and security of supply concerns. Offshore sites are, not surprisingly, more expensive than onshore, while depleted oil- and gas-fields generally offer somewhat lower costs than aquifers. The potentially high costs for EOR and ECBM in Europe can exceed that for saline aquifers in the same setting. This clearly indicates that enhanced hydrocarbon recovery options will simply not be pursued if they are more expensive than straightforward aquifer disposal. On the other hand, there should be a spectrum of opportunities in which an early EOR project, sufficient to cover some of the costs of a capture and transport system, transitions into a longer-term pure storage site.

Figure 5.10. Storage costs for a variety of settings[62]

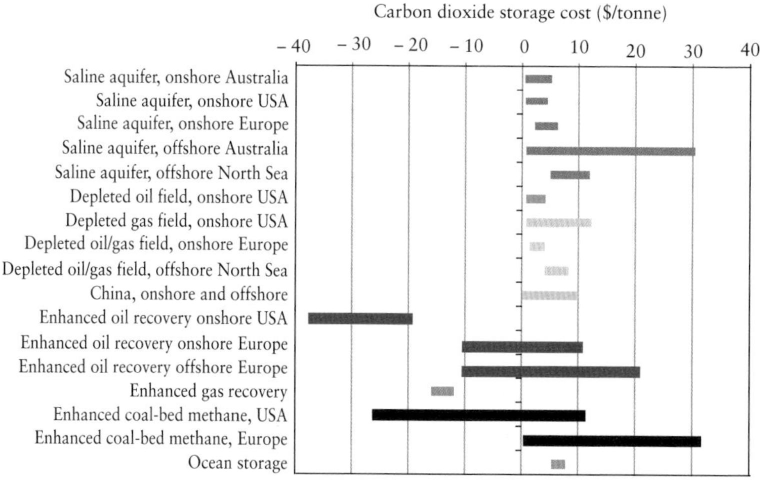

205

It appears that **ocean storage** is relatively expensive, more than onshore aquifers and probably comparable to offshore geological storage. This probably makes ocean storage appear more competitive than it really is; for most nations, deep-water sites are remote and therefore transport costs will be greater than for underground storage. This relatively high cost should be considered with all the other problems of ocean storage (Chapter 3). Carbonate neutralisation, not shown on this chart, has a very wide range of cost estimates, $10–110 per tonne, also rather unpromising compared to other storage possibilities.[63]

Mineralisation is also not shown on the chart. Using industrial wastes such as fly ash, spent concrete, iron and steel slag and bauxite residues, is very cheap, with a suggested cost of just $8 per tonne[64] combined for capture and storage (and no transport, since the work is done on-site). The volumes of such wastes, though, are small and this is a niche application. Large-scale application of mineralisation using silicate minerals will be far more expensive, due to the need to mine and pulverise the minerals, the energy input required, and the long time taken for reaction. Suggested storage costs are $55–100 per tonne,[65] evidently much more than even the most costly offshore aquifers. Some of the *in situ* mineralisation efforts, on the other hand, might achieve lower costs and greatly reduced environmental impacts.

Given the very large projected underground storage capacity, it seems unlikely that mineralisation will be required unless there is a technical leap forward, geological storage falls badly short for unanticipated reasons, or there is a massive amount of fossil fuel burning and associated air capture effort. There should be plenty of time to pursue basic research in mineralisation, and possibly to build a demonstration plant, while at least the first two or three generations of CCS plants use geological storage.

Total Cost

We are now in a position to put the entire chain together.

The total cost of a full CCS system evidently varies enormously depending on the combination of capture, transport and storage method—particularly the capture cost. For each link in the chain, we can define low, medium and high costs as follows, giving the cost in $ per tonne of CO_2 avoided in each case.

		Low	Medium	High[66]
Capture	Example	Gas processing, hydrogen from coal, pet-coke IGCC	New coal IGCC or pre-combustion gas plant	Coal retro-fit
	Cost	10	30	60
Transport	Example	Storage site next to capture site	Onshore site 100 km distant; 10 Mt per year	Offshore site 200 km distant; 2 Mt per year
	Cost	0	2	12
Storage and monitoring	Example	EOR, onshore USA	Onshore saline aquifer	Offshore North Sea aquifer or EOR
	Cost	– 30	2	10
Total		– 20	34	82

Compared with typical carbon dioxide emissions permit costs of $15–25 per tonne, the implication is that most low-cost capture options will be economic, as long as they have a low- or medium-cost storage option within a reasonable distance. Similarly, value-adding EOR or ECBM projects using carbon dioxide from new coal or gas plants should be economic. The most typical application, for a new fossil-fuelled power station disposing of its emissions in an aquifer within a reasonable distance, is some way short of current carbon prices but, as shown below, these will probably rise sufficiently over the next five years or so. Even the large group of expensive options—coal retrofits on small plants, with a rather distant offshore disposal site—is still considerably cheaper than the ~$130–140 per tonne mitigation costs expected later this century by Stern and the IPCC. However, it will take some time and experience for costs to fall to these levels; the capture cost for first-of-a-kind plants will be substantially higher.

For the USA, one study has derived a cost curve (Figure 5.11) showing the volumes that might be captured (with present-day sources and technology) and the associated cost. This can be divided into four groups of increasing expense:

207

Group 1: Value-added. A fairly small number of negative-cost capture and storage options, typically high-purity carbon dioxide sources near oilfields. For comparison, global potential for carbon dioxide storage with EOR is estimated at about 0.5 Gt per year.[67]

Group 2: Low cost. About 0.5 Gt per year capture at costs of up to $50 per tonne, consisting of high-purity sources near depleted fields, or large coal plants with enhanced hydrocarbon recovery possibilities.

Group 3: Mid cost. The curve flattens out, as the bulk of the CCS possibilities, more than 1.5 Gt per year, cover coal power stations and iron- and steelworks within reasonable distance of saline aquifers.

Group 4: High cost. The curve steepens again, as there are a smaller number of costly opportunities, including capture from gas power plants, cement, and smaller or more distant coal plants.

Figure 5.11. CCS cost curve for the USA[68]

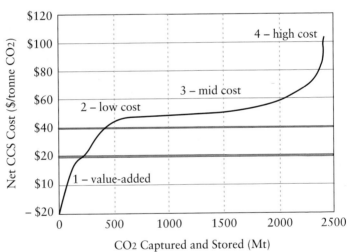

CCS should begin to be widely adopted once the carbon cost reaches the medium costs mentioned above for the full system.[69] Before carbon prices reach that trigger level, there will be some early adoption via low-cost opportunities: for example, high-purity industrial sources, and EOR projects. The bulk of CCS could be implemented with a carbon price around $50 per tonne.[70]

The effect on the wider economy can be understood by examining the cost increases in various products when CCS is included.[71] We can distinguish three classes: electricity; synthetic fuels (hydrogen, methanol and DME); and other products. The increase in the cost of these products depends on two factors: the carbon intensity of their manufacturing, and the ease or difficulty with which their emissions can be captured.

- Electricity is, of course, heavily affected, with increases of between 27% and 62% for coal IGCCs, and more for pulverised coal plants with post-combustion capture, and natural gas plants. This is due to the high carbon intensity of power generation (all the fuel carbon ends up in the flue gas) and the relatively expensive capture techniques required. As noted earlier, though, the retail price to consumers, which includes transmission and distribution costs, is affected much less in percentage terms than the wholesale price.
- Synthetic fuel production (hydrogen, methanol and DME) is moderately affected, typical cost increases being 10–20%. This does not include any carbon cost when the fuels are ultimately burned. Production of these fuels is carbon intensive, but the nature of the process makes capture relatively easy.
- Other products include refined oil (diesel, kerosene, petrol, etc.); paper; iron and steel; and purified contaminated natural gas. Carbon capture in these cases is fairly difficult and expensive (apart from natural gas). However, the products have high value and relatively low carbon intensity. Their costs are therefore barely affected by including CCS: steel, refined oil products and paper see production costs elevated by only 1.5–4%, natural gas even less. This would certainly be enough to wipe out a refiner's margin, but as long as all competing refineries see a similar cost (from installing CCS or from a carbon tax or cap), they should be able to pass most of the price increase on to customers.

Air Capture

Air capture is expensive, but offers an ultimate backstop to climate policy. In principle, it can tackle any emissions source, including very problematic ones such as aviation. But is air capture physically feasible on a large scale? And would it be affordable?

As mentioned, two of the leading proponents of air capture are Professor Klaus Lackner from Columbia University and Professor David Keith from Calgary University. Lackner suggests that 50% of global emissions could be trapped by a structure as large as the Great Wall of China.[72] This might sound ridiculously big (although, presumably, there would actually be many smaller devices in different locations). But when we reflect that the Great Wall was built in pre-industrial times with no more powerful machine than oxen, the task seems quite manageable.

An 'artificial tree' the size of a shipping container can reportedly capture 1 tonne of CO_2 per day. To capture all anthropogenic emissions would therefore require about 80 million such machines, occupying 1,200 km^2, about the area of Denver, or three-quarters of London. Keith's system (which he acknowledges is deliberately very simple) requires about ten times this area. If we were only trying to offset aviation, then the problem is even more manageable: about 1.3 million of these artificial trees. For comparison, more than 50 million cars are built worldwide annually, so the manufacturing task required is small by comparison.

In practice, like wind turbines, air capture units have to be spread widely, otherwise they compete to extract the same carbon dioxide, but, as with wind turbines, the land between them remains free for other uses, such as agriculture. Allowing for this effect, the total area of Lackner's 'air capture farms' would be some 75,000–200,000 km^2,[73] the upper limit being somewhat smaller than the UK, of which only 0.5% would actually be occupied by the machines. This is some 50 times less area than required for the same carbon offset achieved by wind power or biomass[74] (which would fill the whole of Canada[75]).

A large programme (one wedge's worth) of electrochemical weathering to increase ocean alkalinity would require electrolysing some 6,000 m^3 of sea water per second, a flow rate equal to about 100 large sewage plants.[76] Again, this is large but not impossibly so.

Consider also that large-scale CCS would be applied on many stationary sources, and that switching to gas, renewable power and perhaps nuclear, and upgrading efficiency, should make a major contribution to reducing greenhouse gases too, so that air capture is required on only a fraction of projected emissions.

When in mass production, a 10 tonne per day air capture unit could cost as little as $20,000.[77] The main expense, though, is for energy to recover and compress the carbon dioxide. There is a range of estimates for the energy requirement and cost of air capture,[78] but if we compare to a typical American's transport-related emissions of 11 tonnes of

CO_2 per year,[79] the energy required would be about a quarter of annual consumption. This sounds extravagant, but most of it is waste heat. The remainder, which requires electricity, could be provided by cheap off-peak nuclear or renewable power.

Typical suggested costs of air capture are around $130 per tonne for a somewhat improved system, while in the best case electrochemical weathering might also achieve the same level.[80] Such a figure implies about $1,400 per person per year to offset all transport emissions, substantial, but equivalent to a petrol (gasoline) tax of about $1.30 per gallon. It suggests that a New York to London flight, available at the time of writing for around $500 (one-way economy class), would have to increase ticket prices by about 40% to offset its emissions. This sounds like a lot, but is considerably less than the range between the cheapest and most expensive airlines on this route. Such a price hike would reduce long-distance business travel by about 10%, and short-haul leisure trips by about 60%,[81] which would of course also save on emissions.

The air capture approach may well therefore be cheaper than hydrogen cars, or giving up flying. Air capture provides a 'backstop' for climate mitigation (Figure 5.12). Without it, as carbon cuts get deeper, we have to proceed to ever more expensive options and the carbon price rises steeply. With it, at a certain cost, we have essentially an infinite amount of mitigation available, subject only to building enough

Figure 5.12. Air capture as a backstop technology

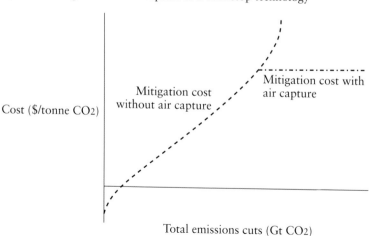

artificial trees (and to the availability of underground storage space or to development of mineralisation methods).

If air capture cost decreases to ~$30 per tonne CO_2, as Lackner believes is possible, then air capture would cost much less over the twenty-first century than the mitigation costs estimated by Stern and the IPCC. Since most of the cost estimates even for today, at a very early stage of the technology, are around the $130 level comparable to Stern and the IPCC,[82] it appears that air capture can have an important role to play in capping the upper level of carbon costs.

Transport and storage costs need to be added to this figure, but since air capture can be implemented anywhere, it will presumably be placed near a high-quality storage site. With a 1 million tonnes per year system spread out over an area of no more than 7 km² (about twice the size of Central Park in Manhattan), transport distances will be small. A pilot version of Keith's machine capturing about 0.3 Mt per year would cost about $115 million to build and $25 million annually to run. Making this relatively modest investment might well be worthwhile to establish the feasibility of air capture.

Once some initial plants are up and running, air capture costs should be far easier to estimate than those of alternative mitigation strategies, since the technology:

- Is simple and modular.
- Requires little adaptation to local circumstances.
- Has no obvious diseconomies of scale (unlike, in particular, biomass).
- And does not demand that we make detailed assumptions about government policies, society and human behaviour.

An alternative to air capture is to extract the carbon dioxide using biomass, burn that biomass for power, and capture the resulting carbon dioxide. With suggested costs of $41–55 per tonne, this certainly appears cheaper than indirect capture, and competitive with CCS on coal-fired power stations. It is, as noted, a very expensive way of generating electricity. However, at high carbon prices, electricity from biomass CCS plants becomes progressively cheaper while all other fossil-fuelled power becomes more costly. At carbon permit prices of $80 per tonne, the cost of electricity drops to zero: i.e. the plant can pay for itself purely on the strength of its carbon dioxide abatement.[83]

Co-firing a CCS coal plant with about 10% biomass makes it carbon-neutral,[84] and has the advantage of allowing larger plants with greater economies of scale. Employing higher proportions of biomass gives carbon-negative electricity. Even at the lower end of estimates, there should be comfortably enough biomass to make all the world's coal power stations carbon-neutral.[85] The ranges of estimates for available biomass are so wide as to be almost meaningless,[86] but, once allowing for the biomass used to make coal carbon-neutral, this method could mitigate between a third and all transport emissions.[87]

Turning this biomass into biofuels would mitigate only a third as much transport emissions, due to the lower efficiency of biofuel conversion. Therefore, from the sole viewpoint of carbon mitigation, a better strategy for transport than biofuels is to continue using oil, and to mitigate its emissions with biomass-fired CCS power plants.[88] Still, unless available biomass is plentiful and used intensively, this still suggests a remaining need for decarbonisation of transport—whether by electricity, hydrogen, offsetting air capture or another method.

Capacity and Cost of Biological and Land-Use Sequestration

The main options for biological and land-use sequestration, as mentioned before, are forestation; sequestration in wood products; preserving peatlands; extinguishing underground coal fires; biochar; conservation tillage; and ocean fertilisation.

Currently, biological sequestration by **forestation** and changes in land-use is one of the cheapest options for carbon mitigation. Forestation costs can be broken down as follows:

- Land costs—either rents or outright purchase—which, in the long run, should represent the 'opportunity cost' of not using the land for other purposes (farming, recreation, urbanisation, etc.). These costs vary enormously from around $600 per km^2 in remote parts of Canada and $1,600 per km^2 in India, up to $800,000 per km^2 or more in parts of the USA. A tropical forest may contain around 30,000 tonnes CO_2-equivalent per km^2, and temperate forests rather less (perhaps 4,000 tonnes per km^2 on typical forestry land). However, unlike CCS, sequestration is not instantaneous, so future removal of carbon dioxide by growing trees has to be discounted versus the up-front costs. This also applies to the cost of forest creation, the next point in this list.
- The cost of establishing the forest, including a failure rate which may be around 15%. These costs cover a broad range but around

$60,000 per km^2 seems typical,[89] or a range of $0.26–4 per tonne CO_2, the smaller projects costing more. Individual projects ranged from 0.07 to 22 Mt CO_2 avoided over their lifetime, so the very largest project was similar in magnitude to the Sleipner carbon storage scheme, and about a quarter the size of CCS on a medium-sized coal plant. For avoided deforestation, of course, this cost is zero. Estimates for preserving peat-lands in South East Asia are comparable, $0.8 per tonne, but these probably do not take into account land and other costs.[90]

- Maintenance costs, including the risk of forest destruction by fires, storms, pests or other disasters, which most researchers have not clearly spelled out.
- Administration, which, even in developed countries, may be as much as 15% of the land, establishment and maintenance costs, and typically ranges from $0.05–1.20 per tonne CO_2. Transaction costs vary due to the difficulty of putting forestation schemes together, particularly in remote or corrupt regions, and the availability of good information about which forests are in danger. It is not necessary to protect all the Earth's forests at the same time, since many are not in danger of being cut even in the absence of protection measures.
- Monitoring, verification and enforcement, which is (as with CCS), a modest part of the total cost, $0.03-$0.14 per tonne CO_2.[91]
- Other environmental benefits from forests, such as providing fruit, wood, fibre and fuel, preventing erosion and floods, and preserving biodiversity, which can be taken as a 'negative cost'.

But unlike carbon capture from power and industry, the costs of which fall or rise only slowly with increasing use, bio-sequestration becomes rapidly more expensive as its scale increases. The reason is that it competes increasingly with other land uses—for food, timber and fuel. The breakdown above shows that, for low land prices, most of the sequestration cost is for establishing the forest, but as land prices rise, they become the largest single component of costs. With land costs being generally lower in developing countries, this suggests that they will be the main host for forestation initiatives, but this may be at least partly offset by population pressures and higher transaction costs.

Drivers for deforestation, and hence costs of mitigation, vary greatly between countries. In Africa, trees are mostly cut down for firewood; in Asia, timber and biofuel exports are more important. The cost of forestation in Asia is therefore generally higher than Africa, a fact exacerbated by much higher population densities. Indicative costs by region to halt deforestation are as follows:[92]

- Africa, $11 per tonne CO_2
- Central America, $35
- South America, $40
- Asia (excluding Japan, China and India), $77

But, as with CO_2-EOR, there are some low- or negative-cost opportunities, when co-benefits are included. For instance, agro-forestry such as growing mango and tamarind in South India may have a negative cost (i.e. a benefit) as great as– $40 per tonne CO_2; miombo (a kind of tree typical of tropical and subtropical savannas and grasslands) in southern Africa about– $25 per tonne; and cedar in Mexico about – $15.[93] In Costa Rica, a very successful campaign against deforestation pays landowners about $8 per tonne CO_2 to preserve forests.

Some OECD countries, such as the USA and Canada, have significant potential for increasing forested areas. Others, with dense populations, have only very limited opportunities. For instance, the UK could sequester about 11 Mt CO_2 per year by doubling the area of its forests, but only during the period of tree growth.[94] Fitting CCS to a single coal power station, Drax in North Yorkshire, would save almost twice as much, and would run indefinitely. This indicates that for many developed countries to make use of bio-sequestration, some form of international offsetting or carbon trading is essential (as covered in Chapter 6).

Short-term costs appear to be low. For instance, McKinsey's study estimated that saving 3 Gt CO_2 per year by 2030, almost equal to one of Pacala and Socolow's wedges (and two decades early), could be achieved by spending $17–28 billion per year to reduce deforestation in Africa and Latin America. This equates to a cost of no more than $21 per tonne;[95] 3–5 Gt per year savings appears to be a practical maximum, falling gradually through the century.[96] A carbon price of this level would, all other things being equal, increase wood prices by about a quarter.[97]

However, in the long term, forestation costs rise significantly. If we look ahead to 2100, to get a full wedge of emissions savings, $400 per tonne is required,[98] well in excess of likely costs for CCS on biomass, or air capture. This is because deforestation is predicted to decline naturally anyway as the century wears on, and because there will be more competition for land. This would require about 12 million km^2 of forests to be established or preserved, almost the area of India plus the entire USA, and equivalent to more than four-fifths of the world's entire arable land. As with CCS, a 'crash programme' of forestation would be likely to suffer escalating costs due to shortages of accessible land, equipment, forestry personnel and so on.[99] It is likely that the costs quoted above include neither this effect, nor the positive non-climate side-effects of forestry.

For carbon **sequestration in wood products**, about 1.2 billion m^3 of harvested wood is converted to long-lived items annually. This amounts to some 1.1 Gt per year sequestered for more than twenty-five years. Of course, this carbon will eventually be returned to the atmosphere: typically three-quarters is released within a century. But since wood demand is growing by almost 3% annually,[100] the size of the sink is increasing too. New Zealand studies suggest that building wood use could be increased by 17%,[101] but the cost is not indicated.

Preserving peatlands offers another large source of savings, though it is not clear whether it may already have been included in available deforestation estimates. About 2 Gt CO_2 per year are released from peat, mostly in South East Asia,[102] and in the worst years of peat fires, such as 1997, emissions may have been as much as 3–9 Gt.[103] Given that peat is being cleared for similar reasons to tropical forests, and that the physical costs of preserving it are small, mitigation costs would probably be similar to those given above for avoided deforestation.

Emissions from permafrost and high-latitude peat are significant too, around 1.8–3.7 Gt CO_2-equivalent annually.[104] However, as this is mainly caused by the warming climate and occurs from many small points in a remote area, it is not really obvious how it could be stopped.

As for **underground coal fires**, recent estimates suggest that these are much smaller sources of carbon dioxide than previously thought, probably around 30 Mt per year in China and a maximum of 75 Mt worldwide (older figures suggested 200 Mt per year from China alone). One estimate is that it would cost over $650 million to put out all US fires, which suggests that a global total might be about $2 per tonne CO_2 discounted over the twenty-first century.[105] The carbon dioxide saving is not very significant, but the cost is low, and there are large ancillary benefits to people living nearby.

For **biochar**, if we consider mixing biochar into all arable land,[106] this would amount to the equivalent of 400 Gt CO_2. If this were carried out over the next century, it would amount to a bit more than one wedge. Pastureland covers about two-and-a-half times the area of cropland, but for practical reasons it could probably not be filled so intensively with biochar, so the overall opportunity here may be of a similar size, say another wedge.[107] This may be optimistic since biochar may cause loss of soil carbon in temperate regions, and anyway has only limited fertility benefits in soils that are already fertile.[108] However, biochar does not only have to be used to enrich agricultural soils: it could also be used on fields that are then forested, or for landscaping, filling sea defences, raising land above floodplains and so on.

Can we actually generate this much biochar? Potential supply from crop residues plus forest waste could amount to some 10.5 Gt CO_2-equivalent per year. Secondary forest harvest and specially grown energy crops could increase the total further. So it should be possible to meet the target mentioned above of somewhat more than two wedges totalling some 800 Gt CO_2 sequestered over the twenty-first century. One more ambitious study sees biochar production potentially increasing to 20–35 Gt per year by 2100, enough to soak up most or all emissions.[109] However, a large part of this comes from replacing slash-and-burn agriculture with 'slash-and-char', so overlaps with avoided deforestation. It also assumes very optimistic levels of agricultural yield and ignores other competing uses for waste biomass.

Biochar sequestration in combination with energy from pyrolysis is estimated to be economic at carbon prices of \$37 per tonne CO_2 and above,[110] a somewhat lower price than for a major forestation or CCS programme. The total carbon dioxide savings would be greater than just those of sequestering the biochar, since the bio-oils and other useful energy produced would displace fossil fuels and potentially reduce deforestation. However, costs might well rise steeply for a very large biochar programme, which would be competing with others for usable plant matter.

Conservation tillage can be most effectively used where it already overlaps with good agricultural practice. One US example calculated that up to 40% of farmland could be used for conservation tillage at no net cost, and almost all of it around \$55 per tonne. Assuming this applies to arable and pasture land globally, then this could sequester 50 Gt CO_2, probably overlapping to some extent with biochar.[111] The tillage opportunity is therefore about half the size of that available from forestation, at similar cost levels. Both biochar and conservation tillage should increase soil fertility, thus freeing land for forestation or energy crops.

Ocean fertilisation has been costed at \$1–9 per tonne CO_2, which would indeed make it a very cheap mitigation option.[112] A massive fertilisation effort could potentially absorb around 3 Gt per year of CO_2, on the scale of one wedge. However, as noted in Chapter 4, it is far from clear that it even works, or would function on a large scale without unacceptable side-effects.

So, in very rough order of increasing costs, we have:

- Agro-forestry
- Extinguishing underground coal fires
- Ocean fertilisation (albeit probably ineffective)
- Forestation in Africa
- Biochar
- Forestation in Latin America
- Conservation tillage
- Forestation in Asia

We can see that some low-cost bio-sequestration clearly beats CCS, and most other carbon mitigation technologies. However, bio-sequestration costs increase sharply as we scale up our efforts. Given competing land uses, forestation and agriculture are a vital component of fighting climate change, but not large enough in scale to substitute for technological carbon capture.

Costs at the Macroeconomic Level

So far, we have considered costs primarily in terms of single facilities or systems. At a national, regional or global level, we need to answer:

- What cost of carbon is required to make CCS attractive, and when will this be reached?
- Is carbon capture competitive with other low-carbon technologies?
- How much contribution can carbon capture make to meeting climate targets, and at what cost?

Carbon Pricing for Capture to Take Off

As with all assessments of climate change mitigation economics, this requires many assumptions, some of them mentioned in Chapter 1. For instance, what would the base-level emissions be in the absence of climate change policies? The IPCC's scenarios, considering various combinations of economic growth, population, fossil fuel availability, social and technological change and environmental awareness, cover a range of emissions from 2,900 to 9,200 Gt CO_2 over the twenty-first century.[113] This is obviously a gigantic range, requiring drastically different mitigation approaches.

On top of these 'business-as-usual' scenarios, we have to overlay assumptions about the stringency of carbon limits; the cost of the various mitigation measures and how they will change with technological progress and implementation at scale; and which policies will be brought in (for instance, whether trading of emissions cuts between regions is allowed). In general, CCS will be more prominent in a world with more baseline fossil fuel use, stricter curbs on carbon, slower development of alternative climate-friendly techniques, and less trading of emissions cuts between regions.

In most of the world, carbon has no cost and therefore no CCS will be implemented, except for pilot purposes or when a value-added opportunity is present. It is no coincidence that two of the world's four longest-running CCS systems are in Norway, which introduced a carbon tax in 1991, and a third is the value-added Weyburn EOR project. The EU's emissions trading system does set a carbon cost for large emitters, and a similar 'cap and trade' arrangement may eventually come into force in the USA. In principle, the economic effects of a tradable cap and a tax are very similar (although the implementation is different).

Therefore, as in Figure 5.13, we can compare predictions for future levels of carbon taxation or permit costs with the costs of CCS. This shows a band (blue solid) for a likely range of CCS costs for the bulk of projects. This falls with time, due to learning effects, assuming government support for early projects in the time before CCS becomes competitive. These costs are then compared to proposals for carbon taxes from various economists, and for forecasts of future carbon prices under two 'cap-and-trade' systems, those of the EU and USA. The wide range of values for future carbon costs mirrors the economic debate mentioned in Chapter 1, Nordhaus in particular being a long-term advocate of introducing a carbon tax but at a very low level.

When carbon prices reach the lower part of the blue band, then we could expect large-scale CCS implementation to take off. Prices under

Figure 5.13. Carbon tax/permit levels and CCS costs[114]

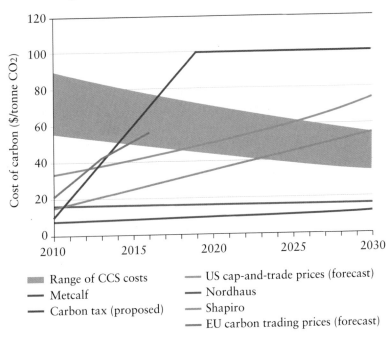

Range of CCS costs — US cap-and-trade prices (forecast)
— Metcalf — Nordhaus
— Carbon tax (proposed) — Shapiro
— EU carbon trading prices (forecast)

the EU's trading scheme were, at the time of writing, around US$18 per tonne,[115] and may soon move into the territory where CCS becomes increasingly attractive. This depends on the pace of economic recovery in Europe, and future policy decisions about the level of carbon cuts to target. The Carbon Tax Center's proposed path is rather similar and would make USA CCS attractive around the year 2015, practically today given the lead-time of projects.

On the other hand, US cap-and-trade proposals would take until 2022 before medium-cost carbon capture comes into play. The very low-priced proposals will not encourage significant take-up of CCS for decades to come, except in some value-added niche opportunities.

In practice, cost estimates will remain uncertain until a number of plants have been built, while carbon prices will be volatile, as has been the experience of the EU trading scheme. This will probably further deter investment, as utilities will wait until they have ample confidence that carbon prices will be high enough to justify a decision to install CCS. In a classic chicken-and-egg problem, the capture costs will not fall significantly until numerous plants have been built. Carbon prices may reach the $70 per tonne or higher level required for 'first of a kind' plants around 2015–20 in the EU, but under current proposals, this will probably not happen until 2030 in the USA. Of course, if early take-up in the EU is significant, then the United States will be able to build on that experience. But this does suggest that significant government support, of the kind already extended to various renewable technologies, is necessary if the first generation of CCS plants is to proceed swiftly. This is explored further in Chapter 6.

Mid-range CCS costs would add noticeably to the cost of petroleum, electricity and ammonia, and greatly to the cost of coal.[116] However, these cost increases would be considerably less than those caused by rising energy and commodity prices generally during the 2000–8 period. Electricity prices easily vary by 3 ¢ per kWh or more between different countries, US states or seasons of the year,[117] so an increase of 1–2 ¢ per kWh caused by CCS implementation is not excessive.

Competitiveness Against Other Low-Carbon Approaches

CCS, of course, has to compete with other greenhouse gas abatement approaches. One possible 'cost curve', from work by McKinsey, is shown in Figure 5.14.

The curve shows the size of the abatement opportunity up to the year 2030 as the width of the bar, and its cost as its height. On the left are some negative-cost opportunities, mostly involving energy efficiency, which save money. On the right is a longer tail of higher-cost options, up to €40 per tonne CO_2 abated, considered enough to stabilise atmospheric carbon dioxide at 450 ppm. More expensive options certainly exist above this level, and would be required for stabilisation at 400 ppm or lower (or if some cheaper approaches are not feasible, for various reasons), but were not considered in this study.

It can be seen that carbon capture is not one of the cheapest options (value-added EOR not appearing here), but it certainly forms a competitive part of the solution. CCS on new coal power plants falls somewhere in the middle of opportunities; coal retro-fits and industrial CCS come in at the high end. Note that solar power and offshore wind are estimated to have costs greater than €40 per tonne CO_2 and so do not feature at all. This level of CCS costs is consistent with Socolow's estimate for most potential stabilisation wedges; wind power, biological sequestration, nuclear and energy efficiency.[118]

Figure 5.14. Cost curve for greenhouse gas abatement[119]

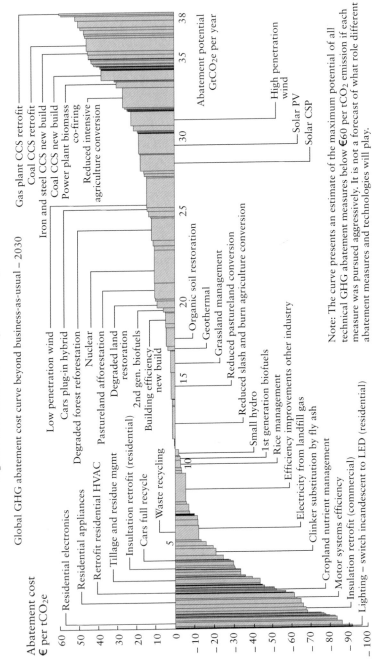

Note: The curve presents an estimate of the maximum potential of all technical GHG abatement measures below €60 per tCO₂ emission if each measure was pursued aggressively. It is not a forecast of what role different abatement measures and technologies will play.

Similarly, Figure 5.15 shows the cost of electricity for new power plants in the USA, which can be compared against wholesale electricity prices of US¢ 3–9 per kWh during 2008–9.

This confirms that carbon capture is competitive against renewable alternatives. Solar[120] and offshore wind appear extremely expensive, and even onshore wind is somewhat more costly than carbon capture on natural gas or coal-fired plants. Geothermal and hydroelectric power are fairly competitive but have limited room for expansion, while nuclear power has its own problems of public acceptability. The competitive position of CCS plants may change somewhat if carbon capture equipment inflated more than renewable technologies during the volatile construction market of 2008–9.[121]

Figure 5.15. Electricity generating costs for various technologies in the USA[122]

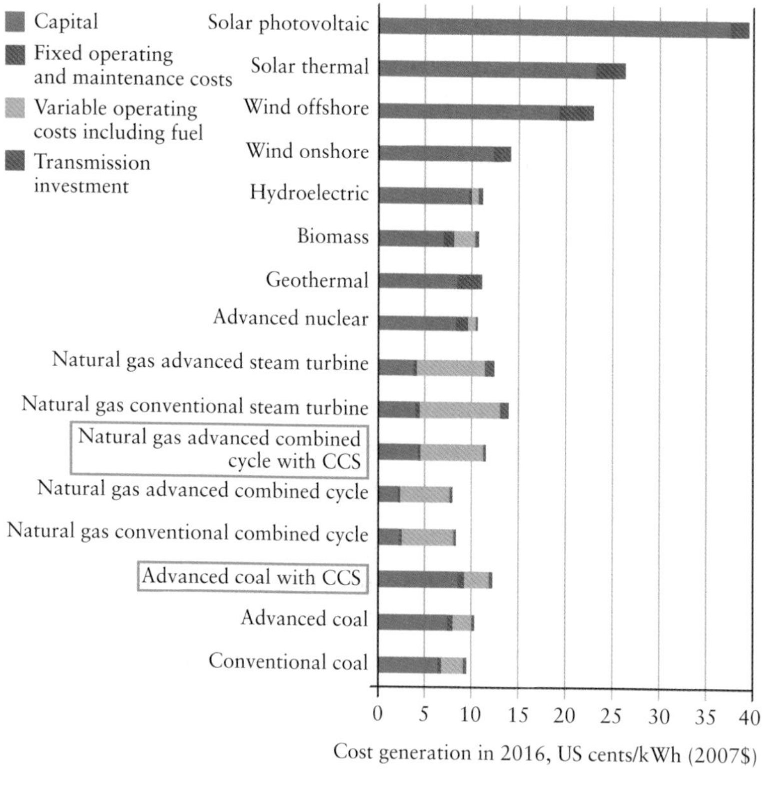

Cost generation in 2016, US cents/kWh (2007$)

Long-Term Costs and Scale

In order for CCS to play its part in reaching climate goals, substantial investment is required. In the early phase, up to around 2025, $70 billion would be required to begin fitting a small proportion of OECD power stations with carbon capture.[123] This compares to some $24 billion committed so far ($7 billion from the EU, up to $14.7 billion in the USA and $2 billion in Australia, plus additional funds from South Africa and Canada). This is encouraging, given the fairly early stage of development. Another $50–65 billion would kick-start industrial CCS up to 2030.[124]

With such investment, CCS can supply at least 3 Gt CO_2 per year of reductions by 2030, nearly 12% of the total identified opportunity. This is impressive, given gloomy prognostications by others that CCS cannot contribute at all before 2030.[125] In contrast, additional renewables (mostly wind) contribute only 5% of mitigation by this date.

Beyond 2030, CCS is likely to become increasingly important, as the technology advances, and as the pool of low-cost opportunities in energy efficiency, non-CO_2 greenhouse gases and forestry becomes exhausted. During 2030–50, the phase of large-scale deployment, investment and additional operating expenses, mostly from extra fuel consumption, average $200 billion annually.[126] This represents a bit more than a tenth of the total carbon mitigation spending. It is, of course, a huge amount of money but, in the long term, even complete decarbonisation through CCS would cost only 3–4% of the total expenditure on energy infrastructure.[127]

For carbon costs above $55 per tonne (likely to be reached in the period 2020–30, if we consult Figure 5.13), biomass CCS could also be a major contributor, helping to offset transport and other fugitive emissions. With strict climate targets, net carbon dioxide emissions may even be negative from 2070 onwards.

Figure 5.16 shows how CCS and bio-sequestration might help in emissions reductions in a scenario of moderately heavy fossil fuel use. The red area shows permitted emissions, and the various blocks above show the contribution of various mitigation options.

Initially forestation and agriculture, which can be implemented relatively quickly, take up the burden of emissions reductions. Bio-sequestration might be more than half the required mitigation in 2020 (most of the rest being simple, low-cost measures such as energy efficiency,

Figure 5.16. One scenario for emissions reductions to 2100[128]

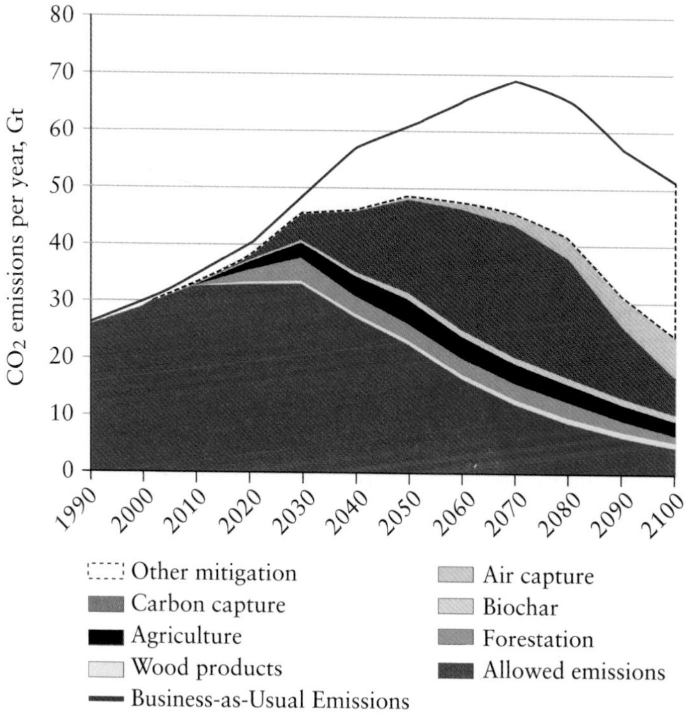

fuel-switching from coal to gas, and some measure of renewable energy). The contribution of sequestration in wood products is minor.

In the longer term, with a reasonably aggressive forecast for implementation, CCS grows to the largest single contributor, with about half of the total reductions in 2040–2080. Starting in 2030, but remaining small until 2050 and beyond, air capture begins to build up to mitigate aviation emissions, offsetting them completely by 2100. Biochar may also play a significant role, although at this stage its realistic potential is very uncertain. Forestation continues at a steady level but becomes proportionately less significant over time.

From 2070 onwards, fossil fuel use begins to fall and so CCS on stationary sources gradually starts to be phased out. Some of the existing storage infrastructure is taken up by a growing air capture industry.

A CCS contribution on such a major scale greatly brings down the cost of meeting climate targets, by removing the need for very costly, marginal mitigation measures. For instance, if we consider the International Energy Agency's scenario for strong action on climate change,[129] carbon mitigation in general costs about $1.8 trillion per year by 2050, about 0.6–1.9% of the predicted global GDP.[130] But if we rule out CCS (or assume it does not work, for some reason), annual costs balloon by 70%—an extra $1,280 billion. Consider that the cost of reaching the Millennium Development Goals (eradicating extreme poverty, improving education and health, etc.) by 2015 is 'only' about $50 billion per year,[131] and we can understand the vast increase in human and environmental welfare that is possible if CCS works.

For very low stabilisation targets (say 350 ppm CO_2 in the atmosphere), biomass carbon capture reduces the overall costs dramatically, typically by 50%, but is only a marginal improvement for less strict targets. Indeed, achieving 350 ppm may be completely infeasible without biomass capture,[132] Contradicting the often-heard argument that non-CCS technologies are 'enough' to avoid dangerous climate change, it appears that climate targets can be achieved as follows over the twenty-first century:

- Stabilisation at 350 ppm with CCS on fossil fuels and biomass: $6.1 trillion.
- Stabilisation at 450 ppm without CCS: $ 4.3 trillion.
- Stabilisation at 350 ppm without CCS: $26.0 trillion.

In other words, for a 40% increase in overall cost, we can go from a world with the likelihood of dangerous climate change to one with less carbon dioxide than today. But to stabilise at low levels of carbon dioxide without using carbon capture requires eye-watering levels of spending, something like half of today's GDP—completely unaffordable.

Similarly, without carbon capture, the marginal cost of mitigation virtually doubles, to almost $400 per tonne CO_2. This higher carbon cost would triple oil prices, taking them well above record levels. Two-thirds of the consumer price of oil, and 90% of that of coal, would be carbon costs.[133] These very high carbon costs are reached mainly because, without CCS, we are forced into making deep cuts in transport emissions, very difficult to do cost-effectively.

These costs can be compared to those for land-use changes and carbon sequestration in the biosphere, often advanced as an alternative (not a complement) to CCS. Tropical deforestation from 2000 to 2200 is estimated to cause damages of $12 trillion, in the mid-case.[134] A policy to cut deforestation by 90% would save 4.5 Gt CO_2 per year and cost 'just' $1.5–7 trillion.[135] Starting the policy later considerably reduces the net benefit; hence, as also with CCS, it is imperative to begin now. Promoting forestation is always a good strategy, regardless of the amount and trends of other carbon dioxide emissions[136]—therefore forestation is not an optional extra, but an integral part of climate strategy.

Conclusion

CCS can be a major contributor to emissions reductions by 2030, perhaps the second-largest single option, after energy efficiency. A substantial demonstration programme could be in place by 2020. The size of the effort required is very large, but not inconceivably so, compared to analogues from the oil and waste-water industries.

There are significant uncertainties in cost estimates, due to the lack of existing full-scale systems, the wide variety of options and assumptions, and the recent volatility of raw material prices. IGCCs appear probably the leading choice for new coal-fired power, but this conclusion could change with more experience and perhaps further development of oxyfuel systems. Retro-fits are costly, so the first priority is to apply CCS to all new plants. Some industrial sources have low capture costs and could be early adopters of CCS, but other large emitters such as cement are expensive.

Capture costs are likely to be high for the first generation of plants but will fall substantially with experience. Long-term capture costs for new coal plants may be around $20–40 per tonne CO_2.

Transport is a small part of overall system costs. Storage is a larger cost, but still cheap compared to capture. Enhanced oil and gas recovery can pay for at least the cheaper capture options. For pure geological storage, $1–10 per tonne seems typical. Mineralisation options are far more costly. A complete carbon capture and storage system would therefore probably cost from– $20 per tonne (with value-added EOR), to $34 (mid-cost case) to $80 per tonne (high cost, e.g. coal retro-fit with offshore storage).

These capture costs result in sharp increases in the cost of electricity, wholesale rising 30–50%, but modest to negligible rises for many industrial products. Air capture, or CCS on biomass plants, can reduce atmospheric carbon dioxide. Costs are high, but this is a viable approach to mitigating transport emissions.

Bio-sequestration is a large and low-cost option initially, but is ultimately limited by competing land uses. Including this factor, carbon costs range from $11 per tonne in Africa to as much as $77 in Asia. Biochar can potentially supply very large-scale mitigation, at costs probably somewhat lower than CCS and forestation, but feasibility remains highly uncertain.

CCS is not the cheapest mitigation option, but it is competitive, more so than solar or offshore wind power. Carbon capture is likely to expand substantially from 2030 onwards, particularly in the second half of the century. It dramatically reduces the overall cost of meeting climate targets. Very strict limits, such as 350 ppm, can only be realised with application of carbon dioxide capture from air and/or biomass. Forestation is an essential complement to carbon capture, not an alternative. For all these options, it is much cheaper to start as soon as possible.

6

POLICY

'*Let us be realistic. CCS is inevitable.*'
Maria van der Hoeven, Dutch Minister of Economic Affairs[1]

'*Politically, you can't pass legislation without support from coal states.*'
Daniel Weiss, Director of Climate Strategy, Center for American Progress[2]

The possibility of capturing carbon has been recognised since the early days of awareness of climate change, dating back to the 1970s. Carbon capture has been a prominent part of the climate change agenda since at least the late 1990s. Yet there are still few operating projects, and no full-scale CCS power plants. Public awareness remains low; where there is awareness, the subject can be highly controversial. As with renewable energy and efficiency, carbon capture needs the right policies—from government, and also from business and non-governmental organisations (NGOs)—to take off.

CCS policy can be divided into the short and the long term. In the short term, specific governmental and industry financial support is required to test the key concepts, demonstrate their safety, practicality and cost-effectiveness, and lay the physical and institutional infrastructure for a major CCS industry. In the long term, CCS will have to fit within the panoply of carbon mitigation options—which requires a consistent carbon price, whether by trading or a tax. Governments will have to ensure that laws, regulations and international treaties cater for carbon capture, transport and storage. And they will have to work with industry and civil society to ensure that CCS is managed safely and well, gains public acceptance and receives necessary non-monetary help.

Carbon Abatement Policies

Climate policy is a fast-evolving and complicated subject, on which a vast amount has been written, and which remains the subject of intense political debate, at national and international levels. I therefore give just a brief overview here of climate policy as it relates directly to carbon capture and storage.

At the moment, there is (in general) no charge made for the 'negative externality' of carbon dioxide. That is, polluters reap the full benefit of polluting, but suffer only a small fraction, if any, of its consequences. The damage borne by others is, as we have seen, the 'social cost of carbon': the harm inflicted globally by emitting 1 additional tonne of carbon dioxide. Unlike most pollutants, carbon dioxide is a global problem; a tonne of CO_2 is equally damaging whether emitted in Albania or Zambia. And many goods, carbon-bearing or made with carbon-fuelled energy, are traded internationally: coal, oil, gas, steel, aluminium, food, chemicals, plastics, cars, silicon chips. It is therefore difficult for one country or region to make policy unilaterally.

One solution to such externalities, which has been followed for sulphur dioxide in the USA, is to impose an industry-wide cap on emissions. Facilities emitting more than the allowed amount have to purchase permits from others. Any emitter who goes over their cap without purchasing permits suffers a fine set at punitive levels. The cap can, indeed must, be tightened progressively over time. The result is that those who can cut their emissions most cheaply do so. This should be the lowest-cost way to achieve a given target.

The other major alternative to a cap-and-trade system is a carbon tax, set at a certain level and progressively increased over time. Some examples of this are given in Chapter 5. The carbon tax does not have to add to the overall burden of taxation; its revenues can be used to cut, say, income or corporate taxes. Polluters who can cut their emissions more cheaply than paying the tax will do so.

In the case of a cap, we choose a limit on emissions, and the market determines the price of carbon. In the case of a tax, we choose a tax level and the market determines the amount of emissions. In principle, though the operations are very different, the emissions cuts and carbon cost should be identical under either system. However, new taxes are politically unpalatable, and there appears little faith that they would, in fact, be used to offset existing taxes.

The political momentum appears to be moving strongly in favour of cap-and-trade. The European Emissions Trading Scheme (EU ETS) has been running since 2005, covering almost half of EU emissions. In January 2009 ten north-eastern US states created a similar scheme, and Japan has had a voluntary carbon market since October 2008. Further proposed schemes are progressing in the USA under the Waxman-Markey/Boxer-Kerry bill[3] (to start 2012), Australia (the Carbon Pollution Reduction Scheme, mid 2011), eleven US states and Canadian provinces (2012) and New Zealand,[4] which is in talks to synchronise its scheme with Australia's.[5] However, both the US and Australian plans have run into political battles and their immediate future is doubtful.

Eventually, these schemes may be joined up with those in Japan and other countries, to approximate a global cap on carbon dioxide emissions. Of course, other big emitters will have to be brought into a worldwide plan in some way; adding China, Russia and India would bring coverage over 80%. The Copenhagen Climate Conference in December 2009, though, was unable to reach such a binding agreement.

Either carbon taxation or cap-and-trade should impose a cost on carbon which, rising over time, will eventually make CCS attractive. There are some legal issues to do with the qualification of CCS for carbon credits, and how they are accounted for. The main question, though, is how soon carbon prices will be high enough to support CCS projects, and whether—given the high volatility seen on carbon markets so far—project developers will have enough confidence to go ahead with costly, long-lived and inflexible investments. Research and development is also promoted by stable prices. This is true of nearly all energy spending, not just carbon capture.

In this sense, a tax would have the advantage of predictability. It is likely to suffer, though, from regulatory meddling: alas all too likely, especially at times when governments need extra revenue, or when a new government of a different political philosophy takes power. Cap-and-trade has the advantage that we know exactly how great the carbon cuts we get and can therefore match these to the recommendations of climate scientists.[6] A possible hybrid is to include a ceiling and a floor in carbon trading. If prices go too high, extra permits are released to prevent economic damage (which has the disadvantage, of course, of weakening climate change mitigation). If prices drop, some permits are withdrawn or bought up by the government, so giving a measure of protection to investors in low-carbon technologies, whether solar, wind, nuclear, carbon capture or any other approach.

Policies can also be imposed to drive CCS specifically, or to prescribe certain technologies and forbid others. Current approaches include a patchwork of carbon trading, taxes, subsidies, mandates and regulations. For instance, mandates can be set to require new plants to include CCS, and for old plants to retro-fit, at a certain date or when carbon capture is considered sufficiently mature.

In the EU, Australia and USA, in particular, there are already informal 'anti-fossil fuel' mandates, in the form of strong and growing public hostility to new coal plants, demonstrations, regulatory challenges, lawsuits and so on. Will Day, head of the UK's Sustainable Development Commission, warns that coal power stations could become a 'lightning rod' for climate protests.[7]

- In November 2008, a two-day campaign was launched against construction of the Kingsnorth coal plant in Kent in the UK.[8] For various reasons, the project was put on hold in October 2009.[9]
- Also in November, two Greenpeace ships attempted to close Rotterdam harbour in the Netherlands to prevent coal imports and protest against a new coal-fired power plant.[10]
- In March 2009, more than 2,500 activists blockaded the Capitol Power Plant in Washington, DC.[11] (Figure 6.1)

It might be better for the power industry to sidestep such opposition, by having clear legal mandates for CCS that at least put all companies on a level playing field, and clear up regulatory grey areas.

Mandates, though, run the risk of spurring unintended consequences. For instance, they may encourage a power company to keep an old, inefficient coal-fired plant running rather than building a new, efficient one. And, once built, the new CCS plant will have higher running costs than non-capture coal plants, and so will be higher up the 'dispatch curve': it will be turned off when demand falls, even though its electricity is less carbon-intensive than that from 'baseload' plants. Policy also needs to consider partly or wholly biomass-fuelled plants: carbon capture mandates should not make them uneconomic to build,

Figure 6.1. Protesters against the Capitol power plant in Washington, DC
(Photo: istockphoto.com)

but equally it would be desirable for them to use CCS to become 'carbon negative'.[12]

If a mandate is set from a given future date, say 2020, it should be made clear that any plants built before then will be required to retrofit, to avoid a race to pre-empt the tougher standards. There may, though, have to be an exemption from retro-fitting for plants that are close to the end of their life, with a firm commitment that they will indeed be shut down on a given date. Overly-prescriptive rules risk locking in obsolete technologies, especially for a rather new field like carbon capture. For instance, as discussed in Chapter 5, it may be more economic, and have the same carbon abatement effect, to fit partial capture to many plants rather than full capture to just a few. Furthermore, a mandate can only be realistically imposed when it is clear that CCS works safely and economically, but this creates a perverse incentive for the electricity business to delay proving that carbon capture works.

This get-out clause also prompts accusations that CCS is a fig leaf for polluting industries, not a serious attempt at a solution. If mandates are to be used, the government of the day must be tough in enforcing them. Industry will have to be clear that they will not be allowed to continue building and running unabated coal power plants—and, in time, gas power, cement, ammonia, synfuels, oil sands and so on. This will be a powerful incentive to make CCS succeed. If industry genuinely finds carbon capture unviable, then, without hope of a regulatory *deus ex machina*, they will have to have a 'Plan B'. This plan, though, might inflict severe economic damage.

Another, perhaps more desirable, alternative would be to include CCS plants within 'renewable portfolio' standards or obligations (which should perhaps be renamed 'low/zero carbon portfolio'). Under these schemes, utilities have to generate a given percentage of their power from renewable—or low-carbon—sources. If they cannot manage this, they can buy offsets from another utility which has an excess of low-carbon power. Alternatively, an average carbon intensity level can be set across an electricity utility's portfolio, at such a level that coal power without CCS will not qualify. This is analogous to fleet mileage standards for car-makers. The utility will then have to bring in a mix of low-carbon generation: solar, wind, nuclear, carbon capture as appropriate. Generators not in compliance with the obligation face a hefty fine. These limits tighten progressively over time. Such standards, though, have the disadvantage of being a minimum approach; there is no incentive for doing more than demanded.[13] And mandates and portfolio standards may not be economically efficient: they do not necessarily ensure that we get the deepest carbon cuts possible for a given expenditure.

Other alternatives include:

- Direct subsidies for CCS.
- The creation of a trust fund, perhaps financed by a levy on electricity, to manage carbon capture investments.
- Extending 'feed-in' tariffs, which give a preferential price for renewable electricity, to CCS power.
- Allowing regulated utilities to recover CCS costs from their customers.

Subsidies are always problematic: they are much easier to bestow than to take away; they may reward projects that would have happened anyway or go to the politically astute rather than the deserving; and they may weaken the incentives for developers to operate projects efficiently and drive down costs.[14]

Feed-in tariffs, offering a high and stable price which falls predictably with time to encourage innovation and cost savings, have been successful in spurring solar development in Germany and Spain: too successful in Spain, in fact, where they have had to be withdrawn due to the burgeoning overall costs of the programme. This leads to the very regulatory and price uncertainty that they were designed to avoid. Similarly generous feed-in tariffs in France and Italy have not been very successful in encouraging renewable energy, because of bureaucratic obstacles, showing that financial incentives are only part of the necessary package.

It is also difficult to ask consumers in a limited region to bear all the costs and risks of CCS demonstrations that will ultimately benefit everyone, a problem that has arisen with the proposed AEP IGCC in West Virginia,[15] where bills were forecast to rise by 12%. A compromise would be to allow regulated utilities to earn a higher rate of return on investments in new technologies, if those plans meet performance standards.

Whether carbon abatement is achieved by caps, taxes, mandates, regulation or subsidies, it has to be done internationally: at a minimum, on the scale of a large economy such as the USA or EU, but ideally globally. This will prevent 'leakage' of jobs and investment to areas with lower (or zero) carbon prices, which would undermine global emissions reductions. Achieving such agreement is, of course, tremendously challenging. It has proved hard enough to reach global consensus on free trade, from which, in principle, all nations can benefit. For climate change, there will be winners and losers, and some—most likely the wealthy countries, with the largest historical emissions—will have to bear a larger share of the future burden of cleaning up. For CCS, though, just six emitters account for more than 80% of all coal use: China, the USA, the EU, India, Japan and Russia. Adding South Africa, Australia, Canada and South Korea brings this to 90%. So if a 'G10' of coal-users, half of them high-income nations, could reach agreement on CCS, the bulk of the problem could be solved. In turn, coal is the most serious contributor to global warming, and the most amenable to carbon capture.

The Need for Government Action

We might first ask the question: why does CCS need government support at all? Surely it should stand or fall on its own merits?

Firstly, as we have discussed above, carbon emitters should pay for their emissions. Secondly, current carbon costs are insufficient to encourage CCS, and we cannot afford the luxury of waiting until they have risen sufficiently. Thirdly, private companies are unlikely to build the first generation of CCS plants without government support, since they will bear all the risks but only capture a small portion of the benefits. Fourthly, for similar reasons, a variety of other low-carbon technologies are already receiving substantial subsidies.

As Figure 5.13 shows, it is likely to take several years for carbon costs to rise sufficiently to make CCS economic. To spur its development, carbon costs will need to rise to $50 per tonne by 2020 in the OECD and by 2035 in non-OECD. Carbon costs alone will not be enough to achieve the G8's target of twenty full-scale CCS projects by 2010 at a cost of $30–50 billion.[16] The US's tax credit might be sufficient to encourage some EOR projects and perhaps low-cost industrial capture near good storage sites, but not costly power plant schemes.

Driving up carbon costs more rapidly could be very damaging to the economy, and would push many CO_2-emitting facilities to try to fit CCS at the same time, possibly before it is really ready. Carbon costs and fuel taxes are sufficient to spur incremental improvements, such as driving up car mileage and power plant efficiency. But as we have seen with wind, solar power and electric cars, government support is needed to nurse new technologies.[17] Until they are implemented on a large scale, and costs fall through learning, they can never compete with existing energy systems. Research and development (R&D) support is useful but not sufficient. Wind power did not advance to its current stage of near-competitiveness with fossil fuels because of carbon costs. It had to be encouraged by subsidies and mandates sufficient to overcome its initially large cost disadvantage. Even in Norway, which has had a hefty $50 per tonne tax on offshore carbon dioxide emissions since 1991, there are only two operating carbon storage projects, both using low-cost capture from contaminated gas.

Figure 6.2 demonstrates this point. One plausible estimate is shown for the increase with time of carbon permits under a cap-and-trade system (or, equivalently, a carbon tax). Carbon capture is initially very expensive, and so no CCS projects are launched until the cost of emitting carbon rises to more than $70 per tonne, which occurs in 2030. From this point on, CCS costs fall due to 'learning by doing'.

However, if a government programme had been launched by 2010 to accelerate the deployment of carbon capture, then learning will begin at that point. Then, by about 2022, CCS will advance to the stage where the carbon permit cost alone is sufficient to make new projects economic, and further explicit government support is not required. By this point, CCS is already almost $20 per tonne cheaper than it would have been without the government programme. For an emitter the size of the USA, the difference in carbon permit bills by 2030 might be as much as $100 billion every year.[18] This could be achieved by a one-off programme costing perhaps $10 billion.

Figure 6.2. Effect of government support on cost of CCS[19]

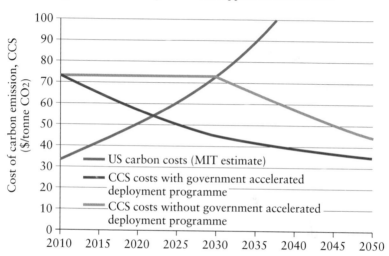

The first generation of CCS plants is likely to be very expensive. Probably all the major approaches will work in a technical sense, but some may prove to be unreliable or impractical. It seems currently unlikely, but from the three main techniques a clear winner might quickly emerge. The 'first movers' face bearing all the costs and risks of a new plant, but will only be able to capture a small part of the benefits. Some parts of the technology may be patentable, but much will boil down to simple choices between existing technologies, to 'prior art', and to intangible knowledge. The companies that wait for the second- or third-generation plants will already know what works, and what doesn't. The orders of equipment, and the training of staff, for the first few projects will have driven down costs for everyone else who follows. In such a crucial area, governments might even force first movers to give up their intellectual property for the common good. In a fast-evolving industry such as computing, a year or even a month's lead over competitors may be vital; for the power industry, where a coal plant may operate for forty years, a year is immaterial.

For similar reasons, governments already pay heavily for research and expansion of other energy sources, usually much larger sums than the $7–12 billion required for 10–12 carbon capture demonstrations (as discussed below[20]). For instance, ITER, the international organisation developing nuclear fusion, has a budget of some $16 billion.[21] For all its promise, fusion faces huge scientific challenges, much greater than those confronting CCS, and will not be ready before 2030–50 at the earliest. In Germany alone, $187 billion will go to solar over the next two decades, if the current guaranteed price is not cut.[22] Ironically, in 2005 German coal mines received $15 billion in subsidies,[23] suggesting a certain incoherence in policy. R&D in nuclear power was supported with nearly $50 billion in the USA over the last thirty years, and in the same period renewable energy and efficiency were subsidised to the tune of $26 billion.[24]

Short Term: Need to Start Soon

The implementation of carbon capture is urgent. The IPCC report, published in 2005, covered all of the key issues and received great attention. A flood of research emerged on all the key issues: capture technologies, transport and storage, economics, law and policy. Around the same time, BP announced the Peterhead-Miller project, and Shell/Statoil the Haltenbanken carbon capture plan. Yet since then both projects have been cancelled, and disappointingly little concrete progress has been made.

During this time, although perhaps obscured by a period of cooler weather, the scientific evidence on climate change is, if anything, growing more worrying. Coal use and carbon emissions have continued to climb (until temporarily derailed by the 2008–9 recession). Lord Stern says, 'If coal is going to be used, the only response—because it is the dirtiest of all fuels—is that we have to learn how to do carbon capture and storage and we have to learn how to do it quickly on a commercial scale...If we can't then it's plan B and plan B will be more expensive probably.'[25]

Between 2005 and 2018, 750 new coal plants are planned (fifty in Europe, fifty in the USA, almost 300 in China and 200 in India). These will mostly be large, modern, efficient designs, which will operate for decades and be expensive and difficult to retro-fit with CCS. On the storage side, there is a window of opportunity for making use of low-cost EOR options, especially in Europe. Most of the big North Sea fields are coming up to decommissioning. It is already too late for two of the largest fields, Brent (UK) and Statfjord (Norway), as well as Miller. On the side of business opportunities, it may be advisable for facilities with a low capture cost to press ahead. They can possibly lock up value-added opportunities in enhanced hydrocarbon recovery now, before the much larger volumes from power plants come into play.[26]

Government support is essential in this early phase. The first movers will bear all the risks: technical, commercial, political, reputational. But they will not be able to reap all the benefits. Once one electricity generator demonstrates that CCS is feasible, others will be able to proceed with confidence, and will have a clearer idea of what works and what does not. The first-of-a-kind plant will, as discussed in Chapter 5, probably cost much more than its successors. There are therefore strong incentives to wait; but if everyone waits, then nothing happens. Only a large-scale, government-supported programme can catalyse action.

We do not know which technology choice will be best. We do not have the luxury of time, waiting to move sequentially, trying first one approach and then another. This is analogous to the Apollo Programme, which was moving too slowly to meet Kennedy's deadline until the bold decision was made to combine and test whole packages of systems together. In any case, different situations have different requirements: gas versus coal, new plants versus retro-fits, aquifer storage versus EOR. The initial CCS programme has to be broad and cover all the main technology choices.[27]

The problem, though, of moving fast is that learning from the first generation of plants will be reduced,[28] and there is a risk of repeating the same mistakes. The first group of demonstration facilities should therefore each test out different aspects of CCS so that there is more learning, and less risk that a single common factor derails all the projects.[29] This also reduces the risk of 'lock-in' to a single technology, which may not turn out to be the best. Even if CCS itself does not take off in a large way, much of the fundamental research can still give economic and environmental benefits, particularly in driving significant advances in power station efficiency.[30]

As we have seen, current carbon dioxide prices, around $25 per tonne under the EU's Emission Trading Systems, or likely US cap-and-trade proposals, are still below those of likely CCS installations, particularly first generation. Australia's begins at an even lower point, $7 per tonne.[31] Costs will come down as experience is gained, networks of transport and storage are established, and a workforce of engineers and operators gains familiarity with CCS. There is a need for some comprehensive studies which benefit all players, such as inventories of storage space, and studies of regional aquifers and potential leak points. Again, the first movers will bear a large part of the costs and risks of such work, while latecomers capture a share of the benefits.

It is not only financial support that is required from governments. In Western countries in particular, but in India too,[32] permitting large industrial installations has become increasingly difficult, due to local opposition and special interests. The challenge is to ensure that safe, well-planned CCS projects can receive planning permission within a reasonable period, recognising the urgency of the climate challenge, without over-riding democratic practices. This is a similar problem to that confronting new nuclear and some wind farms.

Existing regulations and laws may have to be changed to accommodate CCS, especially in the areas of transport and storage. International agreement—clearly a governmental responsibility—is needed on issues such as cross-border carbon dioxide transport and storage, and the transfer of carbon credits or permits. One specific action, which might be helpful in kick-starting value-added CCS, would be to ban the use of natural underground carbon dioxide for EOR projects, or to demand that extraction of natural carbon dioxide should bear the same carbon costs as other sources of carbon dioxide. Existing enhanced recovery operations could then transfer to using anthropogenic carbon dioxide. We have to recognise, though, that much 'natural' carbon dioxide is extracted in combination with methane.

In the short term, there is also great promise from forestation measures. For instance, in the USA, these could sequester about one-third of the country's emissions. Some of these measures are low-cost and, even at current carbon prices, could make a significant contribution. They might help in bridging the gap until the large-scale arrival of carbon capture, renewable technologies and others beyond 2020. No special technology development is required, but support programmes have to be well designed and avoid unintended consequences, such as merely displacing deforestation elsewhere, or denying poor farmers cropland of which they have informal ownership.

Near-Term Policies

A coalition of wealthy countries will have to take the lead; once CCS is established, then those developing countries who are heavy coal consumers may follow.[33] But if Europe, the USA, Australia and Japan do not move first on carbon capture, then it is very unlikely that China, India and South Africa will go ahead independently. As Stern notes, 'We can't ask India and China to use new clean coal technologies if we are not prepared ourselves to demonstrate that they work and share those technologies.'[34] The United Nations Development Programme advocates that the EU and USA should each build thirty demonstration plants by 2015,[35] a considerably more ambitious goal than that outlined below.

Recently there have been significant advances in government support for carbon capture in Europe, the USA and Australia. The UK has announced that new coal-fired plants will have to have at least 300 MW of CCS capacity (net of power needed to run the capture system), and that when CCS is established as commercial, expected by 2020, they will have five years to retro-fit for complete capture.[36] All new carbon-emitting plants will have to be 'capture-ready'—have room for CCS kit, a route mapped to a feasible offshore disposal point, and no identified barriers to implementation. And in October 2009, the Hatfield IGCC in northern England was declared as the winner of a competition to receive EU and UK funding.[37]

The US's current cap-and-trade bill has been criticised for being so prescriptive that it is really top-down regulation, not a trading scheme at all.[38] In order to receive free allowances, electric utilities have to comply with mandates on carbon capture (and others, including renewable energy and efficiency). Power plants have to meet strict emissions standards which will be impossible for non-capture coal plants, and new coal plants have to be capture-ready, and to fit CCS once 2.5 GW of carbon capture has been installed in the USA, or 5 GW worldwide,[39] a target that should be reached before 2020 with existing programmes.

A small levy on fossil-fuelled electricity is intended to provide up to $1 billion annually for CCS. There is also a specific incentive for carbon storage, a tax credit of $20 per tonne CO_2 stored in a geological reservoir, or $10 per tonne for industrial CO_2 safely stored via EOR or EGR,[40] for sources yielding more than 0.5 Mt CO_2 per year (which may exclude some smaller low-cost industrial sites). The incentive is available until a total of 75 Mt has been stored. The injection site has to be located within the USA (which would rule out Weyburn, for one, from receiving the credit, even though it disposes of 'American' carbon dioxide).

The EU is launching a demonstration scheme intended to get CCS to commerciality, including some ten to twelve projects. Modelled on that proposed by the Zero Emissions Platform, a coalition of European stakeholders, it would require about $7–12 billion of government aid. This 'economic gap' is the difference between the plan's total cost and the revenues that would be raised anyway from electricity sales and carbon permit trading at current European prices. Though a large bill, the support is smaller than those given to other new energy technologies, and is modest compared to the economic benefit of accelerated deployment, which as shown above (Figure 6.2) can be in the order of tens of billions of dollars annually. This programme could capture up to 60 Mt per year of CO_2, about five times the current worldwide carbon storage effort.

One possible way to run the scheme would be as a competitive auction, in which companies bid the lowest subsidy they would be prepared to accept for building and operating each plant. The industry, therefore, has to cover the construction and the normal costs and risks associated with a conventional power plant. This set-up also creates the incentives for the operator to optimise the running of the plant as an integrated whole over several decades, and probably steadily to improve the efficiency of capture with post-construction 'tweaks'.

The bidding would probably be based on a premium price per tonne of carbon captured, rather than a lump sum, to ensure that the plant is actually operated, and that it both generates electricity and captures carbon. Alternatively, as long as the plant meets a certain emissions standard, a premium to the standard electricity tariff could be paid. The problem with that approach is that it would not incentivise the operator to maximise capture, although possibly a penalty could be levied for carbon dioxide released above the limit, or existing carbon costs might be sufficient. Total subsidy could be limited, say to the first 60 Mt stored, so that governments are not on the hook for unlimited sums of money, and so the winning companies have an incentive to keep their plants running efficiently.

It is possible that companies' bidding might cover some of the economic gap, in return for the intellectual property, first-mover advantage, reputational gains and future higher anticipated carbon prices. However, the issue of intellectual property would have to be considered carefully: since the whole idea of these demonstrations is to kickstart CCS, it would not help much if a few large power companies accumulate at public expense patents and proprietary techniques that others cannot use. This is particularly an issue for technology transfer to the developing world. Similarly, if consortia are formed to bid, then the competition law/anti-trust issues would have to be thought through. Alliances—for instance, an engineering company, an electricity utility, pipeline operator and oil company with storage facilities—could be very powerful in delivering integrated projects. But several large power companies working together might gain an effective oligopoly over carbon capture. Third-party access rules to pipelines and storage space are needed, as defined for the natural gas industry.

A reasonable demonstration portfolio would cover all the main capture ideas. However, it is important not to over-specify the projects. There should be as much room as possible for innovation, and the private sector should bear as much responsibility as possible for building and running successful plants. This might require careful design, to avoid capture by special interests, who may try to steer the design of the competition to favour a technology in which they are strong. Governments also have to resist the temptation to 'pick winners' prematurely.[41]

The projects would therefore encompass pre-combustion, oxyfuel and post-combustion, with both hard coal and lignite. Natural gas is not essential right now, but Norway will anyway be testing gas CCS. Biomass co-firing is not critical now anyway, but it would gain public support, and be useful for demonstrating the possibility of 'carbon negative' systems. A retro-fit would prove that we can address the backlog of existing plants.

Given the large potential of industrial capture, that is also a priority, but there is a question here: should we focus on a low-cost, straightforward capture option like ammonia or hydrogen, or on a volumetrically more significant, but harder and higher-cost emitter like iron- and steelworks or cement? Important demonstrations of enabling technology on the industrial side are:

- a nitrogen-free blast furnace and smelt reduction for iron;
- an oxy-fuelled cement plant;
- CCS on a black-liquor gasification unit (pulp and paper);
- and for refining, fluid catalytic crackers[42] with high-temperature combined heat and power (CHP) and CCS

But it may be overstretching to investigate all these possibilities in the first wave. Statoil at Mongstad in Norway, and Shell at the Pernis Refinery in the Netherlands, are already investigating refinery and CHP capture, so maybe the most valuable step would be to fit capture to some of the easiest industrial plants, so that CCS here can become standard as soon as possible. Alternatively, an industrial cluster might be able to demonstrate several types of capture and share a single transport and storage system.

On the storage side, there should be investigation of at least one oil- and/or gasfield, and various kinds of saline aquifer. Transport should test a cross-border pipeline, to establish legal precedents, and also one ship-borne method, which can tackle smaller, remote sources and requires less permitting than pipelines. Both onshore and offshore storage should be employed: both are technically proven already, but it would be useful to test legal, permitting and social issues. It seems well-established that storage space would be sufficient for at least the first generation of projects, so further work on mapping capacity should not delay progress.

It is highly desirable to do at least one of the projects in a developing country in order to get China, India and perhaps South Africa involved. The developing-world project could use any of the main technologies, since it is primarily a test of policy and institutions. In general, there should be a wide geographic spread to increase public support and awareness, test a variety of regions and legal systems, and to encourage international cooperation (and financial contributions). The ITER cooperation on nuclear fusion is a possible model. The appeal of hosting a major new international scientific project, and attracting the related jobs and funding, might also help in overcoming local misgivings.

Given all these different requirements, ten to twelve projects would probably be enough to test all the key elements. For instance, we can have a coal pre-combustion plant in China connected to a saline aquifer, a lignite-fired post-combustion system in Germany sending carbon dioxide to an offshore gasfield in the Netherlands, and so on.

Following this crash programme, by 2020 we could have confidence that CCS works in a variety of situations; or, we would at least know when, and perhaps why, it doesn't work. We would know a reasonable upper limit on costs of capture, and would have identified some of the main problems and their solutions. We would have laid the ground-work for capture and storage networks, including pipelines, in key regions: such as the North Sea, the US Gulf Coast, Alberta, northern/eastern India, eastern Australia and north-eastern and coastal China. We would have trained experienced cadres of CCS professionals who could execute the next generation of projects. These achievements would greatly lower the risk threshold for developers to go ahead with new capture schemes. If the first few projects over-run badly on costs, or have poor reliability, then the risk will be absorbed by the private-sector developers, but it will be necessary to persist with the technology and learn the appropriate lessons.

The Commercialisation Phase

The next step would be to build some 80–120 commercial projects in Europe by 2030,[43] with corresponding numbers in other key areas, to test all the main industrial capture methods, and to establish CCS as standard for new coal power plants and many industrial plants. The commercial roll-out should follow as closely as possible on the heels of the demonstration programme.[44] It should refine some of the core techniques, while continuing to progress break-through approaches. Figure 6.3 shows a plausible view of how some core technologies might have advanced by 2030.

These schemes could be executed in 'capture clusters', sites with a dense concentration of polluting industries near good sequestration sites. Economies of scale can be substantial and the 'cluster' idea can greatly speed the adoption of CCS; for instance, Germany's Ruhr area is only 10% of the area of Germany, but contains 75% of its large stationary carbon dioxide emissions (Figure 6.4).

In Europe, possibilities include northern Spain, northern Italy, Netherlands-Ruhr, northern England, eastern Germany-Silesia and eastern Poland. Other such areas are the US Gulf Coast, north-east China and, if suitable storage sites can be found, eastern India. Plans are already being put together for a Dutch 'Energy Valley' in the northern Netherlands, where proposed new coal/biomass-fired power sta-

Figure 6.3. Technological progress delivered by a demonstration programme[45]

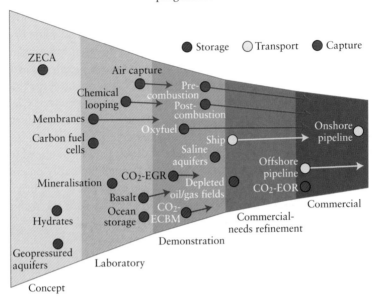

Figure 6.4. Carbon dioxide concentration in the atmosphere over the Netherlands-Ruhr area[46]

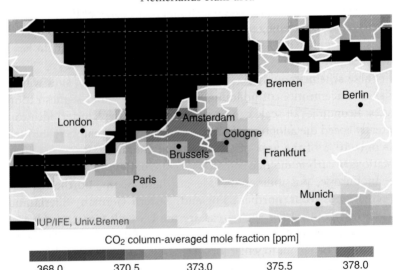

tions are located close to depleting gasfields.[47] Industry can form consortia, probably in combination with the state or national geological survey, to map storage space in the area, perform basin-wide fluid flow modelling, and identify and plug abandoned wells.

These clusters can save costs, couple power and (often smaller) industrial sources, optimise re-use of waste heat, share common pipelines and sinks, simplify permitting and gain greater public acceptability. Installations requiring pure oxygen supply—synfuels, hydrogen generation, IGCCs, oxyfuel power plants, and oxyfuelled cement, steel- and glass-making—could be grouped to share a single oxygen plant. Heat from steel- and glass-making can be used to regenerate CO_2-capture solvents. 'Polyfuels' plants, capable of producing electricity, heat, synthetic fuels, hydrogen and pipeline-ready carbon dioxide, can be particularly efficient and responsive to market demand. If air capture begins to emerge, a few small units could be added to take advantage of available compression and pipelines, in order to test the technology at modest cost. The creation of such industrial clusters may require supportive public policies, and be facilitated by building alliances between industries that historically have had little to do with each other.

Industry, in particular, will also have to consider capacity-building. Geoscientists, technicians and engineers will be in particularly heavy demand, and, for engineering in particular, businesses will have to compete with other 'green jobs' in wind, solar, nuclear[48] and so on. Engineers will have to be able and creative to design highly efficient systems, and to capture the opportunities that exist in integrating heat recovery, building 'poly-generation' systems and combining power plants with industrial capture. For geosciences, most of the skill set already exists in the petroleum business, but the average age there is high, usually estimated to be over fifty, and many people are coming up to retirement. During the recent economic boom, human resources, especially of experienced engineers, were in short supply. If the same constraints come to apply again, during a rapid transition to future energy sources including CCS, then it may prove better to design simple, robust systems rather than highly optimised ones which are harder to build and operate.

As well as these human resources, there is also a need for lawyers, financiers, commercial people and negotiators, biologists (for assessing environmental risks and developing bio-sequestration), social scientists (for building public acceptability), accountants, consultants, regula-

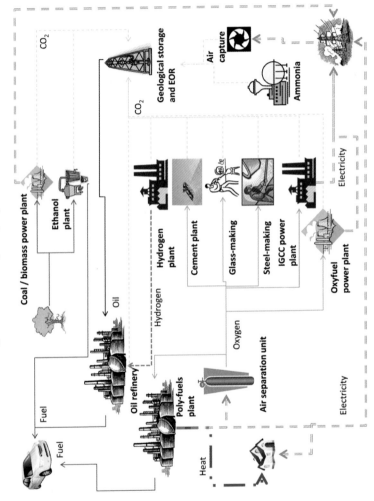

Figure 6.5. Industry/CO$_2$ capture cluster

tors, communications and PR staff, and researchers including chemists and materials scientists. Many of these will have to be people with experience in related fields, who can easily adapt to CCS. For instance, the petroleum consultancy Senergy also offers carbon storage studies. The management consultancies McKinsey and Boston Consulting Group and accountancy firms Ernst and Young and KPMG have published significant studies on CCS.[49]

Increasingly, as the industry grows and matures, university students will seek supporting qualifications. Encouragingly, the University of Edinburgh already offers a Masters Programme in Carbon Capture and Storage[50] and the University of Durham in northern England include CCS in their undergraduate teaching programmes and as part of the training within the newly established (2009) Centre for Doctoral Training supported by the UK's Engineering and Physical Science Research Council (EPSRC).[51]

Growing academic experience in CCS, buttressed by researchers with industry backgrounds, will be important in engineering and geoscience departments, and to some extent in the faculties of life sciences, social sciences, economics and law. The Massachusetts Institute of Technology (USA) and the University of Calgary (Canada), amongst others, have published extensive research on various aspects of CCS; Heriot-Watt University in Scotland is a partner in the Scottish Centre for Carbon Storage Research; Durham University hosts a Carbon Capture and Storage Group; and the University of Melbourne is taking part in Australia's first pre-combustion carbon capture project. Remarkably, in the ten months between August 2009 and May 2010, there were no fewer than twenty-one conferences and workshops around the world on various topics related to carbon capture. This suggests that interest in the subject is surging, and industrial-academic collaboration is expanding accordingly.

Longer Term: Paying the Cost of Carbon

Once carbon capture has been established as commercial, then it should not receive further subsidies. Indeed, neither should other mature low-carbon energy sources. Then all the different energy technologies and industrial processes can compete on a level playing field, with the one caveat being that all economic activities must pay the full cost of their carbon emissions. Large and continuing subsidies for energy in general are undesirable, since they:

- are expensive for governments (many of which are already struggling with deficits);
- lower energy prices and hence work against increasing efficiency;
- favour large consumers of energy, who tend to be more wealthy than average;
- tend to go to the politically connected and to large, capital-intensive 'flagship' projects rather than smaller but sometimes more effective initiatives;
- distort the market. For instance, Germany is far from the sunniest country in the world, yet it has become a leader in solar power due to its heavy subsidies. Those same solar panels could generate much more electricity if installed in California, Australia or India.

Without carbon costs, a utility will rarely run a CCS plant, due to its higher operating costs—particularly fuel. With an appropriate carbon price, they can make an informed decision between, say, zero-carbon wind, low-carbon coal with capture, mid-carbon natural gas, or high carbon non-CCS coal. Having a consistent carbon price throughout the economy will also free us of the very involved and debatable calculations about the 'carbon footprint' of a given product. For instance, how much carbon dioxide is emitted in making a solar panel, or the concrete and steel for a wind turbine? Once every activity pays its full carbon costs, then we no longer need to be concerned about a 'carbon footprint'; all the information is there in the price of the product.

The shape of any future global agreement on climate change will shape the adoption of CCS. Some industries, such as power and cement, are very local—little of their product is traded internationally. So if one country adopts strict carbon limits and another does not, this will not greatly affect business; the company in the country that has the stricter climate policy will largely be able to pass on its increased costs to customers. Except for a few energy-intensive industries, the competitiveness of most of these customers will not be greatly affected by higher energy bills.

For some energy-intensive, globally traded products, though, notably iron, steel, aluminium and perhaps petrochemicals and fertilisers, emissions penalties will form a substantial part of the cost.[52] An incomplete climate deal, not including all nations, raising global compliance costs significantly.[53] Even worse, by raising the costs of some players and not others, a deal which is not comprehensive can cause polluting industries to migrate to non-participating countries. This can easily reduce the expected emissions cuts by half; under plausible assumptions, global greenhouse gas release can even go up:[54] in this case, no deal at all is better than a partial one.

Therefore all players in these industries will need to observe the same rules, or nations will have to shield them in some way. This could be done by reaching a global sectoral deal: for instance, to cut overall emissions from the cement industry by an agreed amount; issuing free permits to such industries;[55] or imposing tariffs on imports to account for the carbon content of products. Though France has propounded it, and former Australian Prime Minister Kevin Rudd warned of this risk to the Australian economy,[56] the tariff option sounds like an administrative nightmare, and anyway would probably conflict with World Trade Organisation (WTO) rules and trigger retaliation.[57]

Oddly, though, CCS may actually see more adoption if a global deal on climate is delayed. Some regions, notably the EU, are clearly going to forge ahead with carbon caps, though they may tighten them if agreement is reached with other nations. Since CCS is, at least initially, going to be more expensive than other options such as avoided deforestation and energy efficiency, it will be used more if EU polluters cannot easily 'offset' their emissions with abatement elsewhere. Energy-efficient economies such as the EU and Japan, with relatively low coal use, have fewer low-cost carbon abatement options than Brazil or China.

In order to allow developed countries to make use of low-cost abatement opportunities in developing nations, the Kyoto Protocol allowed some 'offsets'. The EU Emissions Trading Scheme (EU ETS) also includes this principle, as do voluntary 'offsetting' schemes which allow businesses or individuals to pay to cancel out emissions they cannot easily avoid. The voluntary schemes have so far primarily relied on renewable energy and forestry, and expanded rapidly from $60 million in 2006 to $705 million in 2008.[58]

Such offsets have major implications for the use of bio-sequestration. Kyoto was not only inconsistent in its use of offsets, particularly land-use, but also failed to define and tackle some key issues, for instance what happened if reforested areas were subsequently destroyed, particularly if this was not directly caused by human activities. Kyoto also did not give credits for avoided deforestation. The protocol was thus inefficient and ineffective in driving the required changes.[59] The bill to establish a US cap-and-trade system allows substantial domestic and international offsets from forestry, agriculture and land-use;[60] California's permits forestry credits; while the EU ETS does not allow any bio-offsets, and the Australian scheme credits forestry but not agriculture.

Offsets have been heavily criticised by environmental commentators such as George Monbiot, who sees them as shuffling off a moral imperative to reduce emissions on to others,[61] a cosy way of preserving 'business-as-usual' while salving the consciences of the wealthy global elite, and doubts that many or any of these 'reductions' actually cut greenhouse gases at all. There have been accusations that some Chinese factories were deliberately built to emit large quantities of non-CO_2 greenhouse gases, such as chlorofluorocarbons (CFCs), and then received lucrative credits for some simple re-engineering to destroy these gases.[62]

This comes back to the principle of 'additionality': under the current system, carbon credits can only be earned for a project if it would not be economic in the absence of the credits.[63] For instance, if wind power is the cheapest form of generation in a given location, it would not be additional. If, though, coal power is more competitive, then wind turbines might be able to claim credits to overcome the 'economic gap'. This entails complicated verification procedures and counter-factual projections. Experience in mergers and acquisitions shows me that two parties, with the same data, can reach widely varying views on valuation. And there are other related issues, such as non-financial criteria: what if local protesters are preventing a new coal plant, but would allow a wind farm? Forestry, in particular, is problematic because of the vulnerability of trees to fire, logging and so on, and their rather short lifetimes, as discussed in Chapter 4.

Yet some kind of offset scheme is surely going to be included in the post-Kyoto climate deal. Emissions targets for less developed countries are going to be less strict than those for the developed world, but it makes no economic sense, is not ethical, and helps neither the climate nor poor countries to be spending $50 to cut 1 tonne of CO_2 in Germany if we could save 50 tonnes for the same money in Ghana. Carbon offsets in developing countries are also a way of providing financial assistance, potentially mobilising huge sources of funding.

In this context, carbon capture and storage is extremely attractive as a source of offsets. It is demonstrably always additional: no one would do carbon storage in the absence of carbon costs (excluding, of course, some value-added applications such as EOR). As discussed in Chapter 3, storage can be very secure and monitoring can give good verifiability. The amounts injected can be monitored precisely; whereas with, say, solar power, we have to project what the alternative source of electricity would have been, and how much power would have been generated over the next twenty to thirty years. Most CCS projects, particularly the first generation, will be carried out in developed countries, plus China and perhaps the wealthy Arab states in the Gulf, where it should be easier to arrange verification than for reforestation schemes in Congo or the Amazon frontier. Carbon dioxide is not of any obvious value, so, accidents apart, there is no incentive for anyone to drill into a storage site. In this, CCS differs from forestry. It also differs from the usual environmental prescription of leaving fossil fuels in the ground; there is no guarantee that a government, irresponsible or perhaps financially stressed, might, two or three decades from now, decide to open up coal- or oilfields that were previously under moratorium.

Once firms have adjusted to environmental regulations, they rarely have much interest in seeing them relaxed.[64] They have incurred 'sunk costs' in establishing the technology. They also face reputational risk, both within and outside the firm, by not adhering to procedures that have become 'industry standards'. Indeed, some elements of the energy industry are calling for a swift decision on climate regulations—so that they can move ahead on investments with some certainty.[65] The industry far prefers tough but consistent legislation to laws that are badly drafted, applied irregularly, or vague in scope. So once CCS is proved as a reliable and cost-effective technology, then carbon costs alone should be enough to ensure it expands. One point for environmentalists who oppose CCS to ponder: if carbon capture does become standard, then it will greatly narrow coal's cost advantage over gas and renewables.

Bio-Sequestration Funding

Bio-sequestration (in forestry, agriculture and soils) may expand to a major mitigation method via carbon costs, offsets or directly in countries with carbon caps. Within developed countries, bio-sequestration can be driven by payments (either directly from government, or indirectly from emitters buying offsets) for sequestering carbon, or alternatively a tax on activities such as logging that release carbon dioxide. Voluntary carbon offsets are already a major source of support for reforestation projects, such as ArborCarb's in Ghana.[66] Perhaps most logical would be for agriculture and forestry to be brought within emissions trading schemes, but given the small scale of such operations, and issues of precise monitoring and verification, this might prove administratively costly. In any case, payments directly tied to bio-sequestration are preferable to mandates or subsidies for activities such as planting trees.

Naïve payments for forestation run the risk of repeating boondoggles of the past, rewarding planting the wrong kind of tree in the wrong type of soil. Partly because of the political power of farmers, agriculture is notorious for attracting expensive subsidies that prove impossible to repeal: as experience in the USA, EU, India[67] and elsewhere demonstrates. The current absence of meaningful carbon prices for forestry, combined with mandates and subsidies for biofuels, leads to disastrously counter-productive activity. For instance, in Indonesia, oil palm cultivation earns about $11,400 per km², yet clearing the land for these plantations releases as much as 50,000 tonnes of CO_2, worth $1 million or more at current carbon prices.[68]

As for developing countries, it is likely that the post-Kyoto climate change treaty will include avoided deforestation, with tradable carbon credits, a development fund or some combination of the two.[69] Kyoto did not include carbon sequestration in soils or wetlands, nor any emissions or sinks from land-use in developing countries, and excluded accounting for natural or indirect changes. These will all have to be components of the next global climate change treaty. The Protocol also limited afforestation offsets to 1% of a country's emissions, but in practice only 1% of this small allocation has been taken up, due to transaction costs and 'leakage'—payments made for carbon sequestration that, for various reasons, does not occur. Given political pressures to limit international offsets, carbon markets may not be sufficient to funnel the necessary money to developing countries.[70]

As we have seen (Figure 4.2), a small number of countries account for most deforestation. Brazil and Indonesia, the top two, are middle-income states with reasonable prospects for implementing bio-sequestration projects. Indeed, Indonesian President Susilo Bambang Yudhoyono recently committed to curb deforestation and make his country's forests a net carbon sink by 2030,[71] while President Luiz Inácio Lula da Silva offered to eliminate four-fifths of Brazilian deforestation by 2020.[72] However, corruption and a tenuous presence of government in remote areas of these vast nations are problematic.

Continuing down the 'unlucky thirteen' who account for 90% of gross deforestation, the Democratic Republic of Congo, Nigeria and Myanmar in particular, all look like tough places to make a success of such projects; the others are somewhat more promising.

Bio-sequestration initiatives can also be funded directly, via overseas development aid (which has so far been insignificant in forestry) or via 'debt for nature' swaps. In these deals, a developing country is forgiven some or all of its debt, and in return the money that would have been spent on repayment is allocated to conservation of native forests.[73]

Analogous to logging concessions, 'conservation concessions' can be granted where investors pay for sustainable management of a forest area. In 2008, Brazil launched a $20 billion fund to protect the Amazon, towards which Norway pledged $1.1 billion over the next decade.[74] In the end, the Amazon, and perhaps other rich forest areas, might become giant national parks, with a sparse population carrying out some sustainable industry, tourism, forestry and farming on abandoned land. Such concessions require close coordination with local government and communities, education (as in Kenya, where schoolchildren are taught the value of their wildlife, given the importance of tourism), and policing, to prevent illegal logging and poaching. 'Payment for environmental services' can cover other benefits such as biodiversity and watershed management, although greater expertise is needed to manage such combinations. Forestry companies, food retailers and wood-users, such as manufacturers of paper and furniture, can obtain certification that their production methods are sustainable. This brings them a premium price, and may even be a *sine qua non* of social or government acceptance in developed countries. But, contrary to some rosy expectations, it is not always possible for community-based programmes to be both pro-poor and pro-environment; poverty reduction often takes priority in very low-income countries.

Other possibilities for financing bio-sequestration include private philanthropy, and encouraging commercial activities via micro-financing, specialist credit agencies, and long-term bonds securitised against income from sustainable forestry, eco-tourism and agriculture. Private companies including Marriott Hotels and Bradesco, a Brazilian bank, have provided $8.1 million to 6,000 families in return for ceasing deforestation, while the largest private forest owner in California, Sierra Pacific Industries, recently agreed to preserve redwoods and other trees to sequester 1.5 Mt of CO_2.[75]

Two forestation success stories come from Latin America. Jared Diamond tells the story of the Dominican Republic, whose enigmatic dictator Joaquín Balaguer used sometimes brutal, sometimes far-sighted methods to restore the country's forests, including bulldozing the homes that wealthy Dominicans had built illegally inside a national park. The achievement here is in stark contrast to the loss of trees in neighbouring Haiti, which has only 1% of its original forests.[76] In Central America, Costa Rica has been particularly successful in reversing a trend of severe deforestation since the early 1990s. Its efforts have been funded with a debt-for-nature swap, public donations, USAID assistance and an agreement with pharmaceutical giant Merck to provide it with potentially useful plant and insect samples in return for royalties on commercial products. The country, perhaps a model for future bio-sequestration campaigns, channels the revenues into several policies:

- Direct payments for preserving forests.
- Tax breaks for reforestation.
- Offers of residency status to people who carry out reforestation.
- Purchase of private landholdings for conversion into conservation reserves.
- Integration of neighbouring national parks into entities that can attract international funding.
- Encouraging local communities to develop sustainable economic activities such as eco-tourism.

In general, all these approaches require a major effort in capacity-building: establishing the required skills in law, management, forestry techniques, finance, marketing and value-added industry, for schemes to function effectively at a local level. Community forest management has been successful in Mexico, but experience elsewhere has been less positive.[77] So far, most offsets have gone into China, which offers a secure environment, and indeed progress in reforestation has been dramatic.[78] However, Chinese timber imports have increased in consequence, so it is possible that the problem is just being displaced.[79] In some major potential carbon sinks, such as the Democratic Republic of Congo, private investment is deterred by very high political and security risks, and a weak legal and banking system. But these lower-cost opportunities in Africa and Latin America will have to be taken up for bio-sequestration to deliver its full potential.

Legal framework

CCS is a fairly new concept, particularly in practice, and so some of the key legal issues regarding its use have not been defined, or at least not tested. The main issues are:

- Is carbon dioxide a waste or pollutant, in which case certain restrictions apply to its transport and disposal?
- Who has the long-term liability for the monitoring and safety of carbon dioxide in geological storage?
- How should carbon reductions from CCS projects be accounted for? Does CCS qualify for Clean Development Mechanism credits, whereby a party that has obligations under the Kyoto Protocol to reduce its emissions carries out a CO_2-abatement project in a developing country that does not have such obligations?

Some of these issues were addressed further during the climate negotiations in Copenhagen during December 2009. As then UK Energy and Climate Change Secretary Ed Miliband said, 'We're all agreed that CCS has got to happen, and making sure the world can afford it, through a new financing mechanism, will be a central part of any Copenhagen deal.'[80] In general, regulations that emerge will have to be flexible, have a degree of imagination in covering situations that may arise in the future, and be adapted as CCS practice evolves. However, Copenhagen itself did not reach agreement on including carbon capture in the Clean Development Mechanism; instead, UN working parties will report back at conferences in Mexico in 2010 or South Africa in 2011.[81]

The main relevant treaties for CCS are the UN Framework Convention on Climate Change (UNFCCC) from 1992, and the much better-known Kyoto Protocol, which set binding emissions limits for certain developed countries; the UN Convention on the Law of the Sea (UNCLOS) of 1994, defining nations' rights and responsibilities for using the world's oceans; the London Convention (1972) and London Protocol (1996), controlling marine pollution and dumping; OSPAR, covering environmental protection of the north-east Atlantic; and the Basel Convention, designed to regulate and reduce international transport of waste, especially to less developed countries.

Only parties to treaties are bound by their requirements, and not all nations have signed all the agreements mentioned above. For instance, OSPAR only covers a set of north-western European countries; amongst others, India has not signed the London Convention; and the USA has not ratified UNCLOS. In any case, none of these treaties were designed with carbon storage in mind. For instance, it is not clear if sub-surface (geological) storage is prohibited, or only sub-sea dumping (i.e. if waste, say in drums, is simply sunk to the bottom of the ocean). Some of the treaties make distinctions between supply from land or offshore, and placement from land or from an offshore pipeline. In this case, it might be legal to dispose of carbon dioxide via a pipeline running directly from land, but not one that first connects to a platform, or from a ship. These distinctions are technically meaningless, but legally significant, and would have to be ironed out prior to beginning projects.

At the moment, carbon dioxide is classified as a non-flammable, non-toxic substance. There are therefore no obvious legal restrictions to its transport under treaties such as Basel. This could change if it is defined as a pollutant by the US Environmental Protection Agency (EPA).[82]

Ocean storage is legally problematic because it may move across international boundaries, or from a country's territorial waters into international waters. This might conflict with regulations that disposal activities should not damage neighbouring states. Such migration is less likely for geological storage, and, where countries share a common aquifer (as, for instance, the Utsira aquifer in the North Sea between the UK and Norway), it should be possible for them to work out a framework for its joint use. Otherwise, countries might feel that 'their' aquifer is being unfairly filled up. Cooperation in the North Sea is probably straightforward, given decades of similar collaboration on oil and gas; in the Persian (Arabian[83]) Gulf, Caspian or South China Sea, it may well be more problematic. Rules may need to be agreed in case carbon dioxide injected by one country starts leaking from the seabed in the territory of a neighbour.

The inclusion of CCS in the Clean Development Mechanism (CDM) has been deferred several times, most recently at the December 2009 discussions in Copenhagen. Under the CDM, a developed country that has obligations under Kyoto to reduce its emissions can do so by sponsoring a carbon abatement project in a country (usually a developing one) without such a responsibility. Only three carbon capture and storage projects have been proposed for CDM so far: the White Tiger field in Vietnam, where carbon dioxide from a gas power station was to be used for EOR; capture of carbon dioxide and hydrogen sulphide from an offshore field in Malaysia, with storage in an aquifer; and a small-scale ocean storage project with added alkalinity. It has been proposed to restrict CDM to only the least developed countries, so ruling out India, China and the wealthy oil-producing states, in particular.

There are also accounting issues. The Kyoto Protocol does not contain a methodology for CCS, but the 2006 National Greenhouse Gas Inventory Guidelines do. One question is whether carbon capture is considered as a sink (i.e. as removing carbon dioxide from the air, in the same way as forestry) or simply as a reduction in emissions. In the case of international projects, this will have an impact on where the carbon credit occurs. If the carbon regime is uniform worldwide, this is not a problem, but if, as is more likely, different countries have different obligations, it may be problematic.

Imagine if Spain, a country with emission reduction obligations under Kyoto (and no doubt under successor treaties) were to ship carbon dioxide to Morocco, a country currently without such obligations. If CCS is counted as an emissions reduction, there is no difficulty. If it is classified as a sink, though, then Morocco will receive the credit while the emissions will count against Spain. This can be partly solved using the Clean Development Mechanism (or its successor), but there are restrictions on how much of their emissions developed countries can offset using the CDM. Similar problems emerge if the cost of carbon is not the same in neighbouring countries. Also, agreement will have to be reached on which country (in this example, Spain or Morocco) is responsible for any leakage, or if carbon dioxide migrates from one country into another (for instance, as noted, the UK and Norway share the Utsira Aquifer, into which goes the Sleipner carbon dioxide).

As we have discussed, carbon storage carries risks of leakage. Carbon sequestration in soils, forests and so on is even more vulnerable. After all, a tree planted last year can be burned down this year. An accounting system for carbon sinks needs to allow for leakage.[84] Some possibilities include:

- Temporary credits, that can be withdrawn if leakage occurs.
- A time limit, so that the owner of the permit has to renew it after, say, ten years, either by demonstrating that the carbon is still safely stored, or by funding a new abatement project.
- Discounts, whereby storing 1 tonne of carbon only earns, say, 85% of a credit, thus giving a margin of error for leakage. Possibly the

remainder of the credit could be granted after some years once safe storage is demonstrated. This would, though, have a negative impact on the economics of CCS. The accounting scheme would have to be consistent for all other carbon sinks, and be tied to their relative security of storage/sequestration. For instance, mineralisation should probably not have any 'leakage penalty'.

Rules will have to be agreed, probably internationally, for carbon dioxide pipelines, setting common standards for carbon dioxide purity, assigning responsibility for any leakage during transport, and ensuring 'third-party access' on reasonable terms. Analogous issues have already been encountered and, in the OECD at least, mostly solved, for natural gas pipelines. Regulations should be defined to cover storage, including minimum standards for monitoring. It is important that these regulations are strict and well-written, but not inflexible, given that this is an evolving technology. The rules ought also not to be so unreasonably tough that carbon storage becomes uneconomic. Carbon dioxide is not plutonium. It would make no sense to be so risk-averse on carbon dioxide storage that we have to continue emitting it to the atmosphere, with much worse consequences.

The legal liability for carbon dioxide storage has also to be defined. During the injection phase, it seems reasonable that the operating company (or possibly, if different, the emitter of the carbon dioxide) would bear liability. If the storage site started leaking, the company involved would have to seal the leak, and make compensation for any consequent emissions. A limit will have to be set for what constitutes 'acceptable' levels of leakage, and the associated confidence that will have to be demonstrated before a permit is granted. In consensus-based political systems, this may be problematic, since some stakeholders will demand extremely stringent regulations, possibly beyond what is scientifically or economically justified.[85]

It is not practical or reasonable for the storage corporation to bear liability forever. After all, present-day polluters do not bear any liability for the carbon dioxide they emit, 100% of which ends up in the atmosphere. The storage company may not even exist fifty or hundred years in the future, if leakage starts then. Insurance contracts may run for no more than thirty years.[86] A requirement to set aside escrow funds dedicated to a potential clean-up might be excessively onerous, since in all or most cases this will never be needed. So there will have to be a defined handover point, at which the site is declared to be safely closed, and monitoring determines that the carbon dioxide plume is no longer moving significantly. All the injection wells will have to be thoroughly sealed, probably with permanent measuring devices installed. As with nuclear waste, a detailed inventory of sites will have to be maintained, so that future societies do not accidentally drill into them and re-release carbon dioxide.

Once the site is closed, the government then takes over responsibility. A publicly owned entity could be established to carry out ongoing monitoring and remedial work, funded by a levy on the original operator. This fee could be varied depending on the track record of each storage company—those who build demonstrably secure sites paying less. A similar principle has been agreed for nuclear waste disposal,[87] and the government of Western Australia and Australian federal government have agreed to take on long-term liability for carbon dioxide stored at the Gorgon LNG project.[88] This will inevitably raise complaints that the government is taking on unlimited liabilities, and hence subsidising investor-owned companies. The company that built the site, though, will still be liable for any losses caused by its gross negligence or wilful misconduct. The government would also only take on responsibility for sites that meet a set of stringent standards.

There may be complicated issues if numerous operators are injecting into the same formation and, after some time, the sources of carbon dioxide become mingled or indistinguishable. Alternatively, carbon dioxide injected by one operator might start emerging through an old well in an area held by another company. Regulations on storage integrity will have to be strong, enough to ensure high confidence of centuries without significant leakage. Unlike oil extraction, poor modelling and sub-surface understanding does not immediately lead to any economic loss, so the operator's incentive is to inject as much as possible, and possibly to conceal or downplay any problems.

Most legal systems assign ownership of sub-surface minerals to the state, but sometimes, as in the USA, the mineral rights accrue to the landowner (who can sell them on separately from surface ownership). But few legal codes define the ownership of sub-surface pore space. This may become a valuable resource, if CCS takes off, particularly in well-characterised oil- and gasfields. Some legislation has been enacted in Wyoming, Texas and Illinois on the subject, and Australia has launched a bid-round for the rights to carbon dioxide storage.[89] In Ireland, the Malaysian national oil company Petronas and another oil company, Providence, were awarded the rights in August 2008 not only to look for petroleum in an area offshore Dublin, but also to assess the potential of saline aquifers there for carbon dioxide sequestration.[90]

It remains to be settled, though, how pore space ownership interacts with other property rights—to land, minerals, hydrocarbons and water.[91] For instance, in a CO_2-EOR project, does the existing operator of the oilfield automatically inherit the (possibly valuable) rights to storage space? If the owner of the storage space is different from the owner of the rights to oil extraction, then how practically could a CO_2-EOR project operate? In the Australian proposal, existing petroleum rights-holders can apply for storage space in the same area. There may also be commercial sensitivities regarding data for rights-holders in neighbouring areas—the type of data useful for petroleum exploration, and for carbon dioxide storage, is very similar. This issue may arise if hydrocarbons are accidentally (or, in a Machiavellian way, 'accidentally') discovered while searching for storage space.

How long can the rights be held for, if there is no injection? (In the Australian example, the licence endures if carbon dioxide is expected to be available within fifteen years). Since governments extract a whole range of taxes and royalties from oil and gas production, will they attempt to do the same if carbon storage proves to be a lucrative business? If carbon dioxide from a neighbouring storage project migrates into pore space owned by someone else, then this could be thought of as a negative 'rule of capture', the old American system by which oilfield operators could drain oil from neighbouring blocks underground. Would compensation then have to be paid for filling up valuable pore space? What if carbon dioxide contaminates an existing oil- or gasfield, or a potable aquifer or geothermal resource? Presumably the guilty party would have to make recompense.

The insurance and financial industries will have to develop methodologies for handling CCS projects. Insurance companies will have, with so far rather limited experience, to assess the risks of leakage or disasters at pipelines and storage sites, and the possible costs of remediation. Banks already know how to determine the financial capacity of power stations and oilfields, and hence how much they can reasonably lend. Similar approaches may be taken for storage projects, assessing the available pore space and the fee paid for each tonne of carbon dioxide stored. Carbon capture on power stations may, once technical dependability is assured, be a more straightforward matter—simply treating the carbon capture equipment as one more element of a standard power plant, and taking the benefit of saving on emissions permits. However, at present, equipment manufacturers will not provide a guarantee that CCS equipment will perform; banks cannot finance the carbon capture investment until the technology is mature enough for such a warranty to be given. This is another reason for government support of the first generation of projects.

Accounting issues also arise for forestation and carbon sequestration in soils. Here, measurement of the original amount sequestered is more difficult than for CCS, and it is also harder to verify how much remains after some time has elapsed. There is a trade-off between increasing accuracy of measurement while trying to minimise costs. In addition, for forestation and conservation tillage, which are inherently vulnerable to carbon loss due to forest fires, changes in agriculture and so on, the issue of accounting for leakage is even more significant than for CCS. Liability for long-term storage has to be assigned—in the case of biochar, this may be thousands of years. The rules also have to include changes in non-CO_2 greenhouse gases, since land-use changes affect the amount of methane and nitrous oxide released. Regulations need to ensure that carbon dioxide emissions are not simply shifted from one area of forested land to another area which undergoes deforestation. Finally, changes in land-use may be affected by other international agreements on human rights, such as the United Nations Declaration on the Rights of Indigenous People.

International Perspectives

The success or failure of CCS can be a matter of national competitive advantage. It has the potential to shift the balance of power in the energy world. Countries vary in their ability to make use of carbon capture, as shown in Figure 6.6. Here, a number of leading contenders are ranked out of 100, on short and long term, based on technical and regulatory factors. Technical factors include installed fossil fuel generation, age of plants, and availability of storage capacity. Regulatory factors cover market legislation, access to finance, and the track record of implementing large-scale energy projects. The USA is the first country on both timescales, but nations such as China, India, Russia and South Africa, which have limited readiness to make use of carbon capture today, move to the top of the list when considering long-term potential.

Figure 6.6. Readiness of countries for CCS, short and long term[92]

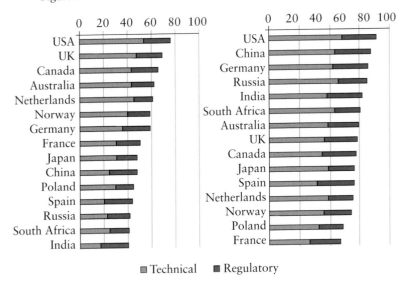

■ Technical　■ Regulatory

Nations rich in coal but relatively lacking in gas can use carbon capture as a vital plank of energy security. This applies particularly to China, India and South Africa, where coal is important too because of its role in domestic employment. Similarly, the USA has large coal reserves, but there is political pressure to ban imports of high-carbon unconventional oil. Carbon capture, by making coal- and gas-to-liquids, oil shales and oil sands more environmentally acceptable, can anchor North American energy security in the face of any threats to Middle Eastern oil. Underground coal gasification, combined with CCS, can unlock large coal reserves with minimal environmental problems.

Europe is currently growing increasingly concerned about energy security. Indigenous gas production is falling;[93] nuclear is being phased out in Germany (though it may enjoy a renaissance elsewhere); renewables, albeit growing fast, are still only a minor part of electricity generation. Pipeline disputes between Russia and Ukraine, and the Russian war with Georgia in summer 2008, have made the EU acutely aware of its over-dependence on Prime Minister Putin's authoritarian state.[94] Coal is one obvious solution: Europe has abundant resources, particularly in Eastern Europe, and could easily increase imports from a diversified group of suppliers, but has been held back by concerns about pollution, particularly global warming. CCS therefore offers a partial solution to these energy security fears. Even if there is no coal resurgence, the prospect that the EU could switch massively from gas to coal may have some effect in compelling Russian good behaviour. CCS offers the additional potential benefit of enhanced gas and coal-bed methane recovery from Europe's domestic resources.

Countries that implement CCS successfully should see lower energy prices, a boon for the entire economy. They will also find it easier to hit whatever global climate targets are then in force, so burnishing their reputation. If a global trading scheme is established, they will earn credits that can be sold to other, more CO_2-intensive countries, or at least reduce the amount of credits they need to buy. This will have a positive impact on their current account balance and general competitiveness. Nations with good geology for storage can be paid for receiving carbon dioxide from other countries, possibly generating additional income from enhanced oil and gas recovery. A caveat is that public opinion may revolt against becoming a 'dumping ground' for other people's waste.

Some oil producing nations—Norway in particular—have very successfully leveraged petroleum extraction to grow a domestic oil service industry. Even now that Norwegian oil output is declining, these companies continue to provide employment and generate exports of leading-edge technologies. To some extent, Aberdeen, Houston and Calgary have done the same. South Korea and Japan similarly turned a dependence on liquefied natural gas imports into a near-monopoly on the construction of specialised LNG tankers.[95] This works for renewable energy too: Germany and California have become leaders in solar technology. As discussed below, CCS can offer similar opportunities for countries to build future energy industries. This is one aim of Abu Dhabi's Masdar initiative, with its carbon capture plans, and is no doubt a factor in Norway's similar enthusiasm. South Africa would find it easier to export its world-leading coal-to-liquids technology if it could combine it with CCS.

Not all countries will benefit from carbon capture. Large-scale use of CO_2-EOR will probably be unfavourable for the major OPEC oil exporters, at least in the medium term, since it will compete with their low-cost oil. The relatively immature and low-cost fields in countries such as Iraq are not immediate candidates for CO_2-EOR, although Saudi Arabia is planning trials in its largest field, Ghawar, on the scale of about half Sleipner, for environmental rather than oil production reasons.[96] Abu Dhabi and perhaps Qatar, though, due to special local circumstances, may employ CO_2-EOR much earlier (as might non-OPEC Oman and Russia).

A more subtle effect is that some major unconventional sources of liquid petroleum—oil sands, heavy and extra-heavy oil, coal-to-liquids and perhaps oil shales—are very carbon-intensive. There is growing recognition, for instance, of the risk this poses to the public acceptability and cost structure of the Canadian oil sands.[97] CCS would ameliorate this carbon footprint.[98] Since the resource base of unconventional oil is enormous,[99] it competes with OPEC oil when prices are high. OPEC would be very happy to see the oil sands torpedoed by carbon taxes. A flood of unconventional oil would force the major exporters to accept lower prices, volumes or both.[100]

In the longer term, CCS may make it environmentally acceptable to continue using fossil fuels. By eliminating large stationary emissions, it can open up a 'carbon space' for oil (and perhaps gas) to continue as a transport fuel, as noted by Shell director Malcolm Brinded.[101] Ultimately, emissions from oil for transport can possibly be tackled by air capture. Carbon capture can therefore extend the viability of OPEC nations' oil far into the twenty-first century.

Since coal with CCS will compete with gas, major gas exporters such as Russia and Qatar will initially lose out, though later in the century they may benefit if carbon capture on gas-fired power enables it to continue growing.

The real world being what it is, of course, the situation is much more complicated. Venezuela, for instance, a leading OPEC member, is also a large holder of extra-heavy oil. Abu Dhabi is championing carbon capture, partly as a drive towards 'future energy' as a pillar of its economy, partly because it has specific local advantages in saving on natural gas that may outweigh increased competition for its oil. Qatar and Iran's rather mature oilfields may gain from CCS, even if their gas exports to some extent lose out. Russia may lose some revenues and geopolitical influence if its 'gas weapon' is blunted, but it is itself a large coal-user and exporter.

The successful implementation of carbon capture would lower the global cost of carbon credits. For countries who expect to be able to make their carbon cuts easily, and to have surplus credits to sell to others, this may be bad news. At the Copenhagen conference, Brazil blocked legislation to establish 'offsets' for CCS, fearing it would undermine their forestry credits.[102] In retaliation, Saudi Arabia opposed eligibility of forest projects.[103] However, cheaper credits would also encourage us to make deeper, more rapid cuts in emissions, with a corresponding (if uncertain) impact on climate.

These are examples of the highly unpredictable effects that widespread adoption of CCS may have. It is a transformative technology. It affects the price relationship between the carbon fuels coal, oil and natural gas, their competitiveness against other energy sources, and the very price put on carbon emissions. It can, via EOR and EGR, enhance hydrocarbon supply. It can change international patterns of energy trade, and even the geopolitical balance of power. Some countries win, some may lose, some lose in the short term but can gain in the longer term (Table 6.1).

Current Projects

Progress on full-scale implementation of CCS has been slow. If ambitious demonstration schemes, like those described above, are to be launched soon, then it is important that, in the popular phrase of the moment, 'shovel-ready' projects exist. At least, there must be CCS schemes that already have a reasonable amount of work done on design and feasibility, which could be ready to bid in such a competition in relatively short order.

Fortunately, in most big emitting regions, a large number of such projects, ranging from research and pilot to demonstration scales, are already in planning or underway. These projects cover multiple capture and storage options, mostly focussed on the more mature technologies but including some 'game-changer' possibilities. A selection of storage projects are shown in Figure 3.13.

Europe

Europe is probably the most advanced region in tackling CCS. As well as the Sleipner and Snøhvit projects, a wide variety of research is in progress. A bewildering range of acronyms covers enhanced coal-bed methane recovery, carbon dioxide storage in coal seams in Poland, mapping storage sites, transport networks, aquifer storage in Germany, pre-combustion capture, post-combustion, gasification and hydrogen

Table 6.1. Countries gaining and losing from widespread adoption of CCS

USA, Canada	Gain from using their large coal reserves and from EOR
Australia	Gains from using its large coal reserves; may be partly offset by lower prices for its exported gas
UK	Gains from greater diversity in energy mix, lower gas prices, and perhaps some EOR and selling storage space to EU neighbours[104]
Norway	May lose somewhat from lower European gas prices and from expensive CCS on domestic gas power, but gains from EOR, selling storage space to the EU, and exporting technology and expertise
France	Gains, but less than other European countries, due to low use of fossil-fuelled power, and possibly reduced opportunity to export nuclear technology
Germany	Gains from using domestic coal, avoiding over-dependence on Russian gas, and potential for technology exports
Ukraine, Poland	Gain from using domestic coal, and avoiding over-dependence on Russian gas
Russia	Loses from lower European gas prices and reduced geopolitical power. Somewhat compensated by gains on its own coal reserves, and perhaps some EOR
Middle East and North African oil exporters	Lose in the short term from greater competition from unconventional oil, and reduced gas prices; gain in the longer term from EOR, and continuing oil use
India, South Africa	Gain from using their large coal reserves (assuming adequate storage capacity can be found)
China	Gains from using its large coal reserves, from EOR, and possibly from manufacturing parts of CCS systems
Japan, South Korea	Gain from lower gas prices, and from maintaining coal use, but need to find storage space
Indonesia	Gains for using its coal reserves,[105] partly offset by lower prices for forestry credits
Brazil	Loses due to lower prices for its forestry credits; maybe partly offset by EOR

production from low-rank brown coals, pre-calcined feed for cement plants, chemical looping combustion, fluidised-bed calcium combustion, site monitoring, integrated electricity and hydrogen generation, ultra-low-CO_2 steel-making, regulation (to be shared with China) and knowledge transfer generally.

Norway has long been a leader in carbon storage, driven by its introduction of a carbon tax on offshore oil operations. We have already mentioned Sleipner, Snøhvit and the Mongstad refinery project, as well as the ill-fated Haltenbanken scheme. A state-owned company, Gassnova, has been established to fund and manage new CCS projects, including a retro-fit of the Kårstø gas power plant to capture 1.2 Mt CO_2 per year. SINTEF (a leading research organisation), Aker Kværner (energy engineering firm) and the Norwegian University of Science and Technology are cooperating on developing capture systems that can work with biomass.[106]

In nearby **Denmark**, Kalundborg hosts a coal-fired power plant and oil refinery within 15 kilometres of a large storage site, with nearly 1 Gt of capacity. **Sweden** has carried out some small-scale tests, notably of the Sargas process (Chapter 2).

Figure 6.7. Sleipner platforms; the world's first dedicated carbon dioxide storage project (Photo: Dag Myrestrand/Statoil)

The **UK** has a competition for a commercial-scale CCS plant of 300–400 MW with offshore storage. Three consortia ran,[107] led by Powerfuel and Shell (900 MW IGCC at Hatfield), Scottish Power (Longannet in Scotland) and E.ON (Kingsnorth), but the process has been subject to numerous changes and some withdrawals. In October 2009 Hatfield was declared as the winner, and awarded some $500 million funding.[108] Scottish Power is currently running a small post-combustion capture pilot at the Longannet coal power station.[109] The UK is also working with China on the 'Near-Zero Emission Coal' project, and plans to have its first CCS plant running by 2014,[110] with up to four operational by 2020.

Ireland's 915 MW Moneypoint coal plant emits 5.9 Mt CO_2 per year, almost a tenth of their total emissions, and could be retro-fitted with post-combustion capture, or replaced by an IGCC.

The **Netherlands** combines good storage capacity in gasfields with high environmental awareness. Key projects include Nuon's planned 1,200 MW IGCC which can co-fire biomass with coal,[111] and will store emissions in onshore depleted gas fields; capture from Shell's Pernis refinery, with storage in the old onshore gasfield at Barendrecht (public opposition to which is discussed below); EnecoGen's Cryogenic project (freezing flue gases for post-combustion capture); a post-combustion pilot at Maasvlakte in Rotterdam; a novel 68 MW oxyfuel trial at the SEQ Zero Emission Power Plant in Drachten; and the offshore enhanced gas recovery project at K12-B. In addition to the Dutch Energy Valley in the north of the country, the Rotterdam Energy Port plans to capture 20 Mt CO_2 per year by 2025, more than 10% of the country's total emissions.

France is conducting a key oxyfuel pilot at Lacq in the south-west, with sequestration in the depleted Rousse gasfield. Capture started in June 2009.

As well as the previously mentioned Schwarze Pumpe oxyfuel test, **Germany** has trialled injection at Ketzin, near Berlin, in the CO2SINK project, while **Austria** is testing capture from a fertiliser plant and paper mill, possibly important for establishing CCS on these industrial sources. The Austrian oil companies OMV and RAG are assessing storage, RAG in a shallow (500 m) onshore aquifer.

Italy has several plans for capture plants: a 50 MW oxyfuel demonstration by the power company Enel, by 2010; a 2,000 MW post-combustion capture plant at Torrevaldaliga Nord; carbon dioxide storage feasibility near Venice; a 1,320 MW capture-ready plant which has started construction at Saline Joniche in southern Italy; and two enhanced coal-bed methane pilots to be tried in Sardinia. However, given Italy's shaky record on permitting other major energy projects, such as oilfield developments and various LNG terminals, it remains to be seen how speedily these projects may progress.

Portugal aims to have 800 MW of clean coal at Sines by 2020, although the capture technique is not specified. In **Spain**, oil company Repsol is investigating using the depleted offshore Casablanca oilfield to store 0.5 Mt per year from its Tarragona refinery.[112]

Poland is a very heavy coal user. One high-profile albeit small-scale pilot, RECOPOL, tested enhanced coal-bed methane recovery in Silesia, while acid gas disposal has been carried out into an aquifer at Borzęcin, near Poznań in central Poland, since 1995. This activity is of particular interest because of the potential to release dissolved methane from the aquifer waters, a kind of EGR. The most exciting project is at Kędzierzyn, for a zero-emission plant using coal and biomass to generate electricity and chemicals, while the fertiliser manufacturer Tarnow plans to capture its CO_2 for EOR. To reduce the carbon footprint of coal, there are plans to retro-fit Vattenfall's 400 MW plant and the Blachownia plant between 2010 and 2016; to construct a 10–50 MW demonstration unit at the Lagisza circulating fluidised bed (CFB) coal power station, an important step for this alternative coal technology; and for the utility BOT to build two new zero-emissions IGCCs by 2016, 858 MW burning lignite and 959 MW burning hard coal.

The **Czech Republic**, also with significant reliance on coal, is considering CCS demonstrations at 660 MW coal and 105 MW lignite-biomass power stations. **Croatia** captures carbon dioxide at a gas processing plant, which could be employed for EOR in three mature oilfields.

Russia, unfortunately given its huge geological storage potential and heavy carbon dioxide emissions, has no known R&D or demonstration projects, but there have been some CO_2-EOR pilots.

USA

Behind only China as a coal-user, coal-mining and power generation is a key part of the American economy. The country's record of scientific innovation, its long history of enhanced oil recovery, its political and economic leadership, and the importance the Obama administration gives to fighting climate change, all make it one of the two current key CCS players (with Europe).

Along with the EOR and ECBM projects, extensively covered in Chapter 3, saline aquifer storage and monitoring was tested in 1994, in the Frio Formation in the Gulf Coast of Texas. An alliance of West Coast states, including the Canadian province of British Columbia, is carrying out the first US enhanced gas recovery field test, while similar consortia of Mid-West, Plains, South-East and South-West states are trialling enhanced oil, gas and coal-bed methane recovery, saline aquifers and acid gas injection. Each of these partnerships will inject at least 1 Mt of CO_2 into a suitable formation.[113] The total project cost is almost $500 million, of which companies pick up about $160 million and the rest comes from the Department of Energy. The Big Sky Regional Carbon Sequestration Partnership (various north-west states) is testing storage in the Columbia River basalts. Consol Energy, one of the FutureGen partners (see below), has tested carbon dioxide storage in coal in West Virginia.

Amongst capture projects, notable initiatives include:

- The Beulah synfuels plant (mentioned in Chapter 3).
- Plans for a 50 MW oxyfuel demonstration on a circulating fluidised bed power station at Jamestown, New York or Holland, Michigan.[114]
- Proposed capture-ready IGCCs in West Virginia (mentioned above) and Edwardsport, Indiana.
- AEP's proposed Mountaineer power plant at New Haven, West Virginia, for which the utility is seeking $334 million in federal funding to cover half the cost of the CCS system.[115] Mountaineer is already capturing small amounts of its emissions.
- Summit Power's 400 MW IGCC near Odessa, Texas, nicknamed 'NowGen', starting construction in 2010 and supplying 3 Mt per year of CO_2 for EOR in West Texas.[116]
- Recovery Act funding for twelve industrial capture projects totalling about 20 Mt CO_2 per year, including cement, paper, petroleum coke, hydrogen, chemicals and others.[117]
- The high-profile FutureGen project.

FutureGen is a partnership of the US Department of Energy with the power companies E.ON of Germany and Huaneng Group of China, and coal miners including BHP, Rio Tinto and Anglo-American. They plan to build a 275 MW IGCC plant in Illinois, costing $1.5 billion, with carbon storage in a nearby saline aquifer. The project was cancelled in 2008 due to rising costs, exacerbated by an error in overlooking inflation, but then resurrected in 2009. Further progress was made in July 2009 towards securing the required permits,[118] and a final decision will be made in 2010.

Canada

With Canada's substantial EOR potential, hefty coal use, heavy emissions from oil sands developments and proximity to the USA, carbon capture is a prominent part of planning for future energy. Canada has substantial CCS experience via Weyburn and latterly the nearby Midale; CO_2-EOR in the Pembina Cardium field, Canada's largest conventional oilfield; and acid gas disposal. The Heartland Area Redwater Project is examining storing emissions from refineries, petrochemicals and oil sands in the country's third largest oil reservoir, a giant reef formation able to hold two years of all Canadian emissions,[119] while storage in the Wabamun Aquifer is being examined in Central Alberta.

Acid gas disposal, though little highlighted, stores 1 Mt CO_2 per year, as much as Sleipner, spread over forty projects. Zama, operated by Apache, is an acid gas EOR project in northern Alberta. ECBM is also being trialled in Alberta using carbon dioxide, nitrogen and flue gases.

It is likely that carbon capture will be legally required for new oil sands plants from 2012 onwards, and new coal power stations from 2018.[120] The government has earmarked CA$2 billion to store 5 Mt CO_2 per year by 2015 in three to five projects in Alberta, and there is a target of 4,000 MW power with CCS by 2030, equivalent to about 22 Mt CO_2 annually if all derived from coal, or about 30% of current total Canadian coal-fired generation.[121] Most coal generation is in Alberta and Saskatchewan, conveniently the two major oil-producing provinces. Key projects include a 100 MW post-combustion plant at Boundary Dam in Saskatchewan (1 Mt CO_2 per year) and a 500 MW IGCC to be built by EPCOR. Swan Hills Synfuels is planning an underground coal gasification project in Alberta, which would feed a 300 MW power plant equipped with carbon capture; the project is to receive CA$285 million from the province's CCS fund.[122] In October 2009, Alberta also agreed to disburse $745 million from this fund to a consortium of Shell, Chevron and Marathon for their 'Quest' project to store emissions from the Athabasca oil sands.[123]

Asia

China, with its massive coal use, has an abundance of CCS potential, yet the country has been somewhat cool on CCS. The country is working on GreenGen, a CCS demonstration of 300–400 MW IGCC, but Su Wei, director-general of the National Development and Reform Commission's climate change unit, has said, 'Carbon capture and storage, particularly for China, is not one of the priorities—the cost is an issue. If we spent the same money for CCS on energy efficiency and the development of renewables, it would generate larger climate-change benefits.'[124] It is probably true that China has cheaper carbon abatement options initially, and therefore the OECD countries will have to be prepared to shoulder much of the cost for the first generation of Chinese carbon capture projects. Yet the partnership on CCS is not all one way: low-cost Chinese manufacturing could be crucial in bringing down the expense of carbon capture, and churning out the vast amounts of steel pipelines, power plant turbines, amine units and so on that will be required.

China has about 12.5 Mt per year of 'early CCS' opportunities, where sources with low capture costs (fertilisers, ammonia and chemicals) are located close to EOR and ECBM sites. CO_2-EOR has been used in China at Daqing (the country's largest oilfield) and I Subei, and ECBM has been tested successfully on a small scale. China's power utility Huaneng is a member of the US's FutureGen consortium, and China has CCS partnerships with the EU, Japan and Australia, in addition to GreenGen. GreenGen is a joint venture of the US's Peabody Energy with Chinese state companies in Tianjin, to build a 650 MW IGCC with carbon capture. An additional six IGCCs are being sponsored by the government.[125]

Compared to China, **India**'s CCS efforts have so far been small. Given the country's heavy coal-use and rapid growth, it is important that carbon capture succeeds here. Due to their high ash content, domestic Indian coals are probably not suitable for IGCCs; pulverised coal or circulated fluidised bed plants (with post-combustion or oxy-fuel capture) may therefore be preferred.[126] A small (0.16 Mt per year) amount of carbon capture has been trialled at the Aonla and Phulpur fertiliser complexes. Given a possible lack of storage capacity, disposal in the Deccan basalts has also been researched.

Japan's research efforts aim to have capture cost at $10 per tonne by ~2020. Storage projects have concentrated on aquifers (RITE and

Nagaoka) and coalbeds (Ishikari). Sumitomo has trialled capture from a chemicals plant and coal power, while Petronas has a fertiliser facility that uses KS-1 (an amine) to capture carbon dioxide to make urea. **South Korea** will build a CCS power plant by 2020 and is also working on some small-scale pilots.[127]

Malaysia stands out as one of the few countries to have made a CDM (Clean Development Mechanism) application for carbon capture, where carbon dioxide and hydrogen sulphide from a gasfield feeding the Bintulu LNG plant are stored in a deep aquifer. Nearby **Indonesia**, a major oil, gas and coal producer, has ECBM potential in South Sumatra and Kalimantan. Like Malaysia, it has many high-CO_2 gasfields, of which the as-yet undeveloped Natuna D Alpha is the largest. Carbon dioxide from these fields could be stored in aquifers or used in EOR, for instance in north-east Sarawak.

Australia

In Australia, growing environmental momentum since the ratification of Kyoto, and the country's heavy dependence on coal power, provides a supportive environment for CCS. AUS$500 million has been allocated for low-emissions coal.[128]

The Callide A power plant in Queensland is testing capture from a 30 MW oxyfuel unit, starting in 2010, while at Coolimba in Western Australia there are plans to build two 200 MW oxyfuel units for carbon storage by 2012. ZeroGen will demonstrate IGCC capture, storing 0.4 Mt per year in the Denison Trough in Queensland, from 2012 onwards, with plans for a 300 MW IGCC starting by 2017 at a cost of more than $1 billion. As for post-combustion capture, a retro-fit is being trialled at the Hazelwood plant in Victoria, and the Munmorah pulverised coal plant in New South Wales is testing ammonia absorption.

For synthetic fuels, FuturGas will gasify lignite, storing the carbon dioxide in the Otway Basin from 2016, while Monash Energy (a joint venture of Shell and coal-mining giant Anglo American) was working on a coal-to-liquids plant capturing a massive 13 Mt per year by 2015, at a cost of $6–7 billion, but announced in December 2008 that the project would be indefinitely delayed.[129]

Geological storage projects include Otway in Victoria, which has trialled injection into a depleted gasfield. The massive Gorgon LNG project will store some 3–4 Mt per year of CO_2 from contaminated gas, when it finally gets underway around 2014 after years of delays. Moomba Carbon Storage will inject 1 Mt of CO_2 for EOR in the Cooper Basin of South Australia, starting in 2010. There are additional high-CO_2 gasfields in the southern part of the Denison Trough, whose emissions could be stored, perhaps in combination with power plant capture schemes.

Latin America

Brazil, a growing oil and gas power, consumes and produces only modest amounts of coal. Still, it has significant potential storage space in coal, and has carried out some research, mapping and ECBM studies. The national oil company Petrobras plans 4 major CCS projects, has used EOR since 1981 in the north-eastern Recôncavo Basin and is running an ongoing project at the Miranga Field in Bahia State.[130]

Some work has been done on a **Mexican** CCS policy.[131] With the country's rapidly dropping oil production, CO_2-EOR is particularly attractive, although there are severe political and commercial barriers to attracting the necessary investment.[132] Nevertheless, projects are underway at Tamaulipas Constituciones, Activo Samaria-Sitio Grande, and Carmito Artesa. CO_2-EOR is also being considered for the massive offshore Cantarell Field, one of the world's largest producers, though now declining steeply.

Middle East and Africa

The MENA region has evidently enormous potential for geological storage, and for value-added hydrocarbon recovery. However, progress is held back by factors that include the minimal use of coal in the region (nearly all generation is gas- or oil-fired), a high level of risk aversion and lack of innovation and commercial *savoir faire*, political challenges to implementing cross-border projects, and a shortage of skilled engineering and geoscience staff, who are very much in demand in the petroleum business. For instance, **Algeria** has not yet built on the successful In Salah capture project, though it could be technically and economically attractive to pipe carbon dioxide from eastern Algerian gasfields to mature Tunisian oilfields.[133]

Similarly, **Abu Dhabi** faces internal issues in coordinating the CCS projects of Masdar, its future energy vehicle, with ADNOC, the national oil company. Nevertheless, carbon capture has huge possibilities in Abu Dhabi, since the Emirate currently uses large quantities of natural gas, urgently required for power generation, for reinjection in its oilfields. Sources of carbon dioxide include the proposed hydrogen power plant, aluminium smelting, and processing of sour gas from the Shah and other fields. Some four to six opportunities have been identified with 6–8 Mt per year capture potential, so each project would be somewhat larger than Sleipner. The system could be extended to Dubai for EOR. Carbon dioxide could be very useful for replacing natural gas injection in **Iran**'s ageing supergiant oilfields such as Gachsaran and Agha Jari, but the prospect of furthering such projects in the current political and commercial climate seems dim.

Shell is investigating carbon dioxide infrastructure, particularly for its Pearl gas-to-liquids project in **Qatar**. Some of Qatar's oilfields, including onshore Dukhan, are rather mature and may be ripe for EOR, as might the **Bahrain** oilfield, the oldest field on the Arab side of the Gulf. **Saudi Arabia** is planning a 1–2 Mt per year CO_2 pilot and screening EOR possibilities, including at Ghawar, the world's largest oilfield. Some of the mature fields in **Oman** might also be good candidates, particularly if the Sultanate turns to coal to supplement its strained gas supplies.

The low level of income and industrialisation in most of sub-Saharan Africa is unpromising for CCS. Coal consumption is negligible except in **South Africa**; gas consumption is also very small, albeit growing. South Africa, though, has significant CCS potential due to its heavy reliance on coal-fired power, and its large gas- and coal-to-liquids plants which can produce capture-ready carbon dioxide.

Business Opportunities

If, as the IEA suggests, some $70–100 billion is invested in maturing CCS up to 2030, and from then to 2050 a further $600-$1,400 billion is spent on projects, plus about $700–3,000 billion of operating costs, there is obviously a huge opportunity for businesses to benefit.

Four main groups of business opportunities can be identified:

- Defending continuing fossil fuel use.
- Providing equipment and services for CCS.
- Supplying storage and sequestration capacity.

- Creating 'carbon space' to pursue their activities for industries without ready low-carbon substitutes.

The first opportunity is essentially a defensive one, for fossil fuel companies to continue to be a major part of the energy picture. This covers coal mines, gas producers (and probably, to a much lesser extent, oil), and fossil-fuel generators. For coal miners and coal generators, carbon capture seems the only realistic prospect for long-term survival, particularly in the OECD.

As discussed above for countries, natural gas companies may, especially early on, actually lose out from CCS, because successful carbon capture would allow us to continue using coal, the main competitor to gas for power generation. If CCS does not take off, then more gas will be required to replace coal, especially in the short term. But in the longer term, if combined with CCS, gas resources are sufficient for it to be a major fuel out to the end of the century and longer.

For unconventional oil, particularly gas- and coal-to-liquids, oil sands and, depending on the process used, oil shales will probably require progressive implementation of carbon capture in order to be environmentally acceptable.[134] CCS is therefore crucial for big oil-sands developers such as Shell, ExxonMobil, Suncor-PetroCanada, Encana and Nexen[135] and CtL players such as Sasol. Growers of biomass for power generation, and collectors of municipal waste, may benefit from increased demand and higher prices, if bio- and waste energy with carbon capture become popular.

Industries with large carbon emissions—iron and steel, aluminium, cement, petrochemicals, oil refining, ammonia, ethanol, hydrogen, synthetic fuels—will also benefit from CCS, especially where they face competition from less carbon-intensive alternatives.

The second group includes those **companies who provide the kit and services for CCS**. This encompasses a huge range of activities, some of the major ones being pipeline assemblers, power plant engineers, manufacturers and constructors such as Siemens, Foster-Wheeler, SNC-Lavalin and GE, and industrial gases companies, such as Air Liquide and Praxair. Potentially oilfield services (Schlumberger, Halliburton), seismic companies, cementing companies with CO_2-resistant cements, and drillers (Nabors, Parker, Rowan, but probably not deep-water drillers like Transocean) will gain work on storage projects. Miners of trace metals used in high-temperature alloys may see increased demand; such metals include niobium (e.g. Anglo American), molybdenum and rhenium (e.g. FreePort-McMoRan and Rio Tinto).[136] Some specialist products can be developed, for instance:

- A reservoir simulation computer program capable of modelling carbon dioxide and aquifers.
- Turbines that burn hydrogen at high temperatures with long lifetimes.
- Specialist materials such as high-temperature ceramic membranes.
- Chemicals including amines and zeolites for post-combustion capture.
- Modular systems for air capture.

Specialist players may emerge, such as:

- Sub-surface monitoring companies.
- Safety and verification organisations, such as Det Norske Veritas (DNV) and TÜV Nord.
- Banks that are able to evaluate and finance CCS projects, especially if they can combine this with monetising carbon credits and managing carbon market risk.
- Legal firms that develop expertise in carbon law and permitting transport and storage projects.
- Start-ups with new technologies (algae, CO_2-containing materials, fuel cells, high-performance alloys, air capture, genetic engineering for bio-sequestration, novel capture methods).

Universities, professional training and development organisations and recruiters will benefit from teaching professionals the various skills associated with CCS, and connecting them to employers. Venture capital may have some role, particularly in supporting some of the niche technologies mentioned above, and perhaps in funding direct carbon fuel cells, for which a mere $20 million might be enough to get to a prototype.[137]

Pipeline companies, such as TransCanada and El Paso, can extend their natural gas expertise into carbon dioxide; Kinder Morgan is already a large operator of carbon dioxide pipelines, for EOR. Large carbon dioxide 'aggregators' may be able to offer competitive transport costs by gathering many sources into a single large trunk pipeline.

The third set of business opportunities covers the **provision of storage and sequestration capacity**, including sub-surface pore space. Well-characterised, secure, high-quality underground sites near major emissions sources will command an economic rent: they will be able to sell their services to the highest bidder. This is especially true of sites with enhanced hydrocarbon recovery possibilities. Consider that an EOR site may gain $30 per tonne from CO_2 injection, an average site might cost $2 per tonne for storage, and an expensive one $10 per tonne. The lower IEA target is for 5.1 Gt of CO_2 to be captured annually by 2050. If just 10% of this were used for EOR, then more than 1 billion barrels of oil could be recovered, nearly 4% of today's global production. This is more than the output of ExxonMobil or Shell, amongst companies, and Abu Dhabi or Kuwait amongst oil exporters. Yearly pre-tax profits could be $16 billion.[138]

Widespread use of EOR and EGR will probably lower commodity prices to some degree, particularly if carried out on a large scale. Enhanced recovery will be a major potential opportunity for skilled operators, particularly international oil companies, who are relatively constrained in their opportunities in traditional heartlands such as North America and the North Sea. Managing EOR schemes is a rather specialist task, and, even amongst large oil companies, not all have the relevant skills. Companies with significant CO_2-EOR expertise include EnCana (from Weyburn), Apache, Occidental, Hess and the smaller specialist Denbury Resources. Statoil and BP have experience in aquifer storage. Owners of large, mature gasfields, such as Shell/ExxonMobil's Groningen, in the Netherlands, may be able to use EGR. Leading coal-bed methane operators such as BP, ConocoPhillips, Nexen, BG, Santos and some smaller specialists such as the Australians Origin and Arrow (if they survive the current acquisition mania) might, particularly in the longer term, benefit from ECBM, while Linc Energy, from Australia, and the Canadian company Swan Hills Synfuels are pioneers of underground coal gasification.

In a similar way, forestry stands to benefit. Owners of suitable land in the tropics, particularly, can sequester carbon at low cost. This may well, as noted in Chapter 4, raise accusations of profiteering, conflicts over land-use, and concerns about poor or marginalised communities who cannot demonstrate legal title to their land. If done sustainably, though, poor farmers may gain a valuable source of income from sequestering carbon via trees, biochar and low-till agriculture. Lumber companies may face problems with restrictions on felling old-growth forests, but will gain from overall higher prices.

The fourth group is more indirect. Big **emitters that are not likely users of CCS**, such as the transport sector and notably aviation, **will** benefit from reduced emissions and lower carbon costs elsewhere in the economy. Indeed, without major carbon cuts everywhere else, it is difficult to see how air travel can be reconciled with climate targets. Ultimately, airlines, and perhaps ground and marine transport too, will need 'carbon-negative' offsets, through technological air capture, CCS on biomass power stations, or biological sequestration. The airline business, although currently financially constrained, might well wish to invest the $125 million or so mentioned in Chapter 5, less than half the cost of an Airbus A380, to build a demonstration air capture machine.

The Human Angle

One of the biggest challenges to successful CCS implementation is the public reaction. Although it has come more into public consciousness recently, with mentions in some popular books and articles,[139] and policy declarations, only 4% of Americans and 30% of Japanese surveyed had heard of CCS. In general, offshore sub-sea storage is viewed least negatively, raising few NIMBY or safety concerns, and ocean storage most negatively. CCS suffers from its association with the unpopular oil and gas industry. It might, like nuclear power, be a concern because of the idea of an 'unknown danger' lurking beneath people's feet, low though the actual risks are. For instance, some survey respondents worried about explosions, which are, of course, impossible.

However, interestingly, surveys also show suspicion of environmental groups. In general, there was low awareness of the trade-offs required between different energy sources, with renewables and energy efficiency being favoured as solutions to climate change (and energy security), but with an accompanying unwillingness to pay higher electricity bills.[140] Indeed, there were variable levels of general awareness of, and concern about, climate change. Unless the threat of climate change generally is better known, then all the tough choices and expenditure needed for solutions will be difficult.

Public support is required on a local and a national and even supranational level. Locally, CCS plants have to manage any concerns and opposition from residents. They have to secure permits and rights of way to build the plant, pipelines and storage facilities. The risk of opposition probably relates more to the transport and storage side than to capture. Carbon capture, especially if installed on an existing industrial site, does not seem to raise the same safety concerns.

There are already some signs of NIMBYism.[141] For instance, 1,300 residents protested against a proposal to store emissions from Shell's Pernis oil refinery in the Netherlands in a depleted gasfield under the town of Barendrecht.[142] Concerns also surfaced about Vattenfall's studies of storage sites at Nordjyllandsværket in Denmark,[143] while local politicians opposed plans to capture carbon dioxide at an ethanol plant in Darke County, western Ohio, USA.[144]

In Germany, a campaign has emerged against utility giant RWE's plans for carbon dioxide storage beneath the North Sea island of Sylt, supported by some North German politicians. Chancellor Angela Merkel, reportedly convinced of the safety of CCS and aware of its importance for Germany's future energy supply, was said to be furious with her colleagues.[145] RWE board member Rolf Martin Schmitz stated, '... if it doesn't work here and we still need to develop it, then we'll go somewhere else...The acceptance on the other end of the pipeline, where you have to pump the CO_2 into the ground, just isn't there.'[146]

We also should not overstate the opposition to carbon capture: in a survey of 500 key European energy decision-makers, three-quarters said that its large-scale use would be crucial to achieving major carbon reductions by 2050.[147] Half of Canadians surveyed were in favour of carbon capture, with only a quarter opposed.[148] Research in Japan, the Netherlands, Sweden and the USA has suggested general support for CCS, with the US being least in favour, and respondents being uncomfortable with storage facilities being located near to them.[149] More than three-quarters of Michigan residents surveyed supported plans for two new clean-coal plants, and two-thirds agreed that coal power could be environmentally friendly.[150]

It helps when the area is already familiar with oil and gas extraction. In Weyburn, Saskatchewan, the town's mayor said, 'It has put Weyburn, Saskatchewan, Canada not just on the provincial or national map—it has put us on the international map...We are blessed.' The province's premier observed, 'I would say modestly we are the leaders internationally in carbon capture and sequestration because of Weyburn.'[151]

On a national and regional level, CCS needs enough public support, or at least lack of opposition, for large sums of taxpayers' or utility customers' money to be spent in support for the early projects. This investment is comparable to, or often less than, that given to renewable energy and some 'boondoggles' of questionable environmental benefit such as corn-derived ethanol. But it may well come in large, discrete chunks, and be vulnerable to accusations that it represents a hand-out to large and profitable energy companies. Scientists, researchers, entrepreneurs and venture capitalists have to feel enthused about CCS, feel convinced that they are playing their part in fighting climate change and delivering affordable energy. This will enable the nascent industry to recruit the people it needs.

There also needs to be some consensus—and practical evidence—that carbon capture is not any of the following: just an easy way out; perpetuating an outdated reliance on fossil fuels; blocking the way to a clean energy future; 'sweeping the problem under the carpet'; coming too late; too expensive; impossible to implement on a worthwhile scale; or less effective than renewables, forestry or other favoured solutions. All these objections are frequently raised in conversation, in blog comments and other debate.[152]

This suggests a few key messages that advocates of carbon capture need to get across. I have done my best to state these messages clearly in this book:

- Climate change is a very serious issue.
- We all have to do our part in fighting climate change.
- We cannot pick and choose solutions; all viable solutions have advantages and disadvantages, and all will be required.
- Carbon capture can deliver, in the near future, large amounts of environmentally friendly energy at an acceptable price.
- The risks of carbon storage are low and can be managed.

It is very important that a self-reinforcing cycle against CCS does not commence. The danger here is that early CCS projects meet public opposition, perhaps due to real or exaggerated problems of safety or leakage, and that arguments from one site then receive media attention and are imported to other areas. In this case, as with nuclear, CCS might face lengthy delays and unreasonably strict regulation that makes it uneconomic. As we have seen, the scientific arguments all suggest that carbon storage should be very safe, certainly safer than other widely accepted activities such as natural gas storage and acid gas disposal. But scientific arguments, especially when presented by unpopular companies perceived to be acting in their self-interest, are often ignored.

There are some approaches to tackle local opposition. Early, honest and sustained communication with host communities is essential.[153] There are now numerous websites giving information on CCS, of which those of Bellona[154] and Vattenfall[155] are notable for addressing the non-specialist. Publications such as the *Carbon Capture Journal* are more targeted at the engineering community. Garnering public consent has to be proactive, not reactive; planning procedures have to be seen to be fair, and the developer has to build trust. Each case will be different, and has to be treated on its own merits. Customers might, for instance, be informed via their electricity bill that they are purchasing 'low-emission power', or communities given a stake in local storage projects, as has been suggested for wind turbines.[156] A 'hotline' can be set up for residents to communicate their concerns; if so, it is very important that these concerns are answered. It has been argued that 'NIMBYism' is not a bad thing, since it forces companies to examine their plans strictly.[157] It is certainly much better than some of the cavalier environmental practices of the past, which left toxic legacies.

Assuming all goes well with the first generation of CCS plants, it should then be much easier to build new ones and to secure public acceptance. If the technology is successful, then it is likely that there will be growing pressure for all new plants to include capture, and for existing ones to be retro-fitted or closed down.

It is entirely possible, as with nuclear, that some countries will adopt CCS enthusiastically: those either with less of a tradition of public participation in decision-making (China, UAE), or those that are familiar with the petroleum or coal businesses (UAE again, Norway, some US states and Canadian provinces, Australia). Those areas with strong environmental lobbies may well oppose carbon storage, for instance California and Germany. Assuming CCS is economically viable, then this would impose extra costs on regions that reject it. That is an acceptable democratic choice for them to make, as long as they meet their climate-change obligations in another way. If CCS is successful elsewhere, they might eventually rethink their opposition; if it fails, they will congratulate themselves on staying clear of it.

In the end, we should recognise that if it is difficult or impossible to site CCS facilities, even if they have been demonstrated to be safe, then it will probably also be very hard to build other key bits of infrastructure that are required for tackling climate change: solar thermal plants, tidal barrages, wind farms, mass transit schemes, nuclear power sta-

Table 6.2. Position of various environmental and other organisations on carbon capture and storage

Organisation	Position on CCS	Key thoughts
Greenpeace[158]	Opposed	'...the technology is largely unproven and will not be ready in time to save the climate... CCS wastes energy...storing carbon underground is risky...CCS is expensive...'
Friends of the Earth[159]	Opposed	'Promoting CCS as the answer, it just sort of pushes the day of reckoning for fossil fuels down the line...Coal, from the mining of coal, is a very dirty fuel....CCS is not going to come on line fast enough, nor is it going to be deployed fast enough to really make a difference.'
Green Party (UK)[160]	Opposed	'CCS is the wrong technology for the UK...Carbon capture projects wouldn't start delivering either emissions reductions or jobs for the next decade. But existing renewables technologies and energy-conservation programmes [sic] could do both.'
Institute for Energy Research[161]	Opposed	'...not currently commercially available. The costs of CCS are an assumption rather than a reality...the consumer will be charged for the new plant every time electricity is used.'
Cato Institute[162]	Opposed	'If the Chinese aren't willing to pay the costs today or tomorrow, what good would it be to hand over blueprints for technology that won't ever be deployed?'
WWF[163]	Opposed/ neutral	'Whilst CCS may play an important role in reducing atmospheric carbon dioxide concentrations in the future, there are currently too many unanswered questions for it to be considered an immediate solution.'
Worldwatch Institute	Neutral/ opposed	'...solutions for reducing emissions by carbon capture in the energy sector are unlikely to be widely utilised for decades and do not remove the greenhouse gases already in the atmosphere'
Sierra Club[164]	Neutral	'carbon sequestration is a potentially important tool...the retention capabilities of geologic formations are uncertain at best....we should not unwisely depend on geologic sequestration to solve all of our problems'

Organization	Stance	Quote
Natural Resources Defense Council[165]	Neutral/in favour	'...not our favorite greenhouse gas reduction solution, but we think it has an important role to play and should be part of the mix'
Bellona[166]	In favour	'The strategy for reducing global CO_2 emissions must therefore be a combination of: • Increased energy efficiency • More renewable energy production • Wide implementation of CCS'
Environmental Defence Force[167]	In favour	'A cap on carbon will create the market for this technology....CCS is ready to begin deployment today....The fact that EDF supports the deployment of CCS does not mean that we are champions of coal...CCS is an important part of the solution but it is only a part.'
Green Alliance (UK)[168]	In favour	'Green Alliance supports the development of...CCS...as part of the transition to a low carbon energy system. However [it] must not be seen as an alternative to, or substitute for, the continued development of renewable energy, energy conservation and demand reduction measures.'
American Petroleum Institute[169]	In favour	'...challenges, including the cost of capturing CO_2 and the lack of regulations governing long-term storage and potential liability, must be addressed. If these hurdles can be surmounted and supporting policies put in place, CCS can help meet the energy needs of the world's growing population with far lower CO_2 emissions.'
United Nations Development Programme[170]	In favour	'a key breakthrough technology.... CCS technologies could be developed and deployed more rapidly'

tions, long-distance electricity cables. Even more so, it will be very hard to impose sweeping lifestyle changes or efficiency mandates. We will then face an unpalatable choice between economic stagnation and decline, or catastrophic climate change—and probably a combination of both.

Environmental groups and think tanks also need to stake out their positions on CCS. Indeed, many have already done so (Table 6.2). Interestingly, environmental groups range from strongly opposed through neutral to strongly in favour; many industry organisations are pro-CCS, not surprisingly, but some are against. There is a rather high degree of consensus, though, that if CCS is pursued, no new non-capture coal plants should be allowed in developed countries.

This is an important step, since it may be difficult for them to change course later. One key analogy is the battle many environmental groups fought in their formative years, during the 1970s and 1980s, against nuclear power. They are now unable to reverse their position, even when it becomes clear that nuclear, though problematic, could be one of the best large-scale hopes for fighting climate change. Pioneer climate scientist James Lovelock is particularly scathing about what he sees as an unreasoning rejection of nuclear power[171] (it's fair to mention that he's positive on biochar, but doesn't believe CCS will ever be implemented on a large enough scale to make a difference[172]).

Once environmental groups are committed to a particular stand on a certain issue, they find it difficult, just as political parties do, to make a volte-face even if contrary evidence piles up. And they naturally attract as members only people who already hold similar views, or they convince would-be members of the correctness of their views. They therefore face the danger of 'policy lock-in' just as governments do. Given the dogmatic conclusions of some groups that CCS will not work, if carbon capture does emerge as a serious contender, which it may well do in less than a decade, they risk losing credibility on other issues. That would be a pity, since on subjects such as deforestation environmentalists have taken an important, leading role.

Some purvey clear misinformation: as one writer dedicated to exposing 'greenwash' writes, 'there currently is no such technology....It is at least two decades...away...don't take my word for it. Check out the Massachusetts Institute of Technology's study on the matter. Or this study by the International Energy Agency.'[173] On the contrary, as we have seen, the technology does exist. And the IEA's report puts commercial CCS happening around 2020, not 2030, while the MIT report describe CCS as the critical enabling technology for reducing carbon dioxide emissions.

Similarly, the key Greenpeace report on CCS,[174] whose title, 'False Hope', gives a clue to its impartial and even-handed approach, is guilty of selective quotation when it cites the United Nations Development Programme as saying 'CCS will arrive on the battlefield far too late to help the world avoid dangerous climate change.' What the report actually says is, 'CCS is widely acknowledged to be the best bet for stringent mitigation in coal-fired power generation... Encouraging as the demonstration project results have been, the current effort falls far short of what is needed....*At this rate, one of the key technologies in the battle against global warming* [my italics] will arrive on the battlefield far too late to help the world avoid dangerous climate change.' The UNDP report is actually arguing for more investment in CCS, rather than rejecting it as Greenpeace imply.

Of course, this is not to advocate blind acceptance of CCS (or any other environmental technology). Environmental groups can play a valuable role by challenging weak points in carbon capture, and letting energy companies know that they are under scrutiny. Concerns for reputation may drive corporations to behave responsibly, more effectively than regulation. As I argued in Chapter 1, energy companies bring tremendous resources of skills, finance, capital assets and political clout. A marriage of industry and environmentalism holds the best promise for delivering not only CCS, but climate change goals generally. A possible manifesto for an environmental group sceptical on CCS would be:

- No new coal plants should be built without carbon capture (so forcing the industry to 'put up or shut up').
- Emissions from existing coal (and perhaps other fossil fuel) plants should be progressively reduced, potentially by retro-fits.
- Coal mining, particularly open-cast, should be subject to tough controls on environmental impact, and should be banned or shut down in sensitive areas.
- CCS should not receive government support in excess of that given to other low-carbon technologies, and CCS funding should not detract from other environmental budgets.
- Environmental and safety regulations should be realistic but stringent.

My viewpoint may be too idealistic. Sections of the corporate world have contempt for environmentalists; many environmentalists bear an inveterate suspicion of companies, and hostility to modern capitalism. But environmental groups might reflect that, in opposing CCS, they are fighting the petroleum, coal and electricity businesses, as well as cement, chemicals, iron and steel. They are already battling nuclear, aviation and auto-makers. They are telling China and India, as well as American coal-mining states, that they cannot continue to use their low-cost domestic fuel, on which millions of jobs depend. Because of these severe restrictions on what they consider 'acceptable' energy sources, they then propose radical lifestyle changes for everyone, including vegetarianism, a near-total ban on flying, washing in a sink rather than showering, and growing our own food. Then environmentalists wonder why they are not getting political traction.

Long Term

During the course of the twenty-first century, and even beyond, the evolution of CCS raises some fascinating policy questions.

Firstly, we can imagine a range of scenarios for the evolution of carbon capture and storage.

1. A dud, not deployed to any significant extent, employed perhaps only in a few value-added options such as EOR.
2. A niche technology, playing a small role in a renewable- or nuclear-led future, perhaps on low-cost capture options only.
3. A distraction, which diverts valuable attention and investment from better approaches.
4. A patchwork, which fails in some countries due to poor economics or (more likely) unsupportive policies or public opposition, but is highly adopted in others.
5. A bridging technology, which plays a big role in saving the climate, soaking up emissions from coal until low-carbon technologies arrive in a big way.
6. A disastrous wrong turn, which is adopted on a large scale but later suffers severe leakage (particularly if ocean storage is widely adopted).
7. A long-term approach, increasingly applied to gas, which sustains fossil fuels as the backbone of the energy economy towards the twenty-second century, perhaps including enhanced hydrocarbon recovery from shales, hydrates and aquifers.

8. The key component of a very long-term strategy for atmospheric management, including capture from air or biomass.

Which one of these scenarios eventuates depends on factors such as the long-term availability and cost of fossil fuels; the competitiveness of CCS versus other low-carbon approaches; the technical success and security of geological storage; the trends and rigour of climate policy; and unpredictable factors of public opinion. Indeed, CCS might well take off in some regions and not others.

Carbon capture offers one way to slide down the 'environmental Kuznets curve'.[175] This curve states that environmental damage initially increases with income. But at some point a level of wealth is reached at which the society feels able to afford environmental protection.[176] This level is different for different pollutants and ecological problems. Clean water is one of the first priorities. At an annual per capita GDP of about $4,600, net deforestation ceases.[177] Air pollution comes next: carbon monoxide, sulphur dioxide and so on. Some research suggests that carbon dioxide emissions per person may peak at an annual income of around $30,000.[178] If so, about twenty countries worldwide would have reached this happy level.[179] As wealthy countries also tend to have low or negative population growth, their absolute carbon dioxide output should also be declining.

The very existence of the environmental Kuznets curve is controversial.[180] It may be an artefact of wealthy nations' tendency to export

Figure 6.8. Simple illustration of a possible Environmental Kuznets Curve for carbon dioxide

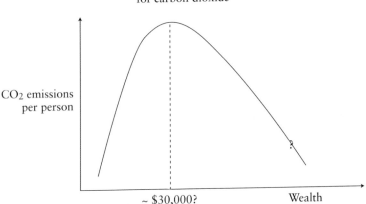

pollution to developing countries (though this is clearly not the reason behind reductions in automobile emissions). It may not apply to carbon dioxide, which does not cause obvious, immediate harm to the emitting nation. Reductions in carbon dioxide might be a side-effect of increasing use of low-carbon fuels (natural gas and electricity). For even higher levels of income, pollution might start increasing again. And the turning point is reached much earlier in countries which enjoy more democratic political systems, secure property rights, and an absence of distorting subsidies.[181]

But if the 'carbon Kuznets curve' does apply, then CCS is one way to hasten decarbonisation. Clearly wealthy nations are more likely to pay for costly energy systems. This suggests that a 'hair shirt' approach to climate change is unlikely to succeed. If we abandon economic growth, we are unlikely to feel, or to be, wealthy enough to afford costly environmental technologies. Lest this sound like First World greed, reflect that a Cameroonian living on an average of $2 per day has other priorities. So, for the long-term evolution of a sustainable and resilient energy system, it is vital that every major option is given a fair chance of playing its part.

As decarbonisation of the economy approaches completion, probably later in this century, indirect capture of carbon dioxide from air might become significant, particularly in Scenario 8 above. Indeed, if the major leakage of Scenario 6 comes to pass (unlikely though that seems, based on Chapter 3), then air capture might be crucial in cleaning up the mess: in other words, turning Scenario 6 into Scenario 8. Of course, we would have to find a more secure form of geological storage, or perhaps implement a massive biochar or mineralisation effort. Similarly, an emergency air capture programme might be necessary if catastrophic global warming begins to take hold and we realise that the climate is much more sensitive than we had thought.

This is illustrated in Figure 6.9. Even in the extreme case of complete decarbonisation by 2020, by 2060 we are still only back at 2005 levels of atmospheric carbon dioxide. If the climate system turns out to be very sensitive, this might be disastrous. But with negative emissions, by 2060 we could be back with a pre-industrial atmosphere. These very rapid rates of implementation are probably practically unfeasible, but the concept does illustrate that we have some more flexibility in climate policy than often imagined. Of course, we must not abuse that flexibility by doing nothing until it is too late.

Figure 6.9. Effect of 'negative emissions' technologies[182]

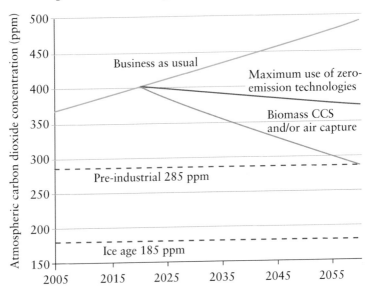

Air capture could allow us to tune the atmosphere to whatever level is desirable. Air capture represents an interesting dilemma for policy.[183] It addresses some otherwise very problematic emissions such as those from air travel. Unlike virtually every other system—carbon capture on power stations, biomass for power, biochar, reforestation, solar power, giant mirrors in space—it has no benefits or side-effects other than reducing atmospheric carbon dioxide. It thus forces environmentalist, anti-capitalist and 'neo-Luddite' groups to confront any hidden agendas they may have.[184] It does not require unanimous international agreement; at some cost to themselves, 'pro-environment' countries could mop up the emissions of polluters.[185]

Air capture allows reversibility of carbon dioxide emissions, which would discourage us from taking tough action on climate today. But if we pass a climatic tipping point, we might not have time to implement capture—it might take less than a decade for disastrous climate change to take hold. It is rather like putting off a programme of exercise until suffering a heart attack. The promise of air capture creates 'moral hazard': it could be used to postpone drastic action today, becoming a convenient get-out clause for those opposed to strong climate action.[186] It is very risky to make policy and delay action on the assumption that air capture will arrive one day, when it may prove unfeasible. Demonstrating air capture on a small scale is therefore desirable, just to know that we have the tool in our toolbox, if required. But although it is fairly easy to prove technical and economic feasibility, there is little way to show large-scale social and environmental acceptability, and sustained policy commitment, without actual implementation.

By 2100 a major air capture scheme, costing some $300 billion per year, might be within the reach of a coalition of non-governmental organisations (NGOs). It would, after all, only represent perhaps $30 per person, in a world perhaps four times richer per person than today.[187] It might only be about 0.25% of OECD GDP, an expensive but far from unfeasible commitment.

But whose decision would it be as to the 'right' level of atmospheric carbon dioxide? A northerly country, or one that has adapted well to a warmer world, might have a very different opinion from one that has become desert. The Greenlanders might not appreciate having their country iced over again, the Tuvaluans might not want to return to their once-flooded, now re-emergent islands, but Indians might welcome the return of the monsoon to an arid land. And, although our climate modelling capabilities will undoubtedly have advanced enormously by then, the effects of playing with the Earth's atmosphere are always likely to be unpredictable. Increasing carbon dioxide to, say, 550 ppm, and then reducing it back to a pre-industrial 280 ppm, may well not land us with the same climate we enjoyed or endured in 1750.[188] Who would make such a decision? Probably the most powerful nations—if nation states are still the favoured form of political organisation in 2100. Perhaps Greenland might desperately pump out greenhouse CFCs to counteract the Indian air capture scheme.

Such thoughts about managing the planetary atmosphere bring us closer to the concept of Gaia, of Earth as a self-regulating organism, much like an enormous beehive. In contrast to most existing Gaia literature, though, where humanity is disrupting nature, humans would have to be working with Gaia—indeed as part of her. As the century progresses, we may increasingly need a global system of sensing, monitoring, certifying and adapting, relying on satellites, pilotless aerial vehicles and a network of ground-based sensors and human researchers. This would enable us to manage and optimise the complex and interrelated issues of climate, carbon in geological storage and biological sequestration, deforestation, energy, water, biodiversity, food production and long-term nuclear waste handling.[189] As with the bees, this does not require any central intelligence or 'world government'.

7

RISKS

'It's become clear that there is no public acceptance for carbon capture and storage in the boundaries of this community.'

Simon Zuurbier, Alderman of Barendrecht, Netherlands[1]

'the first CCS project that is done badly is the last CCS project that will be done...In this respect, it is very similar to nuclear power.'

Mark Brownstein, managing director in the climate and air program, Environmental Defense Fund (EDF)[2]

There remain significant barriers in the path of carbon capture. It is quite possible that it will not take off. Quite apart from the over-riding risk—that the world does not manage to take effective action against climate change at all—such CCS-specific risks cover technical issues, costs, environmental concerns, safety, institutional barriers and public acceptability.

The entirely opposite risk also exists: that is, CCS may be too successful. It may perpetuate a dependence on fossil fuels, with their attendant environmental damage, reliance on big corporations, geopolitical insecurity and problems of ultimate depletion. The investment that goes into carbon capture may mean that we fail to unearth breakthroughs in renewable energy. That is not a view I share, but it is one that may be held by many. This, though, is not the immediate problem: carbon capture first has to overcome several barriers to full-scale implementation.

Technical

We have seen that, technically, it is hard to envision why CCS would not work. Transport and large-scale storage of carbon dioxide is already demonstrated. The basic science and engineering issues of capture are well understood. Both pre- and post-combustion capture techniques are already in use in various places. Oxyfuel is also conceptually straightforward and now working at demonstration scale in two locations. All these systems would require scaling up for large power plants, which always creates an element of risk, but fundamentally there is no obvious barrier. Various breakthrough technologies, such as carbon fuel cells, still require large amounts of research and development, but they are not essential for carbon capture to work in the medium term.

The main technical problem that CCS on power stations may experience is unreliability. Dependability and high uptime is absolutely crucial for electric power, even more than cost. Some of the first generation of IGCCs suffered from reliability problems, as may the oxygen unit of oxyfuel plants. A major power outage blamed on a CCS plant's going down might have a knock-on effect on public opinion, and the confidence of engineers in the technique. CCS on some industrial processes, such as iron, steel and cement, is technically less mature. It requires some basic redesign of the plant, and so might result in unacceptably poor performance.

Carbon dioxide injection for enhanced oil recovery is a proven technique, but it does not always result in significant increases in oil production, mainly due to failures in properly understanding the reservoir. It has, though, been more generally successful than most other EOR methods. Enhanced coal-bed methane recovery is still at very early stages and may not succeed, due to swelling of the coal and hence permeability reduction. Similarly, enhanced gas recovery is an immature technique and may fail due to carbon dioxide breakthrough at producing wells, or an insufficient increase in recovery. If such value-added carbon dioxide storage options are unsuccessful, this would significantly raise the cost of early CCS applications.

Costs

Cost is a much more significant problem. As we have seen, capture costs are generally the largest expense, and the first few full-scale CCS systems are likely to be very expensive. If costs do not fall fast, then CCS may not become competitive with other low-carbon technologies. Heavy cost overruns on one of the first projects may deter all future investors. This has already happened with some nuclear power plants (as in Finland[3]), and Shell's Pearl gas-to-liquids plant in Qatar has also suffered. The sharp increases in construction and operating costs during the early part of the twenty-first century might recur if the global economy booms again. The US's flagship FutureGen programme was, as we have seen, first delayed, then cancelled due to rocketing cost estimates, and then brought back to life.

The depressed power demand accompanying the 2009 economic crisis has also discouraged companies from pressing ahead with costly new initiatives, as with the postponement of Kingsnorth in the UK. Many of the European power companies are carrying heavy debt loads from earlier expansion, and raising new financing remains difficult. However, the economy and electricity demand will presumably recover eventually, particularly in successful developing countries—and a slow economic recovery would cut emissions anyway.

In the slightly longer term, an attempt to build up CCS rapidly might compete for personnel, steel, concrete, turbines and so on with other power plants and industries. This may be particularly acute during the early stages of CCS implementation, when suitably experienced contractors, project managers, engineers and geologists will be in short supply.

Carbon capture has to be implemented on a large scale to be economically viable. It is therefore unlike solar photovoltaics and wind power, which were able to start small. Solar PV, for instance, has niche applications in remote locations away from the electricity grid, which provide a useful testing ground. Solar and wind are modular—a hundred solar panels together function much the same as one. But some problems with CCS emerge only when the technology is scaled up; this was a key concern of Shell's when planning Pearl, the world's largest gas-to-liquids plant, which uses many of the same technologies as a pre-combustion CCS power plant.

It is unlikely, but possible, that inventors tinkering in their garages might produce a new CCS approach. But they would not be able to afford to build a commercial-size prototype. This issue is not unique to CCS: a similar problem has bedevilled the promising solar thermal power.[4] Venture capital (VC) firms have been successful in funding solar, fuel cell and biofuels research. But because most of the skills reside in large, established companies, and due to the problems of scaling-up, covering the extra expense of carbon-capture power, and hanging on to the intellectual property, they have done little in CCS. Carbon capture is therefore reliant on big corporations, mostly not renowned for innovative approaches or moving fast.

Since CCS requires more energy than conventional fossil-fuelled power, it suffers particularly when coal or gas is expensive. Natural gas prices were high in the USA in 2005–8, and rose globally in 2008 before falling back in 2009. As I have mentioned, I do not believe that we are anywhere near 'peak gas' or 'peak coal' (or even 'peak oil') but if I am wrong, then prices of these fuels might soar. A belief that fossil fuels are being exhausted anyway would inevitably undermine support for investment in carbon capture. More plausibly, price rises might be triggered by a dash for gas to meet climate commitments, or instability in a major gas or coal exporter. Operators might even run CCS plants in 'non-capture' mode if energy prices rose much more than carbon permit costs. For countries that import coal and natural gas, fossil-fuelled power may be less secure than domestic renewable energy, or uranium which can be easily stockpiled.

On the other hand, rising oil and gas prices would make it more attractive to use carbon dioxide for enhanced hydrocarbon recovery. The widespread use of CO_2-EOR, and hence the contribution it could make to kick-starting CCS, could be undermined by a period of low oil prices, by high upstream development costs, or by a number of failed implementations. Offshore EOR is particularly high-cost and therefore vulnerable.

Conversely, a rapid expansion in natural gas output around the world, for instance by repeating the North American success in exploiting unconventional reservoirs, would lead to lower gas prices. This would make coal less attractive, and would substantially delay large-scale CCS, until capture from gas plants became economic.

Similarly, if other environmentally friendly options make rapid progress (solar, say, or nuclear), then CCS may fall too far behind to catch up, or may simply not be needed. Or, unlikely as it seems today, new climate science might demonstrate that carbon cuts are not as urgent as now thought. Of course, as long as we achieve our overall environmental goals, this would be a good thing. CCS might also lose out to other approaches that have higher costs, but offer greater energy security benefits.

Environment

On the environmental side, CCS suffers from a continuing reliance on fossil fuels. This gives rise to four problems. Firstly, CCS typically only captures 85–95% of carbon dioxide. If carbon curbs become very strict, this will add an additional cost (and public relations) penalty. The capture rate can be increased further, but only at steeply rising expense.

Secondly, coal-fired plants, in particular, produce other pollutants in addition to carbon dioxide, such as mercury, sulphur and nitrogen oxides, particulates and water contamination. CCS plants tend naturally to be much cleaner than conventional facilities, for good technical reasons. Nevertheless, this local pollution may still be increasingly unacceptable. This might hamper coal-fired CCS plants, but should be less of an issue for carbon capture on industry and natural gas power. In some regions, notably India and perhaps the western USA and southern Europe, water shortages may become increasingly problematic, and CCS plants, with higher water consumption than conventional fossil fuel power, may therefore be undesirable.[5]

Thirdly, extracting fossil fuels generates pollution. Due to its higher energy consumption, CCS will, all other things being equal, lead to greater upstream environmental damage. Coal mining causes water pollution and some carbon dioxide and methane emissions. With open-cast mining, serious landscape degradation can occur, including removal of mountain-tops and the damming of streams. Even underground mining produces spoil heaps. Mining is also hazardous, many miners dying around the world each year, although safety varies enormously between countries and mining methods, open-cast being much safer.

By contrast, natural gas extraction is generally much more environmentally benign. There is concern in the USA, apparently largely misplaced, about the threat to drinking water supplies from hydraulic fracturing, used to extract gas from coals, shales and other low-permeability formations. Coal-bed methane inflicted damage, particularly in its earlier days, from discharging waste water into surface watercourses.

Of course, if we are worried about such problems, we should tackle the problem at source by banning certain kinds of mining, or impose much stricter environmental regulations on fossil fuel extraction. Underground coal gasification with carbon capture can expand coal reserves greatly at minimal environmental cost, and might become a standard for coal use. Nevertheless, some environmental groups will use opposition to CCS as a back-door way to attack coal mining, and will try to ban new techniques such as hydraulic fracturing regardless of the scientific evidence. One possible compromise might be to allow new CCS coal-fired power only if it replaces existing non-capture coal. Or instead, a cap could be set on total coal consumption or on the total amount of electricity generated from coal plants.

Fourthly, storage may leak. As Chapter 3 shows, leakage from good sites should be negligible or small, manageable and open to remediation. But it is always possible that one of the early storage sites may be badly chosen or operated, or our geological understanding of carbon dioxide movement or the integrity of plugged wells might be badly wrong. Escape from storage would tarnish the whole image of CCS, as Three-Mile Island and Chernobyl did for nuclear—indeed, it might kill CCS altogether. Most insidiously, CCS might proceed for many years before large-scale escape begins, presenting us with a major, intractable climate problem, somewhat analogous to, but probably more serious than, dealing with long-lived nuclear waste. This prospect seems unlikely, but, to guard against it, it would be useful to have demonstrated remedial methods; and biomass-CCS, large-scale biological capture such as biochar, or air capture.

Safety

Related to these environmental concerns is safety. As with cost, a high-profile early project might suffer a serious accident. Even if not related to the CCS equipment, this could blight the technology's public image. Again, we have seen that the safety issues of carbon dioxide transport and storage are small and should be manageable, but even a single leak could be problematic, particularly if causing death or injury. A more subtle threat is that disproportionately strict standards may be applied for safety and leakage, making carbon storage uneconomic. Again, this would be analogous to problems faced by the nuclear industry.

Public Acceptability

Public acceptability is closely tied to environmental, safety and to some extent cost concerns. Any accidents or ecological problems could cause an upsurge of public and/or NGO opposition. Anti-CCS campaigners such as Greenpeace already play on fears of leakage from storage, the red herring of the Lake Nyos disaster,[6] and so on. Further trials of ocean disposal are dangerous in this regard, since they could easily become a cause célèbre that, in the public mind, tars all of CCS with the same brush. Such PR attacks could be closely linked with attempts by other lobbies (mostly obviously solar and wind, and perhaps forestation, since nuclear has its own PR problems) to head off the competitive threat posed by CCS. Opponents of CCS might also play up energy security or 'peak oil/gas/coal' concerns, and hence paint it as 'unsustainable'.

Of course, it is always hard to prove causation or demonstrate conflicts of interest: do solar power executives attack fossil fuels in order to further their own businesses, or do they work on solar power in the first place because they believe fossil fuel use is unacceptable? But some groups have a vested, ideological agenda which is impervious to any scientific arguments. The spokesman of the large natural gas producer Devon Energy, speaking to a member of Congress's legislative director about proposed restrictions on his business, was reportedly told, 'What you're saying may be true, but it doesn't matter, because I'm forty-seven, and I want to have this country off fossil fuels by the time I'm seventy.'[7]

CCS already labours under something of a public relations disadvantage, due to its association with the unpopular petroleum, coal and electricity industries. It needs only to attract support from politicians, lawyers and real-estate agents to be completely condemned. CCS might suffer from its promotion by the Bush-era initiative on the 'Asia-Pacific Partnership on Clean Development and Climate', widely (and rather accurately) perceived as a literal and metaphorical smokescreen for polluting countries and industries to escape mandatory carbon curbs[8] and dismissed as 'a nice little PR ploy' by none other than former presidential candidate John McCain.[9] The debate is further clouded by 'clean coal', a term trotted out by industry groups such as the American Coalition for Clean Coal Electricity. Indeed, coal has become vastly cleaner in recent years in terms of non-greenhouse pollutants such as sulphur dioxide. But to be meaningful at all, 'clean coal' has to include carbon capture on at least 85–95% of its emissions. Otherwise, as in Joel and Ethan Coen's satirical adverts,[10] 'clean coal' becomes a byword for hype, empty spin and evading environmental responsibility.

Such bad press leads the public to be suspicious of carbon capture's environmental and safety credentials. There is a natural cynicism when industry proposes a solution so convenient to itself, however solid the scientific arguments. Scrutiny is intensified when the oil and coal industries take the lead in campaigning against climate change bills, as during August 2009,[11] and score PR own-goals such as forging letters opposing environmental legislation. Part of this lobbying is a reaction to elements of the proposed legislation, rather than to the idea of limiting carbon dioxide emissions per se, but the subtlety of this message can easily be lost.

Carbon capture may come to be seen—indeed, is sometimes already seen—as just one more tactic from the energy industry to delay or avoid taking real action on climate change.[12] The major elements of the fossil fuel industry, particularly in the USA, were so slow to acknowledge the reality of climate change, denied the science at every turn, and still continue to spread doubt and misinformation, even allegedly generating fraudulent grass-roots campaigns.[13] By doing so, they set themselves up to be the villains of the piece. To some extent, the global debate over carbon capture (and, indeed, over climate change legislation) is now being held hostage by the ideological clash in the USA between left and right. In Europe, a few mavericks apart, business and environmentalism agree much more closely than they might realise on the science of climate change, and the key solutions.

Such public opposition can lead to lengthy delays, lawsuits, planning inquiries, permitting challenges and direct protests, against new CCS power plants, carbon dioxide pipelines and storage sites. A backlash from taxpayers or electricity consumers might be caused by perceptions that heavy subsidies or rising power prices are being used to support carbon capture. The substantial government aid being given to renewable energy in many developed countries may be more popular. Government programmes, as with America's FutureGen, may be more vulnerable to cuts amid the fickle winds of political fortune than those led by companies planning for their future. Recovery from the financial crisis will, at some point, have to be paid for by spending cuts and tax increases, and this may crimp funding for new technologies, however environmentally vital.

Big businesses, never popular at the best of times, are particularly under the cosh today. Anti-capitalist groups therefore oppose CCS[14] as one way of undermining such corporations, and ushering in a utopia of decentralised, small-scale energy. Organised labour may dislike CCS because it appears to create relatively few jobs[15] (although coal miners' unions should support it, and they are politically powerful in South Africa and Eastern Europe). The really radical 'neo-Luddites' wish to drive down energy use dramatically in order to dismantle industrial society,[16] going along with plans for a radical reduction in population, much more drastic (if, they hope, more humane) than that caused by the Black Death or the 'Great Leap Forward'. They therefore oppose any realistic large-scale energy options.

Institutional

Probably the major group of blockers to CCS implementation, though, is institutional.

These institutional barriers come on the side of both government and industry. On the **government** side, there are capability problems at national and international levels. One of the first problems is the general energy illiteracy of most governments; perhaps this is a general complaint of experts in any field, but energy policy often seems to be conducted with a minimal regard for facts, and a yearning to repeat failed policies.[17] A long-term issue such as energy (and the environment) is perhaps peculiarly vulnerable to political short-termism. It is also so technical, sometimes counter-intuitive and often controversial that non-specialists have difficulty distinguishing good from specious or self-interested arguments. As the Royal Society notes, 'Climate change is a field in which policy disagreements continually find their expression in surrogate disputes about science...the positions taken by scientists and other analysts may interweave policy preferences with technical judgements.'[18] Environmental groups and energy firms are both guilty.

Because of the admittedly huge challenges in agreeing on global climate policy, a general climate change treaty may be delayed, weak, dysfunctional or all three. This would be a problem for the whole fight against climate change, not just CCS. Indeed, as I mentioned in Chapter 6, CCS might, oddly, expand more under regional climate deals than a worldwide one. Specifically, though, failure to agree a suitable successor to Kyoto might mean that costs of carbon (whether through emissions caps or a tax) may be too low or inconsistent to spur carbon capture. Alternatively, trading schemes or regulatory inconsistency may deliver volatile carbon prices that do not provide a suitable basis for long-term planning. It might prove impossible to negotiate suitable support mechanisms (cost of carbon, and eligibility of CCS for credits, and treatment of cross-border carbon dioxide shipments). Specifically, some countries may block CCS eligibility because they have other credits which they wish to sell into a global market, or because they wish to enhance the competitiveness of certain domestic industries.

Because of such problems, the required government support to get the first generation of CCS plants up and running may be delayed or never forthcoming. This might be triggered by cost overruns on an early project, or perhaps due to budgetary constraints. This demands a certain persistence from government and a deafness to industry lobbying. For instance, reducing acid rain would be a highly cost-effective policy,[19] yet we do not yet even have best-in-class control of conventional pollutants on fossil fuel plants in the developed world, let alone in China and India. This is cause for pessimism that carbon controls will be tough enough.

Legislators may also be caught between special interests domestically, and tortuous negotiations internationally. They may therefore be unable to pass the laws required to, for instance, permit cross-border carbon dioxide transit, classify carbon dioxide as neither a waste nor a pollutant, allow offshore disposal (at least underground), and establish long-term liability. As Simon Zuurbier, the Alderman of Barendrecht in the Netherlands, who is fighting Shell's plan to store carbon dioxide under his town, says, 'Nobody wants to be a waste basket. They [the Dutch government] also have ambitions to store emissions from Antwerp, and the Ruhr area. If you ask a Dutchman if he wants to do that? He will say no.'[20] Some environmental groups are adept at using litigation to achieve their policy aims, or at least to frustrate activities they oppose.[21]

On the **industry** side, corporations are under intense pressure to deliver returns to their shareholders while still meeting environmental goals. Building a power plant of any type is a lengthy, complicated and expensive business, and reliability is a key concern of both generators and regulators. Indeed, regulated utilities may not even be able to recover the costs of new technologies from their customers. Power companies are therefore cautious and conservative, preferring tried and tested approaches, and slow, incremental improvements.

CCS today also lacks a natural owner. In this, it is unlike many renewables, such as wind, which slot relatively naturally into power companies' portfolios. CCS is challenging because it spans several different businesses which are not normally combined: electricity (or industry, like cement, steel, chemicals, paper), gas transport and underground storage.[22]

- Start-ups and VC-backed firms have the innovative, risk-taking culture, and some of the smart ideas that might drive 'second generation CCS', but are too small to put a full-scale CCS project together.
- Oil and gas companies mostly do not own power plants; they dismiss the business as 'utility'-type returns (low, but stable and predictable profits).[23] Although they are certainly interested in running the geological operations when this involves enhanced oil recovery, the long-term, low-return pure carbon dioxide storage operations, with their attendant liabilities, are not very exciting for them. Indeed, in the short term, without CCS, gas may gain over coal, which would be good for most petroleum corporations. With oil prices still rather high by historic standards, the business of oil production looks much more profitable than that of carbon storage, and oil companies tend to deploy their limited capital, people and assets in areas where they feel most comfortable. Interestingly, no oil company is a member of the FutureGen alliance.[24]
- Coal companies recognise the threat to their business, which is why they support FutureGen and other consortia, but they are not really directly involved in carbon capture unless they also own electricity generation.
- Power companies would like to use CCS—indeed, as we have seen, the Swedish utility Vattenfall, the Chinese Huaneng and others are leading the way—but they are neither comfortable with, nor readily capable of, managing the sub-surface risk of storage. They also usually have other options for power generation—natural gas, renewables and nuclear—so carbon caps are not an existential threat in the same way as for coal mining.

Every part of the chain will have to understand its role, have a chance of a reasonable return for the risks it takes—and what is more, understand those risks and be able to manage or at least to accept them. The oil company running an EOR operation may have to understand that, due to a warm winter, the power plants have not been running much and it therefore will receive less than the planned amount of carbon dioxide. The power generator will have to plan for the fact that geology is not an exact science, and that the storage reservoir it planned on filling may not be as big as thought. That demands an insight into very different business models, which may not always be readily available.

This set-up is, in many ways, no different from the natural gas business, in Europe and North America at least. It is not at all unusual for ExxonMobil to extract gas in Canada, sell it to a natural gas trader like Duke Energy, which uses TransCanada pipelines to get it to New York where the New York Power Authority burns it to make electricity. But the natural gas business has grown up over a century, and for much of its history was run as a vertically integrated whole. Indeed, this is still the model in much of the Middle East and former Soviet Union. It may take a long time for specialist carbon capture companies to emerge.[25] The nascent CCS industry may not be able to develop suitable commercial models and rules of behaviour quickly enough.

This assumes, of course, good faith on the part of the various corporations involved. This is not a given. The promise of future carbon capture is a useful get-out clause for polluting industries. They can continue to run coal plants and build new ones, with the promise that *mañana*, 'when it is proved', CCS will be implemented. Of course, without serious attempts at implementation, that day never comes—or, at best, the growth of carbon capture is haltingly slow.

More likely, there may be some generalised willingness on behalf of industry to try CCS, but these good intentions may be weak enough that they are repeatedly frustrated by the costs and risks involved, and the short-term demands of the markets. Industry may repeatedly use its lobbying power to negotiate exemptions, delays and special treatment. The US natural gas business has been notable recently for campaigning against what it sees as unfair support for coal in the US's climate change bill.[26] Industry is skilful in using its lobbying clout, and its mastery of arcane technical issues which are not easily explained to politicians or public, to shape legislation. Because of its high upfront costs and risks, carbon capture might even be used by big industries to keep out new competitors.

In an even more Machiavellian way, the fossil fuel businesses may not wish seriously to make CCS work, for fear that it will then be made mandatory, at heavy cost to them. But this outcome seems unlikely. The oil business will be happy to implement storage where it is profitable (via credits or EOR). Power is a rather local business, so as long as carbon prices are consistent at least at a national or supranational level, utilities should be able to pass the costs on to consumers. Anyway, most power companies realise that we are already approaching an era of investment in costly low-carbon generation of various kinds.

Risks to Biological Sequestration

Bio-sequestration has a radically different risk profile from carbon capture. The technology is simple and requires no particular scientific advances. The costs are low, at least at first. The approach is popular with the general public in developed countries, and could offer major benefits (financial and environmental) to the poor in the developing world, but faces some big social and organisational challenges. And the permanence of sequestration varies by technique, but is generally low: most bio-sequestration methods are highly vulnerable to short-term leakage.

Bio-sequestration faces five main risks: sequestration integrity; verification; social challenges; land-use competition; and climate policy.

Of these, **sequestration integrity** is perhaps the most serious, at least for forestation. Forests are vulnerable to the very thing they are trying to prevent: climate change. A drought, forest fire, storm, flood or pest infestation can destroy the trees. Or, 'climate refugees' driven off their farmlands by climatic change may turn to clearing forests. As we have seen, geological storage, even in a badly chosen site, should not leak more than 0.1% annually, probably much less. A forest can easily leak 100% within hours. And a carbon-rich soil, established by years of conservation tillage, can be degraded by a single season of conventional ploughing.

Unlike underground carbon dioxide, a forest represents a valuable resource, a permanent temptation. The land it stands on also has potential other uses, unlike saline aquifers. The forest has to endure for centuries at least to make a meaningful contribution to fighting long-term climate change. Of course, destroyed forests can always be replanted, but this demands a very long-term commitment that we cannot guarantee, as well as extra costs to be incurred, and paid for by somebody, at some point in the near or distant future.

Biochar appears more secure, but it has still not been tried on a large scale and for long periods, unless we include the poorly understood *terra preta*. Much more research is needed to determine how to maximise carbon sequestration in biochar, whether it releases black carbon dust when spread, how it affects other soil carbon, its implications for fertility, and how long it takes to break down. Biochar is already suffering attacks on various other environmental grounds, damaging public support before it can take off.[27]

The issue of sequestration integrity is linked to that of **verification**: determining how much carbon has been removed from the atmosphere. This is a more complicated task than measuring underground injection of carbon dioxide. Forests and agriculture take up carbon dioxide over long periods, and they store it in soils and roots as well as above-ground vegetation. Some assessment has to be made of the value of less than immediate carbon sequestration; since many climate groups maintain that emissions have to peak by 2015, is it valuable to plant Ponderosa pine that reaches maximum carbon dioxide uptake only after seventy-five years? For conservation tillage, it is hard to be sure how much extra carbon has been absorbed without a control plot where, for comparison, conventional tillage is used. This, though, is hard to apply over a large area where farming practices and climate may vary greatly. Very detailed monitoring is possible but at increased cost. Reliable measurement will be essential if a global trading scheme is to have confidence that bio-sequestration offsets are reasonably accurate. Fraud is always a danger, and could undermine the credibility of such a scheme.

Social challenges cover the difficulty of implementing changes in land-use in often conservative rural societies. Contrary to claims that bio-sequestration (specifically, biochar) is an 'easy way out',[28] I know from my own experience that it is likely to involve days' journeys on jolting mud roads, crossing hippo-infested lakes on flimsy canoes, sheltering from rainstorms, catching malaria, all to convince sceptical villagers of the merit of some farming method that goes against their traditional practices and is not at all obviously effective. This is not just a problem of the developing world: agriculture in the USA and EU is notorious for the strength of the 'farmers' lobby', and for wasteful and irrational subsidies. About a quarter of the Brazilian Congress is made up of agriculturalists who support further exploitation of the Amazon and who eventually led environment minister Marina Silva to resign.[29]

The poor in developing countries, lacking formal legal title to land,[30] political power and the money to pay bribes, may suffer from forestation schemes. Only 14% of private land in Brazil has secure ownership, and 1% of the population owns 47% of the country. Land-use schemes may displace the informal inhabitants, causing hardship, or driving them to cut down trees elsewhere. Policing plantations to keep starving people out does not sound very sustainable, ethically or in public opinion. Even if forests are established or preserved on otherwise little-used land, it would still be a missed opportunity to alleviate poverty. Anyway, 'marginal' or 'degraded' land may be under use by pastoralists or hunters. In a middle-income country like Brazil, the government's writ barely runs in remote areas of the Amazon;[31] imagine how much weaker is the state's hold over the Democratic Republic of Congo. Ineffective and corrupt legal systems make forestation programmes much more costly.

Land-use competition may grow increasingly stringent, as a growing population has to find food, fuel and now carbon sequestration from a limited pool of land. The prospects are not entirely gloomy, since increasing agricultural yields and investment in restoring degraded land to productivity (through, for instance, irrigation, and perhaps conservation tillage and biochar) may ease these pressures, and most projections show population growth gradually levelling out. But a massive programme of set-aside for forestation may just displace agriculture elsewhere, or may lead to unacceptably high food prices. Indeed, China had to suspend its reforestation campaign in mid 2009 to protect agricultural output,[32] while there were complaints by poor farmers that they had been forced to give up their land for tree-growing.[33]

Land-use pressures could be exacerbated if high oil prices combine with a lack of progress in second-generation biofuels to lock up large amounts of land growing inefficient energy crops such as palm oil and corn ethanol. There is also potential for cross-border impacts, affecting rainfall, watersheds and fertiliser run-off, though the effects here could be positive as well as negative. The apparently low costs of small forestation programmes could balloon when they are carried out on a large scale and begin to compete seriously for land.

Climate policy is, as we have seen, a risk for all carbon storage and sequestration methods. Bio-sequestration is vulnerable because poor countries may not be in a position to pay for large forestation or biochar schemes, and most of them are not big emitters of industrial carbon dioxide. They will therefore be reliant on offsets or other funding from wealthier nations, but this financing is hostage to domestic politics, which may be averse to sending tax dollars (or euros, sterling or yen) overseas. Increasing aid to poor countries brings the risk of dependency, while climate change itself harms their ability to cope on their own. So just trying to counterbalance climate damage with aid is unsustainable.

Combining the Risks

On their own, these various risks, though problematic, would probably not be enough to sink carbon storage and sequestration. The technical and cost issues certainly appear soluble, with a resolute research and development effort and some large-scale pilots. Solar power, by comparison, started with much higher costs. Nuclear has (mostly) overcome sterner technical and safety challenges. The residual environment impacts of carbon capture are much less than those of conventional power. CCS is not outstandingly popular with the public, but they are also not very familiar with it, and it still ranks better than nuclear power. So the individual challenges of CCS are not insuperable, but putting all these very diverse issues together creates a complicated puzzle.

A central problem is that carbon capture has mostly lacked a powerful champion—be that a national government, an NGO leading public opinion, or a strong industry alliance. Industry-government partnerships such as the Zero Emissions Platform, and FutureGen, may now be leading the way. The next few years will be crucial; the key projects to watch are the demonstration schemes in the EU, particularly the Netherlands and UK, and the US's progress on clean coal. Similarly, for bio-sequestration, large-scale forestation, biochar and agricultural schemes are needed to show that these methods are politically and organisationally feasible.

Perhaps the most likely risk is that, through the accumulation of a variety of problems, none of them fatal in itself, CCS will build up slowly and play only a minor role in tackling climate change. This might mean that climate damage is much more serious than it could have been; it might mean that our solutions to greenhouse gases are much more expensive. Marine biologist Peter Brewer, who has experimented with ocean storage of carbon dioxide, says:

'Whenever we sit around over beers at one of these conferences, we talk about where we think we'll be fifty years from now. To a man, the answer is, CO_2 will still be rising. Climate will have changed. We'll have gotten more efficient at energy generation, but our population will have risen. We'll have developed sequestration technologies—and maybe they'll be doing three percent of what's needed.'[34]

8

CONCLUSIONS

'It's the urgency that creates the scale of the challenge.'

Dr Graeme Sweeney, Executive Vice-President,
Shell International Renewables[1]

Carbon capture and storage is a realistic contender for a leading role in fighting climate change. Amongst the various future energy options we have, it is unique:

- It can tackle the existing stock of fossil fuelled facilities.
- It addresses industrial emissions of carbon dioxide as well as power generation.
- Via biomass or air capture, it can achieve negative overall carbon dioxide output, and so head off rapid climate change, even negating otherwise intractable emissions such as those of aviation.

Table 8.1 (p. 314) shows a simple comparison of various future energy options, ranked from ● (best), through ◕, ◑ and ◔, to ○ (worst). Of course, this ranking is necessarily subjective, imprecise and controversial, and nearly all of these technologies have niches where they will be successful, but it does suggest that CCS comes out somewhere in the middle of plausible possibilities. Of the five options ahead of it, only nuclear fission could realistically match fossil fuel power generation in magnitude.

Table 8.1. Qualitative ranking of future energy options

	Technical maturity	Reliability*	Scalability**	Potential***	Cost	Environmental performance	Long-term sustainability	Public acceptability	Ranking
Energy efficiency	●	●	○	○	●	●	●	●	=1
Offshore wind	◐	◐	◐	◐	◑	◐	●	◐	=1
Biomass	●	●	◑	◐	●	◑	◐	◑	=1
Onshore wind	●	◑	◑	◕	●	◐	●	◑	=4
Nuclear fission	●	◐	●	●	●	◕	◑	○	=4
Carbon capture and storage	◐	◐	●	●	◐	◑	◕	◕	=6
Solar thermal	◐	◑	◑	●	◑	◑	●	◐	=6
Geothermal	●	◐	◑	◕	◑	◐	◑	●	=6
Hydroelectric	●	●	●	○	●	◕	●	◕	=9
Solar photovoltaic	◐	◕	◕	●	◕	◐	●	●	=9
Nuclear fusion	○	◕	●	●	◑	◐	●	◑	=9
Wave/current power	◕	◐	◑	◑	◑	◐	●	◐	=9
Tidal power	◐	◑	●	◕	◑	◑	●	◕	13
Biofuels (second generation)	◕	◐	◑	◕	◐	◑	◑	◕	14

* Technical reliability and intermittency.
** How much energy can be delivered by a given project, hence how feasible it is to implement rapidly on a large scale.
*** How much of our energy needs could realistically be served by this technology.

This table highlights the main areas of weakness of carbon capture: less mature and higher cost than nuclear or onshore wind, less environmentally friendly than wind and solar PV, not indefinitely sustainable because of its use of finite fossil fuels (though as I have argued, a lifetime of seventy years or more is comfortably good enough), and less publicly acceptable than most other clean technologies.

It also, though, indicates the key strengths. Carbon capture works on a large scale, and is cheaper and more reliable than many renewable energy sources. It will probably prove more acceptable to the public than nuclear power, and the existing projects strongly suggest that geological storage can be secure enough to meet our climate change objectives. We are now entering a key phase where, over the next few years, some full-scale demonstrations will vastly advance our practical experience, and establish how expensive carbon capture from industry and power plants is.

Yet if the political winds turn against it, if early projects fail or over-run heavily on costs, or if an NGO or media campaign is whipped up by real or imagined safety or environmental problems, CCS may not take off. If industry moves slowly and conservatively, and is not smart or aggressive enough to secure the right government support and seize the available business opportunities, CCS may move slowly and by mid-century, it might be 'three percent of what's needed'.

Carbon capture fits well into the emerging paradigm of 'post-environmentalism', of resilience in preference to 'sustainability'. It presents a plausible solution to what seem intractable problems. It makes pragmatic use of existing resources. It can be very attractive to some key developing countries, China and India above all, and may help swing them behind decisive climate policies. It can be a short-term sticking-plaster or an enduring pillar—or both—depending on how the situation evolves.

To build the will for our global society to make deep carbon curbs, we need to continuing building the scientific, economic and political understanding of climate change—as best we know today, that it presents a massive and immediate danger. Vested interests, certain big businesses, right-wing ideologues, self-appointed 'experts' in the media, ignorant or self-interested politicians, and scientists willing to prostitute themselves to the highest bidder, continue to repeat egregious falsehoods, objections to the idea of anthropogenic global warming that have been repeatedly demolished by science. They cloud the issue, literally, with propaganda masquerading as research. Many in the media, always in search of a controversy, help to fan the flames.

Of course, there should be, if anything, not less debate but more, less reliance on a cosy consensus, and more clear-headed acknowledgement of the uncertainties and value judgements that inevitably attend the climate issue. But on all sides, we need more intellectual honesty and open-mindedness.

Having built the momentum to take action, we then must offer people the hope of realistic solutions. Having established the dangers of rampant global warming, some commentators go on to tell people that biochar will 'pyrolyse the planet',[2] biofuels lead to starvation, nuclear power builds up dangerous waste that cannot be stored safely,[3] and carbon storage will leak and kill people as at Lake Nyos.[4] These Cassandras display a certain arrogance when, in a few lines, they dismiss ideas that expert scientists and engineers have laboured for years to develop.

The first reaction to new technologies is to say they are unproven, too expensive, dangerous and won't work. (Does anyone remember the first mobile phones? They were too expensive, the size of a brick, and are still accused of being dangerous.) Any technology that survives the initial scepticism is then condemned since it will have to be deployed on 'too large' a scale; not surprisingly, given how large the current global economy and energy system is—yet we built it, most of it over the last fifty years. The respondent's favourite solutions, of course, are able to grow at any speed and to any scale, but CCS cannot be delivered until 2030; therefore we should not do anything today, as though its ultimate arrival is a matter of magic rather than hard work.

So we are exhorted to make drastic changes in lifestyle. The developed nations will have to give up many things that have become essential, such as air travel;[5] the poorer countries will never enjoy them. People are even told, 'It is a campaign not for more freedom but for less. Strangest of all, it is a campaign not just against other people, but against ourselves.'[6] Such a clarion call is, perhaps, to be expected in apocalyptic times. It is oddly reminiscent of medieval campaigns against sin with the backdrop of the Black Death, the world of *The Seventh Seal*.

Such doom-mongering works against a cool, realistic appraisal of the situation, a careful weighing of the twin goals of fixing climate and eliminating poverty, an informed democratic debate on the policies to achieve these aims. Even Jared Diamond, prolific author on the collapse of past societies which (in his view) outran their resources, concedes from his Bel Air mansion that he will continue to follow his affluent lifestyle.[7] Roosevelt would not have mobilised the American people to overcome the Great Depression by telling them that they should fear fear, that fixing the economy was impossible and they would never be wealthy again. Equally, though, he did not deny the immensity of the problem. As Lord Nicholas Stern, of the eponymous review, says:

What's the alternative to optimism? Unless we act as if we can sort this out you might as well just get a hat and some sun tan lotion and write a letter of apology to your grandchildren. The only way we can think of going forward is to try to make the best of a bad starting point...The one way of guaranteeing to fail is to assume that we will.[8]

Ideologues who deny anthropogenic climate change, and environmentalists who reject valid solutions, are equally enemies of the climate and of the poor. My own ideology is simple: I believe in what works. I am humble enough to admit that I can't predict the future, and when specialists tell me that something in their field has a reasonable chance of working, then it shouldn't be rejected out of hand. The fight against climate change is such a gigantic, challenging task that we have to cast aside our prejudices.

So let us accept that carbon capture may be a plausible part of the solution to climate change. What does CCS mean for us at a personal level?

If you do not believe in anthropogenic climate change, or think it is not a serious problem, then you've done well to get this far in the book (or you have skipped ahead to the conclusion). There are many excellent books and articles covering climate change issues. They may convince you of the need for low-carbon technologies, possibly including carbon capture.

If you are an environmentalist, then stand up for solutions rather than negativity and doom-saying. Challenge energy companies hard when they propose carbon capture, but be prepared to give it a fair chance. Don't take everything from environmental organisations at face value. If you feel your particular organisation is taking an ideological rather than scientific stand on CCS, or any other issue, then campaign to change its policies, or join someone else. Big corporations may be self-interested, but they often do know what they are talking about.

See CCS as one major part of a portfolio of many complementary solutions. Campaign against deforestation, wasteful energy subsidies, new coal plants without CCS, destructive coal-mining practices, NIMBYism. If you do genuinely decide that carbon capture is not the solution, then at least let others try to make it work; instead of obstructionism, turn your energies to promoting alternative approaches. Protesting is a feel-good activity that requires, at its most basic level, few skills; developing and commercialising new technologies is arduous work requiring years of study and endeavour.

If you are a policy-maker, adviser, regulator or politician, then understand where CCS comes in the portfolio of energy and climate. Scrutinise any government support for CCS demonstrations, and make sure that the private companies involved are taking a fair share of the costs and risks. Try to develop 'smart regulation' that is flexible, that does not impose impossibly tough standards on theoretical risks, but does correctly identify and manage real risks.[9] Think of the potential to develop new businesses and export products and ideas for your region or nation. If you are from a developed country, consider how carbon capture can be shared with poorer countries to bring them on board. If you are from a developing country, do not see CCS as a luxury to be installed later; it may well be cheaper to start now, and you might be able to attract substantial aid and goodwill from richer nations.

If you work in the energy business, or another carbon-intensive industry, then consider CCS. Become an advocate for it in your company. Realise that, for coal in particular, there is no future without carbon capture. If you're working in renewable energy, then you should realise that, though there is room for all realistic technologies, carbon capture is a significant long-term competitive threat; if you can beat it on cost and environmental impacts, then you can rightly feel pleased. Look for profitable opportunities that make use of carbon capture, such as EOR, especially if you're involved in running a plant with low capture costs, such as ammonia or DRI. If you're lobbying on behalf of industry, look for policies that open up carbon capture, rather than seeking special treatment and exemptions. Be prepared to take risks in developing new technologies—and have a plan for how you will capitalise on success. There is an immense mass of research on all aspects of carbon capture, but far too few practical results and commercial projects so far. We need people who can implement projects, as well as research them.

If you're qualified in subjects as diverse as law, finance, public relations, earth sciences, engineering, life sciences or technology development, then there might be a career for you in carbon capture: experts in all these fields will be needed, and at the moment virtually everyone is starting from the ground floor. You may do better in this as yet relatively unrecognised area, rather than scrambling for 'green jobs' in solar power or carbon trading.

If you are involved in agriculture or forestry, recognise that bio-sequestration may be a valuable source of income, both from carbon credits as well as the additional benefits of increased fertility, by-products and so on. Carbon offsetting groups may be interested in working with you on biochar, forestry or conservation tillage, particularly if you are in the developing world. Recognise that short-term, unsustainable land-use will actually leave you, and your community, worse off.

If you're an investor, you should at least be considering the impacts that climate change may have on your portfolio. Specifically, you might wish to think about which companies and sectors will lose or gain if CCS takes off, as discussed in Chapter 6. Carbon capture is, of course, far from the only part of climate change business, and other changes—in greenhouse gas regulation and treaties, renewable technologies and energy prices—will generally be of greater magnitude in affecting company values. It is also hard to invest directly in carbon capture at the moment; there are no 'pure play' investments, as there are with solar or nuclear, say. But that may change over the next decade or two. Particularly if you're a venture capitalist, then you can look for opportunities in some small entrepreneurial companies working on CCS technologies.

If you belong to none of the above categories, then you may be a voter in a democracy, with a voice in public decisions. In this case, you can contact your political representative and give your opinion on climate policy, specifically relating to carbon capture. If a CCS project is proposed near you, then ask the organisation building it some tough and informed questions, but give it fair consideration. Compare CCS's safety profile with that from similar activities, and carbon storage projects elsewhere in the world. In general, learn as much as you can about environmental issues, and make up your own mind.

And, whether you fit any of the above categories or not, if you're an open-minded person who is curious about the future, you can simply reflect on the possibilities that might await us in 2050 or 2100. CCS may be a major activity or it may not. We may have won the fight against climate change, be winning it, or struggling vainly, or have discovered other problems. By the century's end, we (or our children or grandchildren) might find it amazing that people ever discharged carbon dioxide directly into the atmosphere, or that anyone could have been opposed to capturing it. So working on carbon capture today is like researching the internet in 1990 or nuclear power in 1950 or the steam engine in 1750. It might deliver nothing, it might fall short of our high hopes, but, equally, it might shape our world.

Because of this potential, because of the threat that climate change poses to our civilisation and the natural world, and because the future is inherently unknowable, we cannot afford to throw aside a key weapon in the battle against carbon. Within the next decade, we must decide whether our best strategy includes carbon capture, storage and sequestration. We must aim for an energy system that is not just sustainable, but also resilient to whatever the winds of fortune may throw at it. We need not only to devise a Plan B, but to begin pursuing it; in fact, we need to be working on Plans A, B and C right now, aiming to get them to the stage where they can stand on their own feet. Instead of vain denial of the reality of climate change on the one hand, and reflex opposition to anything but a single, pre-conceived, vulnerable plan on the other, we need post-environmentalism. I take that to mean protecting people and planet with innovation, open-mindedness, flexibility, pragmatism and technology—and potentially, capturing carbon.

APPENDIX

FURTHER DETAILS

This section gives additional supporting material, figures and arguments for the main text.

Chapter 1: The Need to Capture Carbon

Page 12: On the overwhelming consensus on anthropogenic climate change, 97% of US climate scientists surveyed by Harris Interactive believed that 'global average temperatures have increased' over the past century; 84% say they personally believe that human-induced warming is occurring, and 74% agree that 'currently available scientific evidence' substantiates its occurrence. Only 5% believe that that human activity does not contribute to greenhouse warming.[1] A survey prepared by the University of Illinois showed that 90% of Earth scientists surveyed believed that the climate had warmed since pre-industrial times, and 82% considered that human activity was a major contributor to changing temperatures. Amongst those scientists with the greatest climate expertise, these figures were 96.2% and 97.4% respectively.[2] The national science academies of the G8 nations plus those of India, China and Brazil signed a joint declaration in 2005 that scientific understanding of climate change was clear enough to justify global countermeasures.[3]

Page 13: Atmospheric carbon dioxide equilibrates with the ocean over 200–2,000 years, and weathering removes carbon dioxide from the air on a timescale of 3,000–7,000 years. This implies rather long-lasting effects of human carbon dioxide emissions.[4] Weathering and hence carbon dioxide removal is faster when atmospheric carbon dioxide concentrations are higher.

Page 17: On total possible emissions, limiting emissions between the years 2000 and 2050 to 1,830 Gt has a 75% chance of keeping warming below 2°C. A total of 234 Gt CO_2 was emitted during 2000–6, leaving approximately 1,600 Gt from now to 2050.

Page 19: Most other cost estimates for the amount of GDP required to be spent to avoid dangerous climate change fall in the range 0–4%.[5] For instance, the consultancy McKinsey[6] calculated spending 0.6–1.4% of GDP by 2030 to stabilise atmospheric CO_2 at 450 ppm. Using a 2% discount rate (around Stern's central case), the IPCC calculates the cost of carbon at $165 per tonne. Nordhaus, on the other hand, puts the value at $10 per tonne.[7]

Page 22: Against the indications that fossil fuel resources are more than enough to cause disastrous climate change, there are some dissenting views.[8] These rely, in my opinion, on absurdly low estimates for volumes of oil, and in particular coal and gas, especially the amount that can be recovered with improved technology. For example, coal gasification can recover some 95% of coal in place (as against ~37% for conventional deep mining[9]). Reported US coal reserves of 270 billion tonnes, already two-thirds of a low estimate for the entire world,[10] are likely almost to double.[11] Indeed, the US resource base is 489 billion tonnes; not all is accessible due to property rights issues, land-use conflicts and environmental and physical restrictions, but presumably many of these issues could be overcome in the case of a shortage of coal.[12] China is estimated to have 5,500 billion tonnes of coal resources.[13] The volumes of potential carbon dioxide emissions are from the IPCC's A1C AIM scenario, which has the highest fossil fuel consumption and cumulative carbon dioxide emissions to 2100.

Conversion from energy to volumetric/mass units is assuming 7.35 bbl/tonne for conventional crude oil, 6.4 bbl/tonne for oil sands and extra-heavy oil, 29.31 GJ/tonne for coal, and 40 MJ/m^3 for gas. Coal reserves are resources of hard coal plus lignite. Lignite has been converted to hard coal equivalent at a ratio of 2 tonnes lignite: 1 tonne hard coal, and figures rounded to the nearest 100 billion tonnes.

For oil, production during 1996–2007 has been removed from future discovery. The estimate includes natural gas liquids, future exploration potential and reserves, plus extra-heavy oil and oil sands potential, but excludes additional conventional oil exploration potential[14] and oil shales. Figures have been rounded to the nearest 100 billion barrels.

For gas, conventional resources are given, plus tight gas, coal-bed methane and future gas exploration potential. Production 1996–2007 has been removed from future exploration potential. The estimates exclude gas hydrates and aquifer gas, and figures are rounded to the nearest 100 Tcf.

The emissions from burning these fossil fuels are calculated assuming coal and oil are 85% carbon, and gas is pure methane. These compare to IPCC figures, which have potential for some 80,000 Gt of CO_2 emissions, almost twice as much.[15]

Page 35: Comparing the area of different generation technologies is done as follows. For the solar thermal plant, I assume that output is 25% of capacity, and that the solar plant replaces coal with 0.97 kg/kWh specific emissions.

The area of the gas plant is based on the literature.[16]

For the area of the coal plant, assume a 1,000 MW coal power plant operating for thirty years, with carbon emissions 0.97 kg/kWh; plant area[17] 1.4 km², assuming that the carbon capture system doubles the plant area (an over-estimate; likely increase is 23–37%[18]); carbon dioxide pipeline of 200 km with 100 m width right-of-way; sequestration site 200 x 200 m surface footprint (if sequestration site and part of pipeline is offshore, this would further reduce land-use conflicts). This gives a total of almost 12 million tonnes sequestered.

A coal power plant fed from surface mining, though, needs a similar area to the solar plant, when allowing for the area of the mines and feeder railways. The surface footprint of gas production and transport is small.

Woodland carbon content converts the carbon in the trees to carbon dioxide equivalent.[19]

Chapter 2: Capture Technology

Page 62: Although many other ASUs and oxygen-fuelled furnaces are in use for non-CCS purposes, such as synthetic fuel production and glass-making. There is also a 30 MW oxyfuel boiler at Babcock & Wilcox's Clean Environment Development Facility, but it is not used for carbon capture.

Page 72: The estimate for capturable emissions from glass-making assumes 0.45 tonnes CO_2/tonne glass for energy-related emissions and 0.2 tonnes CO_2/tonne glass for decomposition of soda ash and limestone,[20] with special glass, domestic glass, mineral wool and glass fibres each amounting to 5% of the output of container glass and flat glass. This may be over-estimated since recycled glass has lower energy requirements and no raw material-related emissions, and raw material related figures are slightly lower at 0.185 tonnes CO_2/tonne glass.[21] On the other hand, glass production figures are from 2001 and 2004, and output has presumably grown since then.

Page 75: The potential for carbonation of industrial wastes is divided as follows: 86 Mt from coal ash; 70 Mt from steel; 17 Mt from waste concrete; 2 Mt from alumina; 1 Mt from oil shale ash.

For coal ash: 500 million tonnes of coal ash were generated annually worldwide in 1995.[22] Global coal consumption increased by 95% from 1995 to 2008. About 80% of this ash was pulverised fly ash, of which approximately 13% (or 35%,[23] but this is not consistent with the figures in the table given) was used for other purposes in 1995, mainly in cement and concrete manufacture, up from 11% in 1992. Assuming that this increase in utilisation continued, then I project that about 36% of total 780 million tonnes pulverised fly ash was used for other purposes in 2008. This leaves 500 Mt to be carbonated. A simple average of compositional figures[24] suggests that pulverised fly ash is 10% Fe_2O_3, 9.6% CaO, 2.2% MgO, 1% Na_2O and 2.1% K_2O by weight. One tonne of fly ash, completely carbonated, should absorb 0.17 tonnes of CO_2, assuming that the aluminium oxide does not react. Potential for CO_2 removal is therefore 500 x 0.17 = 86 Mt/year, assuming that the 20% of other ash cannot be used. However, fly ash can also be used as a substitute in cement-making, and this may prove a more effective way of reducing emissions than mineral carbonation.

For steel: 0.27 tonnes of CO_2 can be captured by carbonating one tonne of steel slag.[25] Global figures for slag production are not available, but US steel slag production in 2006 was 20.3 Mt.[26] Scaling this up to world steel production gives an estimate for global slag production of 258 Mt in 2006.

For concrete: US flow of waste concrete in 1998 was 68 Mt,[27] with capture of 0.06 tonnes of CO_2 per tonne of concrete. Scaling this by cement production seemed inappropriate, since that presumably relates to new building rather than the demolition of existing buildings. Therefore the US waste concrete total has been scaled up by its share of global GDP to give a world total. Perhaps an estimate based on building stock would be more appropriate.

For oil shale: the only significant commercial oil shale mining at present is in Estonia, burning of which produces about 10 Mt CO_2 per year.[28] Of this, about 10% can be captured by reaction with ash.[29] If oil shale mining were to increase, then the capture potential (and emissions) from ash would increase too.

For alumina: the Kwinana alumina refinery produces 2 million tonnes of alumina annually,[30] and now carbonates all of its residue. This has reduced emissions by 70,000 tonnes CO_2/year. Global alumina production in 2005 was 62.4 Mt,[31] implying potential sequestration of (62.4/2) x 70000 = 2.2 Mt/year. However, the total potential for this technique is quoted as 15 Mt/year in Australia alone.[32]

For other mining: the following figures can be estimated from a graph:[33] 120 Mt/year from residues from nickel mining, 90 Mt from platinum-group elements, 20 Mt from chromium, 15 Mt from asbestos, 25 Mt from diamond mining and 5 Mt from talc mining.

However, some of these residues can also be used as additives in cement-making, to reduce carbon dioxide emissions from calcinating limestone. Obviously these carbon savings should not be double-counted.

Page 80: Carbon dioxide capture on a ship can be calculated as follows. Assume the 'Suezmax' class of tanker, 150,000 tonnes deadweight, 30 MW engine of 52% efficiency,[34] burning fuel which is 85% carbon by weight. A typical sailing time from the Middle East to Asia or from Nigeria to the USA is twenty days. For longer routes, if carbon dioxide tankage was insufficient, the ship could dock at an intervening port to unload. Carbon dioxide is assumed stored as a supercritical liquid at 20°C and 150 atmospheres pressure.

Page 81: Contrail formation, and hence additional global warming, can be reduced by rescheduling, particularly avoiding night flights;[35] cutting down pollutants such as sulphur in jet fuel; rerouting to avoid climatic conditions favourable to contrail formation; flying at different altitudes. The residence time in the atmosphere of contrails and the particles that form them is much shorter than for carbon dioxide.

Chapter 3: Transport and Storage

Page 88: The possible impact of substituting ceramics is calculated from global clay mining in 1997, which was nearly 400 million tonnes.[36] It has been assumed that this clay was all used for ceramics, was 50% water, and could be replaced weight-for-weight with calcium carbonate. Other sources of ceramics are ignored. There might also be some carbon dioxide savings from reducing energy used in brick-making, etc.

Page 95: Typically, 1 tonne of CO_2 recovers 2 barrels of oil, which will release about 0.5 tonnes of CO_2 when burned; assuming a typical light oil of 35° API gravity, the type for which CO_2-EOR is most applicable, and a carbon content of 85% by weight; also assuming that all the oil is ultimately burned (some may end up in long-lived plastics and other petrochemicals; petrochemical use is about 10% of oil consumption). Carbon intensity of some unconventional oil sources is up to 30% greater than conventional oil.

Page 108: Possible impact of adding sodium carbonate to the oceans is estimated, given that the total world reserves base of sodium carbonate is 40 billion tons,[37] which, assuming a molar relationship of 0.8 molecules carbonate: 1 molecule CO_2,[38] could sequester about 13 Gt of CO_2 in the oceans, i.e. about half a year's global emissions. Global potassium carbonate reserves are less than half those of sodium carbonate. Even given that more sodium and potassium carbonate deposits could probably be discovered with incentives for exploration, this approach still seems limited in usefulness.

Page 110: On the possibility for by-products to reduce the cost of mineral sequestration, the minerals at Kempirsai contain, at current prices, at least $10 billion worth of platinum-group metals, plus the value (not quantified here) of chromium and iron, which are actually the primary products of this mine. At $75/tonne CO_2 sequestration cost, this volume of ore could sequester about 160 Mt of CO_2 for $12 billion.

Page 111: Emissions from Oman, UAE, Saudi Arabia, Qatar, Bahrain, Kuwait, Iran and Iraq totalled 1194 Mt in 2004,[39] but a substantial proportion of this is not easily capturable (transport and small stationary sources). Iraq and much of Iran are also far distant from Oman and would require long carbon dioxide pipelines or transport by ship. On the other hand, Iran also contains some possibly suitable ultrabasic rocks, such as the well-studied Neyriz Ophiolite,[40] as do Iraqi Kurdistan,[41] Saudi Arabia and Yemen.

Page 113: ECBM potential of 200 Gt of CO_2, assuming every two molecules of CO_2 replace one of methane, amounts to 1,791 Tcf of natural gas at standard conditions. If produced over 100 years, this averages 49 Bcf/day, compared to 2008 global gas output of 295.8 Bcf/day.[42]

Page 113: The breakdown of storage capacity estimates is as follows. Note that none of these studies give a separate estimate for EGR potential, but this may be included under 'Depleted gas fields'. The IPCC's high case estimates for oil and gas include a 30% increase due to undiscovered fields. For Sonde's figure on basalts, I have calculated a total assuming 29 Gt of CO_2 emitted worldwide (2006), and storage only in the Deccan (ten years of global emissions) and Columbia River (1–5 years).

Page 114: For the area of the Netherlands required to store its emissions, I assume a CO_2 density of 0.5 tonnes/m^3 (at the low end for supercritical CO_2), and an effective CO_2 layer thickness of 1 metre. Current emissions, if maintained to the year 2100, would total about 13 Gt, which would occupy 26,000 km^2 in a layer 1 metre thick. The area of the Netherlands is about 41,500 km^2 onshore and 57,000 km^2 offshore.

Page 118: Storage capacities per region are taken from the literature.[43] The total volumes for capture in the period 2010–2100 are calculated using the 2005 figure for carbon dioxide emissions from power and industry, and increasing it by the following annual growth rates: 0.9% for Western Europe, North America, South Korea, Japan and Oceania; 2.3% for Latin America, the Middle East and Africa; 1.2% for Eastern Europe and the former Soviet Union; 1.6% for India, China and other Asia.[44]

Page 119: To calculate the amount of mineralisation product, assume 8,200 Gt CO_2 converted to minerals, producing 4.7 tonnes of minerals per tonne of CO_2; assume average density of the minerals produced is 3 tonnes/m^3. For the comparison with total cement production over a century, about 4 billion tonnes are used annually today; assume this grows at 1% per year to 2100.

Page 125: The number of wells required is based on a 4 Mt/year sequestration project with injection rates equal to the production rates reported for the McElmo Dome natural CO_2 field.[45] Injection rates at

Table 0.1. Underground storage capacity for carbon dioxide (all figures in Gt)

Potential storage space (Gt CO$_2$)	Source							
	Baines and Worden (1995–1999)[46]	IEA (2000)	IPCC (2002)	World Coal Institute (2005)[47]	IPCC (2005)[48]	Global Energy Technology Strategy Program (2006)[49]	Sonde[50]	IEA (2008)[51]
EOR					61–160			
EGR								
Depleted oilfields	40–190	600–1,000	120	126	65–360	120		657–1,200
Depleted gasfields	140–310	200–700	690	800	1,040	700		
ECBM and unminable coal seams	5–40	200–3,000	40	150	3–200	140		148
Saline aquifers	87–2,700	400–9,000	400–10,000	400–10,000	1,000–10,000	9,500		2,000–20,000
Basalt					0	>240[52]	319–435	200–400[53]
Total	272–3,240	1,00–13,700	1,250–10,850	1,476–11,076	1,678–11,100	10,700+		3,005–21,748

Sleipner are more than twice as high (in fact, just one injection well is sufficient for the whole project).

Page 127: Considering the volume of available seals, mud-rocks (including shales) form about 80% of all sedimentary rocks, but are less well-represented on the continents and surrounding continental shelves, since they are deposited preferentially in the deep ocean.

Page 127: This graph is drawn from my own calculations, assuming 0.1% annual leakage of cumulative trapped volumes for each seal, injection period twenty years, leakage starting from each seal ten years after the beginning of injection (or the beginning of leakage from the underlying seal, as appropriate), and that each formation traps 30% of the gas in solution or as residual droplets, so that only 70% is available to leak.

Chapter 5: Scale, Costs and Economics

Page 156: For the potential of CCS by 2030, Stern suggests 1.4 Gt annually,[54] the IEA 4 Gt[55] and McKinsey/Vattenfall 3.5 Gt[56] (compared to 28.4 Gt of global emissions in 2006). Comparison with wind power assumes 2.3 MW turbines.

Page 159: The graph of CCS potential has been compiled from various sources.[57] Note that power generation with petroleum coke is not the actual generation, but the potential if all pet coke produced were used in IGCC power stations, since this has been identified as a particularly low-cost option. The assumption of 35% capture of oil refining emissions is given by the IEA,[58] but may be too low, since the Mongstad refinery in Norway is planning for 80% capture,[59] albeit expensively.

Page 160: These figures bear in mind that the IPCC scenarios have widely varying levels of 'business-as-usual' emissions, which therefore require different numbers of wedges for the same climate target. About 200 of these typical power stations are equal to 100 GW. In the IPCC scenario with very high fossil fuel use, then the rate of fossil fuel power station building accelerates and more than 200 new coal stations are built annually, so many but not all new plants would have CCS.

Page 160: Some of the studies on the potential contribution of CCS in 2050 are summarised in Table 0.2.

Table 0.2. Contribution of CCS to total emissions reductions by 2050

Source	Scenario	Role of CCS in 2050	
		% of emissions reductions	CO_2 captured (gigatonnes)
International Energy Agency[60]	Moderate action on climate change[61]	14%	5.1
	Strong action on climate change[62]	19%	10.4
Intergovernmental Panel on Climate Change (technical potential)	High fossil fuel[63]	56% (of total emissions)	37.5
	Low fossil fuel[64]	21% (of total emissions)	4.7
Akimoto and Tomoda[65]		11	
Bellona[66]		16	

Page 161: Comparing the total build-up to the proposed route for EU adoption, about half of the 2020 total, 100 Mt/year, would be in European projects, with the bulk of the remainder in the USA, Canada and Australia. This rate of progress could be maintained so that by 2030 about 1/3 of a Socolow wedge would be in the EU.[67]

The comparison of carbon mitigation by solar power and carbon capture assumes solar power replaces an inefficient coal plant with emissions of 963 grams CO_2/kWh. Solar statistics are from 1995–2006[68] and 2007–8,[69] assuming that the 2007 and 2008 load factor remained the same as in 2006. The assumption that solar replaces coal-fired generation is the most favourable assumption possible for solar power. In practice, it probably usually displaces gas-fired plants and perhaps covers for some nuclear phase-out, since coal is used as baseload.

Page 164: For a US comparison of oilfield water to CO_2 injection, produced water re-injection amounts to some 49 million barrels per day in the USA alone (compared to oil production of 6.7 million barrels). This is the equivalent of 2 Gt of CO_2 injected annually—a third of all US emissions.

Page 166: For the possible cost reductions from the learning effect, if we take the lower projections of the IEA and IPCC, which have about 5 Gt of CO_2 being captured in 2050, and we assume the first generation of plants comprises 50 Mt/year, then with learning factors of 12–20% per doubling of capacity, costs would decline dramatically by 2050, to just 20–40% of the initial installations.

Page 167: Capture costs quoted in this section are taken from a range of sources, particularly the IEA and IPCC reports.[70] Many of these studies pre-date the great run-up in engineering and materials costs that occurred particularly in 2006–8. On the other hand, recent studies[71] suggesting higher costs now appear unduly pessimistic, with the end of the economic boom. I have therefore chosen to quote estimates as they are, recognising that they may be somewhat biased to the low side. This seems better than an ad hoc adjustment, which is very difficult to do fairly due to the lack of information given about the assumptions underlying many estimates. The costs for renewable energy and other carbon mitigation options should be similarly understated, so the comparison is reasonably fair.

Page 169: On the decision as to what size of plant to build, consider the example of a 500 MW power station without capture. Compare it to a similar plant with a CCS system that reduces its efficiency from 40% to 30%, and will now produce only 375 MW. Instead of building only a 375 MW plant, we would enlarge it so that the available electricity for sale is still 500 MW. The capture cost per tonne for a 500 MW plant is about 10% less than for a 375 MW plant.

Page 172: On the subject of partial capture of carbon dioxide, although most systems plan to catch 85–95%, some plants have been designed which would capture only 50% of emissions, with the aim of reducing costs. The lower capture rate can reduce aggregate capital costs by 30–35%, and brings down the plant's emissions to the same levels as a natural gas power station. Some calculations suggest, though, that the cost per tonne for post-combustion capture decreases when higher capture rates are chosen,[72] and that the 'partial capture' option is therefore not attractive.[73] This seems entirely understandable when we remember the economies of scale of CCS kit.

For IGCCs, though, the situation may be different. Even for a 'non-capture' IGCC, capture of 25% of CO_2 is straightforward; for capture IGCCs, 50–80% can be achieved with a single stage, so costs rise sharply for capture percentages higher than this. The gas turbine is also up to 10% more efficient without full capture.

Particularly for post-combustion retro-fits, partial capture may be more effective, as, amongst other reasons, it can be tailored more closely to the capacity of the existing steam turbine,[74] and reduces cooling water use. It may be possible to eliminate one of the two 'trains' of capture equipment. The option of running in 'non-capture' mode is also valuable, since it allows boosting output (at the cost of higher carbon dioxide emissions) at times of peak electricity demand. It can also allow the plant to continue running even if the capture equipment is experiencing problems, as may well happen with first-generation installations. Partial capture is more compatible with current equipment designs and plant operating procedures, increasing reliability.

When these factors are taken into account, it may be that costs per tonne do not decrease much beyond 60% capture. If this is the case, it could be equally effective to fit partial capture to all coal plants, as to fit 'full' (actually about 90%) capture to two-thirds of plants. The 'partial capture' option also has much lower total capital costs (even if the cost per tonne avoided is higher), so it may be more suited for demonstration projects. More different systems and approaches could be tried out and more experience gained, for the same total bill, before moving on to full capture for second-generation CCS. Partial capture might also be appropriate for plants which have only limited geological storage capacity nearby. However, we would have to be careful that building or retro-fitting partial capture on a given plant today would not prevent our moving to full capture later on. It might also present public relations problems, since CCS opponents would point out that the plant is still emitting large amounts of carbon dioxide.

Page 172: On the effect of plant utilisation on carbon capture costs, in one study reducing the utilisation of a gas power station from 75% to 50% increased the cost of CO_2 avoided from \$49 to \$68/tonne.[75] The lower cost is competitive with carbon capture from coal plants, but the higher cost is not.

Page 172: Costs for industrial capture are shown in Figure 0.1. On the costs for capture from oil refining, note that estimates for capturing

Figure 0.1. Carbon dioxide capture costs from industrial sources[76]

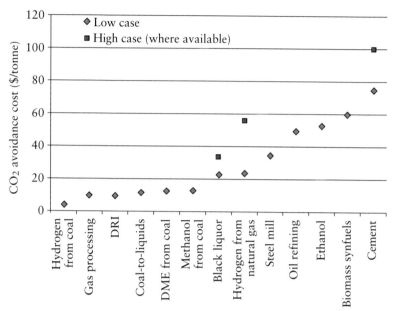

80% of emissions from the Mongstad refinery in Norway were much higher ($185–255/tonne CO_2), albeit these probably reflect the engineering and materials cost spike of 2008; were aiming at a high capture proportion; and were using chilled ammonia, a new technology.[77]

Page 173: Figure 0.2 shows how the cost of electricity generation changes when applying carbon capture to a variety of plant types. Coal without capture is shown in red and with capture in dark green; advanced coal without capture is shown in dark red and with capture in olive green; biomass and gas without capture is shown in orange and with capture in blue.

Page 176: On growing petroleum coke consumption, the increase was 4% per year between 1990 and 2006.[78] Oil use grew 1.4% annually and coal use 1.95% annually over the same period.[79] In addition to increasing demand for light refined products, petroleum coke production was driven by the growing need to refine heavy oils.

Page 177: A recent report,[80] albeit using top-of-the-market costs from 2008, and probably therefore pessimistic, suggests that first-of-a-kind

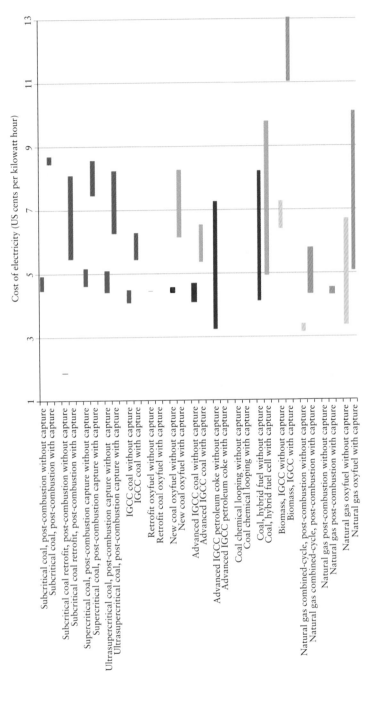

Figure 0.2. Cost of electricity from various power plant types, including present-day and advanced systems, with and without capture

plants will have a carbon dioxide avoidance cost of around \$120–180/tonne and generate electricity at 8–12 ¢/kWh more than a conventional plant.[81]

Allowing for recent falls in input costs as the economic boom evaporates, abatement costs would fall to \$110/tonne for the first plant.

Another recent report[82] suggests costs of carbon dioxide avoidance for first-generation plants around \$87–91/tonne for post-combustion capture on pulverised coal, \$62–70 for oxyfuel coal, \$61 for IGCC and \$112 for natural gas. Cost reductions for later plants appear surprisingly small.

These costs may be pessimistic: costs for capture from a small gas turbine on Alaska's North Slope, remote, with difficult weather-windows, and given that capture from natural gas is generally expected to be more costly than coal, gave \$130/tonne.

Note, though, that the final cost gap shown in my chart of about 1.5 ¢/kWh between CCS-equipped IGCC and non-CCS pulverised coal may be too optimistic; the study from which I take the learning factors has a gap of 3 ¢/kWh, albeit this incorporates the probably transitory high construction costs of 2008.

The learning curves shown assume that installed IGCC capture capacity doubles every four years between now and 2030.[83]

Page 180: For storage costs, as much as 90% of European storage capacity may be available for less than \$2/tonne, while in Australia all onshore sites studied and half of offshore sites cost less than \$5.1/tonne. The cost of the In Salah project, including capture and \$2/tonne of monitoring, was only \$6/tonne. In Salah is onshore, but was probably relatively expensive due to the low reservoir permeability that necessitated drilling several injection wells.

Page 181: In the absence of recent public-domain economic calculations for a CO_2-EOR project with today's oil prices, I have calculated them myself. Figure 0.3 shows in more detail the cashflows from a typical onshore CO_2-EOR project, based mainly on US and Canadian (Weyburn) experience. The prototypical project here captures about 1 Mt/year on average (excluding recycled carbon dioxide), and recovers some 90 million barrels of incremental oil. The main costs are the up-front capital (mostly for injection and production wells) and for the high levels of initial carbon dioxide. This assumes some level of re-use of existing production wells and facilities.

Production costs and profiles are from public sources,[84] scaled to give a 1 Mt/year average injection rate. The number of injection wells is scaled from the Weyburn field example. The oil price assumed is $50/bbl. The figures include transport, site selection and monitoring costs but not capture costs, nor the value of deferring field abandonment or extending non-EOR production.

Note that royalties and taxes have been ignored here. Under a well-designed petroleum fiscal system (such as Norway or the UK), a project that is profitable pre-tax should also be profitable post-tax.[85] Taxes and royalties are at the discretion of the government (and, in the USA, the landowner) and can be waived if required to ensure a project goes ahead.

The carbon dioxide price has been set to the minimum to make the EOR project break even,[86] which in this case, at an oil price of $50 per barrel, occurs at $35 per tonne, fairly consistent with Figure 5.10.

Page 186: The estimated increases in the costs of various products due to adding carbon capture are shown in Figure 0.4. These are calculated

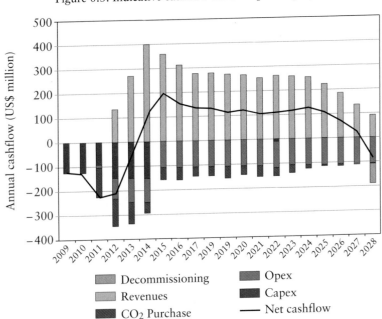

Figure 0.3. Indicative cashflow for a CO_2-EOR project

as follows. Capture costs are mostly from the IPCC.[87] I have added $2/tonne CO_2 transport costs and $2/tonne CO_2 storage costs to the capture costs given, except for natural gas processing where I have assumed reinjection in a nearby formation and hence zero transportation costs. Prices assumed (without CCS): $5/MMBtu for natural gas, $50/bbl average refined oil product price, $700/tonne steel price, $800/tonne for uncoated super-calendered mechanical paper, $1.65/gallon for ethanol. I assume no carbon cost for non-captured emissions. Some cost increases will be somewhat overstated, because only costs per tonne of carbon avoided, not per tonne captured, were available.

Page 188: The costs of enhanced electrochemical weathering are my own calculations, assuming $0.07/kWh for carbon-free electricity; 10% required rate of return; $275 million spent over two years for a 0.5 Mt CO_2/year system, based on chlor-alkali plant costings; energy use 821 kWh/tonne (the low end of available figures[88]); other operating costs 2% of capital costs annually; thirty-year plant life.

For the cost of air capture, I take 2500 kWh/tonne CO_2 as the typical energy requirement.

Page 190: On the potential of biomass to co-fire with coal: global biomass availability is 47–450 exajoules (EJ) per year;[89] global coal consumption was 3,177.5 million tonnes oil equivalent in 2007,[90] equivalent to 133 EJ.

Page 193: The figure of 75 Mt corrects earlier, higher estimates for emissions from underground coal fires. However, it includes waste incineration of fossil carbon such as plastic, so is itself an overestimate of carbon dioxide from coal fires only.[91]

The estimate for global costs to put out these fires given in the text has been made from US figures.[92] This assumes that US coal fires are 20% of the total; the costs for extinction have not changed since the date of the estimate (1998), i.e. that inflation is offset by technological advances in fire-fighting and possibly lower costs in developing countries; and that the 75 Mt of emissions would otherwise continue over the next 100 years. Costs and emissions are discounted at 5%.

Page 193: The total biochar potential is calculated assuming planting the top 20 cm of land with biochar making up 10% of the soil, the biochar having a density of 0.4 g/cm^3, i.e. 30,000 tonnes/km^2. This is significantly less rich than *terra preta* soils, which can contain up to

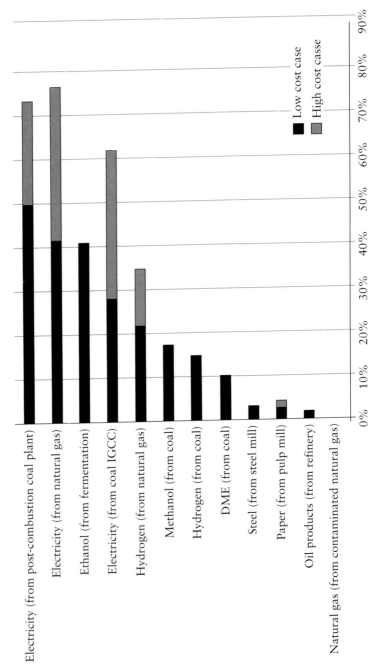

Figure 0.4. Increase in the cost of various products with CCS (including capture, transportation and storage cost)

180,000 tonnes/km².[93] Global arable land is assumed to cover 13.6 million km².

Figures for total biochar generation assume that global arable land on average resembles the USA. For crop residues, global arable land of 13.6 million km² is used with figures indicating about 587 Gt CO_2-equivalent from 1.2 million km² of US cropland.[94] For forestry, about 13.2 million km² of forest have the primary function of production,[95] while 587 Gt CO_2-equivalent is achievable from 2 million km² of US forests.[96]

Page 194: For the potential of conservation tillage, figures assume 1,100 tonnes CO_2-equivalent sequestered per km².[97] Estimates of a much higher potential of 110–220 Gt,[98] using a similar amount of carbon per km², presumably assume soil carbon sequestration over a much larger area of land.

Page 197: Table 0.3 shows the increase in various product prices due to a carbon tax. These are calculated as follows. Industrial emissions factors are derived from various sources,[99] including 0.87 tonnes CO_2/tonne cement; 8 tonnes CO_2/tonne aluminium; 1.75 tonnes CO_2/tonne steel. The ammonia emissions factor assumes no carbon sequestration or generation of urea. Average prices are then determined from various other sources.[100] All are historical average prices in Money of the Day (MOD), i.e. not inflation-adjusted.

Page 196: Carbon tax levels are from the Carbon Tax Center.[101] EU carbon permit predictions are from Point Carbon,[102] assuming €1=US$1.4; USA cap-and-trade proposals are a mid-case estimate by MIT. US figures for 2010–14 are derived by reducing 2015 figure at 4% annually.[103] CCS implementation levels run from $34/tonne CO_2 (discussed in text) to $55/tonne.[104] CCS costs have been scaled using learning curve figures.[105]

Page 200: The chosen emissions path yields atmospheric stabilisation around 500 ppm. Business-as-usual emissions have been taken from the IPCC's A1 MESSAGE scenario. Forestation, soil use and wood products sequestration are assumed to take place due to their relatively low costs and side-benefits. Agricultural potential is taken from the literature[106] and CCS implementation to 2050 from Bellona.[107] Beyond 2050, it is assumed that total CCS can continue growing at no more than the same annual capacity additions from 2040 to 2050.

Table 0.3. Increases in product costs caused by a carbon tax

Product	Increased cost due to $30/tonne carbon tax or CCS cost	Average cost 2000–2008	Price increase caused by carbon tax/CCS cost
Gasoline/petrol	$0.275 per US gallon	$2.06	13%
Crude oil	$13.20 per barrel	$48.26	27%
Coal	$71.50 per US ton	$57.31	125%
Natural gas	$1.65 per thousand cubic feet	$6.05	27%
Electricity from coal	2.42 US ¢/kWh	US¢ 8.03	30%
Electricity from natural gas	1.1 US ¢/kWh	US¢ 8.03	14%
Ammonia	$98/tonne	$250	40%
Cement	$26/tonne	$80	33%
Aluminium	$240/tonne	$2000	12%
Steel	$52.5/tonne	$600	9%

Maximum potential of CCS assumes capturing 90% of plant emissions, applied to up to 99% by volume of large stationary sources. Figures of 5% of oil use, 50% of gas use, 80% of coal use and 10% of biomass use are assumed to be due to large stationary sources. The energy penalty of CCS begins at 20% and drops to 15% by 2070. Biochar grows from 0.2 Gt/year currently to 1.2 Gt/year by 2050 (20% of estimated technical potential, allowing for 20% biochar decomposition and 30% losses during spreading) and remains flat thereafter. Air capture starts in 2030 and builds up to offset 100% of aviation (allowing for contrail effects) by 2100, assuming 4% annual emissions growth from 2004.

Page 201: The future global GDP estimate is made using the various IPCC scenarios, and inflating their GDP figures from 1990 prices to 2008 assuming 2.5% annual inflation. I have quoted market exchange rates since Purchasing Power Parity (PPP) estimates are not given for all cases. Since PPP estimates are generally higher, this approach somewhat overstates climate protection costs as a percentage of GDP. The 2008 global GDP estimate is from the International Monetary Fund.[108]

Chapter 6: Policy

Page 210: The estimate for the varying cost of meeting climate targets to the USA with and without CCS assumes that CCS is the marginal emissions reduction technology and so carbon costs converge to the lower of the CCS cost and the estimated MIT path for carbon permits. US emissions in 2010 are taken as 5.75 Gt CO_2 and fall at 1.3% annually thereafter (in line with targets to 2020 under the proposed cap-and-trade system).

Page 211: The chart showing the effect of a government programme on CCS costs assumes continued learning beyond 2030. Since this will probably be at least partly offset by a need to move to higher-cost CCS projects as the 'easy' opportunities are exhausted, I have reduced the cost decline post-2030. This is counterbalanced, though, by the fact that delaying implementation will lose the chance to make use of many valued-added EOR opportunities, and to replace ageing inefficient power stations with new, efficient CCS units.

Page 240: For the estimate of Canadian coal-fired generation, 467 billion kWh of electricity were generated in Canada in 2006, of which 16.5% came from coal.[109] Assuming 85% utilisation, this gives installed capacity of 11.2 GW.

NOTES

INTRODUCTION

1. Cascio, 2009.
2. International Energy Agency, 2008.

1. THE NEED TO CAPTURE CARBON

1. Quoted in MacKay, 2009.
2. Quoted in Chaffin, 2009.
3. For instance, Monbiot, 2007; Walker and King, 2009; Kunzig and Broecker, 2008; Gore, 2006.
4. E.g. International Climate Change Taskforce, January 2005.
5. See Appendix note to this page 277.
6. Infrared light is an invisible part of the light spectrum, with longer wavelengths than visible light, noted for carrying heat from matter around room temperature.
7. Carbon dioxide was about 280 parts per million (ppm) in pre-industrial times, methane 700 parts per billion (ppb), nitrous oxide 270 ppb, and the various industrial greenhouse gases zero. The 0.03% refers to dry air; the water content of the atmosphere (humidity) is highly variable in time and space, but averages 0.4%.
8. Socolow and Lam, 2007.
9. The other greenhouse gases have also increased; adding them gives an effective carbon dioxide content of 430 ppm. Walker and King, 2009; The Royal Society, 2009.
10. Kunzig and Broecker, 2008.
11. Buchwitz, Schneissing, Burrows, Bovensmann, Reuter and Notholt, 2007.
12. From a graph in Rennie, 2009.
13. Tripati, Roberts and Eagle, 2009.
14. CFCs are also responsible for the 'hole in the ozone layer', an environmental problem entirely separate from global warming

15. A gigatonne (Gt) is 1 billion metric tonnes. Each tonne is 2,205 pounds, and 1 cubic metre of water has a mass of 1 tonne. I use tonnes of CO_2 and $/tonne CO_2 throughout, while many other sources use tonnes of carbon (C). For stocks such as solid carbon in soils, I have still converted them into CO_2-equivalent, as if they were burned. So 12 tonnes of carbon give 44 tonnes of CO_2 when burned, i.e. 1 tonne C = 3.67 tonnes CO_2; $1/ tonne C = $0.27/tonne CO_2. A pyramidal mountain of rock 1 km high weighs about 1 Gt.
16. Edgar, 2009; World Resources Institute, 2009. 1970 figure excludes international bunkers (international shipping and aviation). Figures include land-use change (2005 figure includes land-use change from 2000, the last year for which data is available).
17. Socolow & Lam, 2007.
18. Gerlach, 1991.
19. Lovelock, 2009; Gribbin and Gribbin, 2009.
20. See Appendix note to this page.
21. MacKay, 2009.
22. Italics in the original. 'Very likely' is given a precise definition, as meaning more than 90% probable.
23. Intergovernmental Panel on Climate Change, 2007.
24. However, deliberate creation of sulphate aerosols as a way of reducing global warming is covered in Chapter 4.
25. Viner and Jones, 2000.
26. E.g. Lockwood and Fröhlich, 2008; Lockwood and Fröhlich, 'Recent oppositely directed trends in solar climate forcings and the global mean surface air temperature', 2007; Lockwood, 'Recent changes in solar outputs and the global mean surface temperature. III. Analysis of contributions to global mean air surface temperature rise', 2008.
27. Lomborg B., 2007.
28. E.g. Shenker, 2009.
29. Jolly, 2009.
30. Black, 'Forest fires—a continuing Greek tragedy?', 2009.
31. Though Bjørn Lomborg makes the point that this has to be set against the reduced mortality due to warmer winters. Kunzig and Broecker, 2008. Albeit, this may be at least partially offset by fewer deaths due to cold weather (Lomborg B., 2007).
32. Black, 'Climate "meltdown", yet fusion lags', 2009.
33. Though Tacoli, 2009 argues that these fears are misplaced.
34. Bamber, Riva, Vermeersen and LeBrocq, 2009.
35. Schuur et al., 2008.
36. Burns, 2009.
37. Jones, Cox and Huntingford, 2005; Cox, Jones and Huntingford, 2005.
38. Clement, Burgman and Norris, 2009.
39. Kunzig and Broecker, 2008.

40. Positive in the sense that it amplifies the original warming, not in the sense of being a good thing.
41. Ibid.
42. Black, 2009.
43. Challenor, Hankin and Marsh, 2005.
44. Khadka, 2009.
45. Walker and King, 2009.
46. Ibid.; (Dansgaard, White and Johnsen, 1989).
47. Lovelock, 2009; (Lenton et al., 2008).
48. Of course, some politicians and commentators and a minority of scientists continue to dispute some of these basic points.
49. E.g. Lewandowski, 2006; Michaels, 2009; *The New York Times*, 2009.
50. Chand, 2009.
51. E.g. Stern, 2007; Robinson, April 2008.
52. International Climate Change Taskforce, Jan. 2005.
53. The Council of the European Union, 11 March 2005.
54. Meinshausen, 'On the risk of overshooting 2°C', 2005.
55. Allen M. R., 'Atmosphere: A Reappraisal of Climate Sensitivity', 2007.
56. Of which CO_2 represents 450 ppm. The CO_2-equivalent is reached by adjusting the other greenhouse gases by their relative global warming potential. This is somewhat complicated, since the lifetime of these gases in the atmosphere may be shorter or longer than that of carbon dioxide. Walker and King, 2009; Wang and Watson, 2009.
57. E.g. Meinshausen et al., 2009; Allen et al., 2009.
58. See Appendix note to this page.
59. E.g. Stern, 2007; Lomborg, 2001; Dasgupta, 2006.
60. Garnaut, 2008.
61. Weitzman, 2008.
62. This 'social discount rate' is built up from other parameters, one of which expresses society's level of aversion to inequality, whether between individuals or generations.
63. Stern, 2007. Critics of the choice of discount rate include Dasgupta, 2006; Lomborg, 'Stern Review: the dodgy numbers behind the latest warming scare', 2006; and Nordhaus, 2007. Defenders (broadly speaking) include Quiggin, 2006. Arrow, 2007 thinks the discount rate is wrong but that Stern's conclusion is correct.
64. E.g. Monbiot, 2007.
65. Lomborg B., 2007.
66. E.g. Stern, 2007; Pielke R. J., 2007.
67. Lomborg B., 2007.
68. Mendelsohn and Dinar, 2009.
69. Black, 'Climate protection "to cost more"', 2009.
70. Discussed by e.g. Quiggin, 2006, although he does not consider this fatal to Stern's results. As it's obligatory these days to mention 'black swans',

I should also point out that Taleb, 2008 is very sceptical of long-term forecasts, with some justification.

71. Kalkbekken and Rive, 2005; Socolow and Lam, 2007.

72. Allen M. et al., 2009.

73. This estimate of damage, though, remains highly controversial. See the Appendix note to this page. Strictly speaking, too, we should speak of 'Gross World Product' (GWP) rather than GDP; I retain the more familiar term.

74. Central Intelligence Agency, 2009.

75. Including land-use changes, primarily deforestation; excluding this reduces the figure slightly, to 5.6 tonnes.

76. Prosser, 2006. Excluding additional climate impacts of the flight due to high-altitude vapour trails.

77. Stern, 2007. Year 2000 prices.

78. At a 3% discount rate.

79. Teske, 2008.

80. Lomborg B., 2007; Tol, 2009.

81. Energy sources which are not depleted by their use. Generally overlaps with 'green energy', although renewable energy sources are not necessarily environmentally friendly, and nuclear power, not generally classed as a renewable, has minimal carbon dioxide emissions. The main renewable energy technologies are hydroelectric (energy from moving/falling water, usually by damming rivers); wind; solar (photovoltaic, which makes electricity directly, and thermal, using the sun's rays to heat a liquid); biomass/biofuels (burning plant matter for power, or converting it to liquid fuels such as ethanol); geothermal (using underground heat); and ocean (tidal, wave and currents).

82. As opposed to 'low-hanging fruit' efficiency improvements which, their proponents argue, can be implemented at low or negative cost. Biomass is probably the one exception to the 'high capital cost' rule amongst renewable energies: it has relatively lower capital costs, but high ongoing operating costs.

83. Hydroelectric is competitive too, but most good sites are already taken, and damming rivers causes environmental damage.

84. Mills, 2008.

85. Slaughter, 2007; Heathcote and Fryer, 2008.

86. And perhaps by the Chinese around 1000 BC. World Coal Institute, May 2005.

87. For dissenting views, see Appendix note to this page. E.g. Gibbins, 2008.

88. Nakićenović et al., 2000.

89. Schmidt, Rempel and Schwarz-Schampera, 2008.

90. Ibid.; Ahlbrandt, Charpentier, Klett, Schmoker and Schenk, 2000; and BP, 2008.

91. Schmidt, Rempel and Schwarz-Schampera, 2008; Ahlbrandt, Charpentier, Klett, Schmoker and Schenk, 2000.

92. Using cumulative carbon emissions from Nakićenović et al., 2000. Result is within 6% of that obtained using assumptions in the Appendix note to this page.
93. World Commission on Environment and Development, 1987.
94. As MacKay, 2009 does, in his otherwise excellent book.
95. E.g. Keith, Ha-Duong and Stolaroff, 2005 suggest that the typical inertia of the global energy system is about fifty years.
96. This may seem inconsistent with the much longer timescales required for climate policy. That, though, is a recognition of the protracted periods over which climate change will progress, even if action is taken today. Of course, this does not detract from the formidable uncertainties that accumulate over the course of a century and more, and which pose a huge challenge for formulating robust and flexible policies. Similarly, for some specific cases such as disposing of nuclear waste (and CO_2), secure storage into the distant future is required.
97. Teske, 2008.
98. Note also Aune, Rosendahl and Sagen, 2009, whose modelling suggests no significant rise in long-term gas prices by 2030.
99. Edner, 2009.
100. International Energy Agency, 2006.
101. E.g. Teske, 2008.
102. Nakićenović N., 2002.
103. E.g. Hao, Dong, Yang, Xu and Li, 2007; Hawksley, 2009.
104. Ansolabehere et al., 2007.
105. E.g. Sheffield et al., 2004.
106. Dewees, 2008.
107. E.g. Hao, Dong, Yang, Xu and Li, 2007.
108. E.g. Landais, 2009.
109. Organisation of Petroleum Exporting Countries.
110. Sachs, 2009.
111. Bullis, 2009.
112. Dunnahoe, 2009.
113. MacKay, *Sustainable Energy—Without the Hot Air*, 2009.
114. McCracken, 2Q 2009.
115. Although the UK's grid operator, National Grid, is fairly confident that it could manage about 30% wind generation by 2020 (National Grid, 2009).
116. Lever, 2007.
117. McKillop, 2005.
118. Reuters, 2005.
119. Around €15/tonne (about $21/tonne) of CO_2 in the EU today; zero in most other parts of the world.
120. Dooley et al., 2006.
121. Calling the EU a 'country', for want of a better term, since it does have a common climate policy.

122. Figures for 2008, from BP, 2009.
123. E.g. Rochon, 2008.
124. Quotations Page.
125. Worrell, Price, Hendricks and Meida, 2001.
126. Dooley et al., 2006.
127. E.g. Lovelock, 2009.
128. For instance, Monbiot, 2007; Teske, 2008; Goodall, 2008.
129. E.g. Teske, 2008; Sheffield et al., 2004.
130. McCracken, 2Q 2009
131. REN21, 2009. Assuming 25% capacity factor gives 35 billion kWh generated in 2008. This compares to about 17,700 billion kWh of electricity consumption from all sources (extrapolating 2006 figure from Energy Information Administration, 2008).
132. Including combined heat and power (CHP) in total generation. If it is excluded, the share of PV in generation is even higher, 30%.
133. Allowing for the fact that solar only operates at 25% or less of peak capacity, since the sun does not shine all the time.
134. Mills A., 2009.
135. International Energy Agency, 2006.
136. Wind turbine shortage continues; costs rising, 2007.
137. Lex, 2009.
138. Goodall, 2008.
139. A generous assumption, since at least some was probably replacing gas during a period of high prices, and supporting Germany's phasing-out of nuclear.
140. Assuming $1.4: €1 (approximately the exchange rate at the time of writing; it was both higher and lower during the 2000–10 period); typical coal carbon intensity of 0.97 kg CO_2/kWh; solar installation has produced 1.5×10^{11} kWh.
141. Even given that In Salah, using 'contaminated' gas and with a nearby aquifer, was unusually cheap.
142. Morriss, Bogart, Dorchak and Meiners, 2009.
143. See, for instance, Dikeman, 2009 for solar, and Bryce, 2005 for corn ethanol.
144. UK Department of Trade and Industry, 2003. I have used £1.5: $1 for currency conversion.
145. A point well made by Nakićenović, 2006.
146. Mark Johnston, WWF analyst, quoted in Chaffin, 2009.
147. To be fair, carbon dioxide emissions from these flights are fairly modest. It is about 40 km from the centre of the wind farm to the nearest airport near Esbjerg. Assuming that all flights were made with Uni-Fly A/S's new Eurocopter EC135, which consumes 205 litres of kerosene per hour at its economical cruise speed of 229 km per hour, fuel consumption per flight would be 36 litres (ignoring take-off, landing and hover while winching the worker down to the turbine). Each litre of kerosene emits

about 2.6 kg of CO_2 when burned, so these 75,000 trips were responsible for some 7,000 tonnes of CO_2. This is about 1% of the emissions of a coal power plant producing 72 MW over that one-and-a-half year period (Horns Rev's capacity is 160 MW and its typical capacity factor was 45%, once it was past its teething troubles).

148. Modern Power Systems, 2004; Sweet, n.d.
149. Monbiot, 2007. See also the story relating to goat deaths in Taiwan (BBC, 2009).
150. Wood smoke is the single largest contributor to winter particulate emissions in Sacramento County, California, USA (Faust, 2007).
151. Jaccard, *Sustainable Fossil Fuels*, 2005.
152. E.g. The Royal Academy of Engineering, May 2008.
153. Lima, Ramos, Bambace and Rosa, 2008. Although there are potential solutions (Project aims to extract dam methane).
154. BBC News, 2009.
155. Pearce, 'How a wind farm could emit more carbon than a coal power station', 2009. Though others point out that CO_2 emissions can be reduced by some mitigation measures (Nayak, Miller, Nolan, Smith and Smith, 2008).
156. Glanz, 2009; Kulish and Glanz, 2009.
157. AP, 2010.
158. Renner, Sweeney and Kubit, 2008.
159. Bradsher, 'Outsize Profits, and Questions, in Effort to Cut Warming Gases', 2006.
160. Fossil fuel plants, especially coal, also use a lot of water, but unlike solar power they are not generally located in deserts
161. Woody, 2009.
162. Admittedly, some oil and gas installations may also be vulnerable to rising sea levels and hurricanes.
163. Agrawala, Raksakulthai, van Aalst, Larsen, Smith and Reynolds, 2003.
164. Borenstein, 2009.
165. MacKay (*Sustainable Energy—Without the Hot Air, 2009*) stresses this point particularly.
166. Wood, 2009.
167. Megatonnes, Mt. The world's largest ship carries about 0.65 Mt when fully loaded, and the Great Pyramid of Giza weighs about 6 Mt.
168. For calculations, see the Appendix note to this page.
169. If the Pentagon's power use per employee is similar to other US commercial buildings (instead of being filled with energy-hungry secret weapons), it probably consumes about 30 MW.
170. Fry, 2009.
171. Nugent, 25 April 2009.
172. Obviously since the Earth is finite, so are ultimate resources of carbon fuels and other minerals (although metals, at least, can maintain close to steady-state consumption by recycling). The question is whether we are

likely to approach the physical limits of these resources within a practical time for long-term policy-making, say 50–100 years.

173. E.g. Snow, 2000.
174. Allison, 2007.
175. Metals Place, 2009.
176. E.g. Tahil, 2006. Bolivia, holding 50% of world reserves, is far from friendly to foreign industries who might wish to extract lithium (Day, 2009). Evans (March 2008) is more optimistic.
177. Salameh, 2Q 2009.
178. Osborne (2009) comments on this volatility for indium, another component of some solar cells.
179. Latimer, Kim, Tahara-Stubbs and Wang, 2009; Reuters, 2009; Bradsher, 'Earth-Friendly Elements, Mined Destructively', 2009.
180. E.g. Rochon, 2008.
181. Assuming that the extraction is done in an environmentally acceptable way and that the resulting CO_2 is stored safely.
182. Perhaps 'OSEC', the Organisation of Solar Exporting Countries?
183. E.g. Lucas, 2009.
184. Morriss, Bogart, Dorchak and Meiners, 2009.
185. Maddison, 2009.
186. *Ceteris paribus*, a good thing, since one of the key goals of climate policy is to help poorer countries; but likely to be opposed by anti-globalisation activists, organised labour, and protectionist groups.
187. For instance Teske, 2008 and Monbiot, 2007, which put stress on imports to the EU of North African solar power; and on the development of a grid for North Sea wind.
188. Disclaimer: I have no involvement with the nuclear industry, and the existing literature, on both sides of the nuclear debate, is so contradictory and often self-interested that it is hard even for specialists to reach firm conclusions.
189. For example, Broomby, 2009.
190. MacKay, 2009.
191. Lovins, Datta, Bustnes, Koomey and Glasgow, 2004.
192. Maksoud, 2009. Compared to Greenpeace's study, where oil use has to fall by 12% between 2010 and2020 (Teske, 2008), it rises by 7% in that period in McKinsey's study, even under the assumptions of a severe economic downturn combined with the most optimistic assumptions on energy efficiency and electric cars (McKinsey, 2009).
193. Forecasts used as a baseline, assuming no climate change policies.
194. Vattenfall, 2007.
195. E.g. Teske, 2008; Lovins, Datta, Bustnes, Koomey and Glasgow, 2004; Bressand et al., 2007.
196. Jaccard, *Sustainable Fossil Fuels*, 2005; (Jaccard, Rivers, Bataille, Murphy, Nyboer and Sadownik, May 2006; Gillingham, Newell and Palmer, 2009.

197. E.g. Jaccard, 2005.
198. Lomborg B., 2007.
199. Herring, April 1998.
200. Morriss, Bogart, Dorchak and Meiners, 2009.
201. Gustafson, 2006.
202. Jaccard, Rivers, Bataille, Murphy, Nyboer and Sadownik, May 2006.
203. Or, in Houston or Dubai, sweating in the dark.
204. Edmonds, 2005.
205. Pacala and Socolow, 2004.
206. Socolow R., 'Stabilization Wedges: Mitigation Tools for the Next Half Century', 2006.
207. Romm, 2008. Socolow himself covers the revision of the 'wedges' due to continuing emissions growth: Socolow and Lam, 2007.
208. Each the size of Sleipner in Norway, 1 Mt/year, a rather small project by the standards of likely future CCS.
209. Teske, 2008.
210. International Energy Agency, 2008.
211. Assuming a scheme of the size of BP's now-defunct Peterhead/Miller project, 1.3 Mt CO_2 per year.
212. E.g. Monbiot, *Heat: How We can Stop the Planet Burning*, 2007.
213. As illustrated by the cynicism of Pearce, 'Greenwash: Why "clean coal" is the ultimate climate change oxymoron', 2009.

2. CAPTURE TECHNOLOGY

1. HTC Purenergy, 2009.
2. Quoted in Chaffin, 2009.
3. International Energy Agency, 2008.
4. Converting one kind of fuel into another: mostly oil refineries, gas processing plants, synthetic fuels facilities and hydrogen plants
5. More than 100,000 tonnes (0.1 Mt) of CO_2 per year.
6. E.g. Dahowski, Davidson, Dooley and Gentile, 2008; International Energy Agency, 2008.
7. Dooley et al., 2006.
8. Kohberg, 2009.
9. Of course, this carbon dioxide was extracted from the atmosphere by vegetation in the first place.
10. Quoting efficiencies as 'higher heating value' (HHV).
11. Bohm, Herzog, Parsons and Sekar, 2007.
12. Supercritical carbon dioxide has the properties both of a liquid and a gas. It has the density of a liquid but the viscosity of a gas and, like a gas, expands to fill a container.
13. As does oil extraction, especially heavy oil, but oil use in power stations is small.
14. MacKay, 2009.

15. Though this can be reduced, for instance by Coal Mine Methane (CMM) projects.
16. Kohberg, 2009.
17. Largely based on Rubin, 2008.
18. Amines are a class of organic (carbon-bearing) chemicals containing a nitrogen atom.
19. Herzog, 'A Research Program for Promising Retrofit Technologies', 2009.
20. Page, Williamson and Mason, 2009.
21. International Energy Agency, 2008.
22. OilVoice, 2009.
23. Christiansen; Hetland and Christensen, 2008; Bryngelsson and Westermark, 2009.
24. Nord, Kothandaraman, Herzog, McRae and Bolland, 2008.
25. Klemeš, Cockerill, Bulatov, Shackley and Gough, 2006.
26. Banerjee et al., 2008.
27. National Energy Technology Laboratory, 2008.
28. Feng, 2009.
29. Dortmundt and Doshi, 1999.
30. Metz, Davidson, de Coninck, Loos and Meyer, 2005.
31. Baxter, 2009.
32. Reportedly the target is to remove 20–60% of the CO_2 (Day D., 2005).
33. Johnson K., 2009.
34. Sun, 2008.
35. 50–100 km^2.
36. Salmon, Saunders and Borchert, 2009.
37. Hamilton, 'Capturing Carbon with Enzymes', 2007; Bond, Medina, Stringer and Simsek-Ege.
38. About 10%.
39. Farrell, 2005; Conn, Baum, Mudd and Gunnulfsen.
40. MIT, 2009.
41. Largely based on Rubin, 2008.
42. 95–99%.
43. Lynch, 2005.
44. Men and Women of Wabash River Energy Limited, 2000.
45. Modern Power Systems, 2008.
46. Biello, 'How Fast Can Carbon Capture and Storage Fix Climate Change?' 2009.
47. Hildebrand and Herzog, 2009.
48. Johnson, 2005; Johnson and Pleasance, 'Clean Coal Technology for Brown Coal Power Generation', 1996.
49. Kohberg, 2009; Scottish Centre for Carbon Storage, 2008.
50. Metz, Davidson, de Coninck, Loos and Meyer, 2005.
51. Klemeš, Cockerill, Bulatov, Shackley and Gough, 2006.
52. Giroudiere et al., 2009.

53. Jaccard, *Sustainable Fossil Fuels*, 2005.
54. International Energy Agency, 2008.
55. 24 GW (Sheffield et al., 2004). Electricity equivalent is if utilisation were 100%, and all gasification plants were producing electricity. Electricity generation from BP, 2008.
56. US Department of Energy/Wabash River Coal Gasification Project Joint Venture, September 2000.
57. It was operating 96% of the time. Sheffield et al., 2004.
58. After being cancelled due to cost overruns, then possibly resurrected when, remarkably, these were found to be largely due to mathematical errors.
59. Keesom, Unnasch and Moretta, 2009.
60. Albeit, for a retro-fit at the Nexen plant, there would be some significant engineering issues resulting in a loss of efficiency. These could be avoided in new plants (Ordorica-Garcia, Wong and Faltinson, 2009).
61. *Carbon Capture Journal*, 2009.
62. Krasnodebski, Ziebinski and Zys, 2009.
63. GreatPoint Energy, 2009; Biopact, 2007.
64. Or post-combustion capture.
65. National Energy Technology Laboratory, 2008.
66. A 1 GW coal plant would require a 550,000 m³/hour oxygen plant; Shell's Pearl GtL plant in Qatar has an 860,000 m³/hour ASU (Linsenmaier, 2009).
67. Largely based on Rubin, 2008.
68. Metz, Davidson, de Coninck, Loos and Meyer, 2005.
69. Herzog, 'A Research Program for Promising Retrofit Technologies', 2009.
70. Merour, 2009.
71. Klemeš, Cockerill, Bulatov, Shackley and Gough, 2006.
72. Herzog, 'A Research Program for Promising Retrofit Technologies', 2009.
73. Scottish Centre for Carbon Storage, 2008; Klemeš, Cockerill, Bulatov, Shackley and Gough, 2006.
74. Perhaps from 15–20% with current systems, to 10–14%.
75. Herzog, 'A Research Program for Promising Retrofit Technologies', 2009.
76. Appendix note to this page.
77. Vattenfall, 'Pilot Plant Schwarze Pumpe'.
78. Pronske, 2007.
79. Hurley, Kadrmas and Robson, 2005.
80. Normally 50%; 55% requires a gas boiler to be used to heat the air entering the turbine, which would make carbon capture somewhat more difficult.
81. Kohberg, 2009.
82. Grupo de Combustión y Gasificación.
83. Around 54% (Klemeš, Cockerill, Bulatov, Shackley and Gough, 2006).
84. By 2–3%.

85. Imperial College London.
86. Miracca et al., 2008.
87. International Energy Agency, 2008.
88. As described in National Energy Technology Laboratory, 2008.
89. Sundkvist, Griffin and Thorshaug, 2001.
90. Ausebel, 2002.
91. Imperial College London.
92. Imperial College London; Kohberg, 2009.
93. DirectCarbon, 2008.
94. SRI International, 2005.
95. Cooper and Berner, 2005; Steinberg and Cooper, 2002.
96. Metz, Davidson, de Coninck, Loos and Meyer, 2005.
97. Goodall, 2008.
98. Howell, 2009.
99. Wright.
100. Carvalho, 2010.
101. Mainly based on Metz, Davidson, de Coninck, Loos and Meyer, 2005, with additional information from Kohberg, 2009; Hetland and Christensen, 2008; and International Energy Agency, 2008.
102. 2–3%.
103. Albeit with the caveat that these are the company's own, possibly optimistic estimates.
104. 54%.
105. See e.g. Monbiot, 2007.
106. Gold, 2008.
107. Wasas.
108. International Energy Agency, 2008.
109. Klemeš, Cockerill, Bulatov, Shackley and Gough, 2006.
110. Upstream, 'Shell joins UK carbon race'.
111. Calculation is covered in the Appendix note to this page.
112. Gielen and Unander, March 2005; Kheshgi and Prince, 2005.
113. Blue Flint Ethanol, 2009.
114. Hamilton, 2009.
115. Metz, Davidson, de Coninck, Loos and Meyer, 2005.
116. Assuming 1.3 million bbl/day of 10°API bitumen, with 0.325 tonnes CO_2/tonne bitumen released during upgrading.
117. 86–91%.
118. Ordorica-Garcia, Wong and Faltinson, 2009.
119. StatoilHydro, 2008.
120. Mantripragada and Rubin, 2009.
121. X being C (coal), G (gas) or B (biomass).
122. Upstream Online, 2009.
123. 1–2%. Metz, Davidson, de Coninck, Loos and Meyer, 2005.
124. Seddon, 2006.
125. International Energy Agency, 2008.

126. It also produces substantial amounts of non-CO_2 greenhouse gases, but these can be largely eliminated by refits.
127. O'Sullivan, 2008.
128. Cleanteach Group LLC, 2007.
129. Calculation is given in the Appendix note to this page.
130. Physorg, 2007.
131. With a few exceptions, such as coal-bed methane if produced water is not disposed of properly.
132. Metz, Davidson, de Coninck, Loos and Meyer, 2005.
133. Assuming energy consumption is proportional to the natural log of (final concentration/initial concentration) (MacKay, 2009), air being 387 ppm CO_2, and the coal plant flue gas 15% CO_2.
134. Lackner, Grimes and Ziock, *Capturing Carbon Dioxide from Air.*
135. Institution of Mechnical Engineers, 2009.
136. *The Economist*, 2009.
137. Zeman, 'Energy and Material Balance of CO2 Capture from Ambient Air', 2007.
138. Stolaroff, Keith and Lowry, 2008.
139. If we can refer to such a new technology as conventional.
140. Keith, Ha-Duong and Stolaroff, 2005.
141. Other 'geo-engineering' solutions, involving blocking some of the sun's light, do not address ocean acidification, and may have other damaging and unpredictable consequences (Chapter 4).
142. E.g. International Energy Agency, 2008; Monbiot, 2007; Metz, Davidson, de Coninck, Loos and Meyer, 2005.
143. E.g. Monbiot, 2007.
144. Ibid.
145. Whether you think that the modern, globalised world is a good thing is something else. As this book should make clear, I believe strongly that it is, and that there are ways to sustain this lifestyle without causing disastrous climate change.
146. Or even nuclear, which is entirely possible for large ships, and has even been seriously proposed for aeroplanes (although public acceptability of this is likely to be problematic).
147. Seifritz, 1993.
148. E.g. Green Car Congress, 2007; Page, 2008; GreenBang, 2009.
149. A bit more than 1 litre, or about a third of a US gallon.
150. That said, similar issues bedevil hydrogen cars.
151. Appendix note to this page.
152. The world's largest container ship can carry about 15,000 shipping containers.
153. Aresta, 2003.
154. Metz, Davidson, de Coninck, Loos and Meyer, 2005.
155. Transport Watch UK, 2008; Argonne National Laboratory.

156. Arcoumanis and Kamimoto, 2008.
157. 14–17%.
158. Gielen and Unander, March 2005.
159. CNG Services Ltd, 2009.
160. Appendix, note to this page.
161. Faaij, 2007.
162. Fargione, Hill, Tilman, Polasky and Hawthorne, 2008.
163. Riduan, Zhang and Ying, 2009.
164. Zeman and Keith, 2008.

3. TRANSPORT AND STORAGE

1. American Petroleum Institute, 2007.
2. Biello, 'Storing the Carbon in Fossil Fuels Where It Came from: Deep Underground', 2009.
3. International Energy Agency, 2008.
4. 20 Mt/year (Metz, Davidson, de Coninck, Loos and Meyer, 2005).
5. Ibid.
6. Typically 26 inches or 660 mm (pipeline diameters in the oil industry are generally quoted in inches).
7. Dooley et al., 2006.
8. Metz, Davidson, de Coninck, Loos and Meyer, 2005.
9. About 40,000 tonnes.
10. International Energy Agency, 2008; Kohberg, 2009.
11. About 3.3 Mt per year.
12. *Wired*, 1998.
13. For calculations, see Appendix note to this page.
14. AZO Materials, 2009.
15. Jha, 2008. There appears, though, to be some dispute about intellectual property (Tececo).
16. Dodson, 2006.
17. This is the conventional explanation of petroleum formation. A minority view holds that petroleum is formed by 'abiotic' processes deep within the Earth. The vast majority of petroleum geologists (including myself) do not believe in this theory, which fails to explain a variety of almost universal features of known petroleum basins. Even if the abiotic theory were true, it would make no difference to the issue of carbon storage, except that there would probably be a lot more oil and gas to be burned (and potentially therefore CO_2 to be captured).
18. Down to 13° API, unambiguously a heavy oil (Brown, 2009).
19. After Wilson and Monea, 2004.
20. Mills R. M., 2008; CO_2 injection at Weyburn is predicted to increase recovery from 34.6% of oil in place to 55% (Wilson and Monea, 2004).
21. Rigzone, 2009.

22. *Oil and Gas Journal*, 2009.
23. The horizontal interface in a reservoir, with oil above (being less dense) and water below.
24. Appendix note to this page.
25. Where 99% of the produced fluid is water and only 1% oil; not unusual in the USA and other mature areas. Lifting and treating oily water is very energy intensive.
26. Le Thiez, Torp, Feron, Zweigel and Lindeberg, 2005.
27. Sharma and Dodds, 2008; Turta, Singhal and Sim, 2003.
28. Or, looked at the other way, a lot of carbon dioxide has to be injected to recover a rather small amount of gas.
29. Breunese, 2006.
30. Breunese and van Leverink (2008) give Groningen's gas initially in place (GIIP) as 102.5 Tcf.
31. Kohberg, 2009.
32. International Energy Agency, 2008.
33. *Oil and Gas Journal*, 2009.
34. 7.39 megapascals, MPa.
35. An aquitard restricts fluid flow; an aquiclude prevents it entirely.
36. Not that all drilled structures turned out to be valid traps, but many did, their lack of hydrocarbons being attributable to other factors such as timing of structure formation or absence of source rocks. Data from existing wells is at least valuable in characterising many possible traps at low cost.
37. International Energy Agency, 2008.
38. Hassanzadeh, Pooladi-Darvish and Keith, 2009); Leonenko and Keith, 2008.
39. Nelson, Evans, Sorensen, Steadman and Harju, 2005.
40. CO$_2$SINK Project Team, 2005. The relevant aquifer waters at Weyburn move 0.8 metres per year (Wilson and Monea, 2004).
41. Ibid.
42. Watson, Tingate, Boreham and Gibson-Poole, 2006.
43. Sengul, 2008.
44. Haszeldine et al., 2005; Gilfillan et al., 2009.
45. International Energy Agency, 2008.
46. Shi and Durucan, 2005.
47. Moon, 2008, and Chevron's CRUSH process.
48. Natural Gas Supply Association; Bundesanstalt für Geowissenschaften und Rohstoffe (BGR), 2005.
49. An idea independently proposed by Lubaś, Warchoł, Krępulec and Wolnowski, 2008.
50. Brehm, 2009.
51. Heydari, 2004; Dickens, Castillo and Walker, 2001.
52. Fitzpatrick, 2010.

53. Hyndman and Dallimore, 2001; United States Geological Survey, 2009; Traufetter, 2007.
54. Cohen, 2009.
55. Izundu, 2009.
56. World Oil, 2009.
57. Trap is a Sanskrit word meaning 'step', referring to the layered nature of these basalt flows.
58. McGrail, Schaef and Spane, 2008.
59. Sonde.
60. Pandey, 2009.
61. Doughton, 2006.
62. Kunzig and Broecker, 2008.
63. Goldberg, Kent and Olsen, 2010.
64. Currently about 826 billion tonnes.
65. Green.
66. Demonstrating, incidentally, that well-founded concern about climate change is not new. Marchetti, 1977.
67. Metz, Davidson, de Coninck, Loos and Meyer, 2005.
68. This will happen eventually anyway if we continue to emit carbon dioxide into the air—but that is another reason in favour of underground storage.
69. Ibid.
70. Ibid.
71. See Appendix note to this page.
72. House, House, Aziz and Schrag, 2007.
73. Hare, 2008. That said, leading climate scientist Wallace Broecker is in favour (Broecker, 2008).
74. Lovelock, 2009.
75. Raymo, Ruddiman and Froelich, 1988. However, there are a number of other theories about the contributory factors to the latest series of Ice Ages, including the configuration of the continents and variations in solar output.
76. *ScienceDaily*, 2008.
77. Krevor, Graves, Van Gosen and McCafferty, 2009.
78. Kelemen and Matter, 2008.
79. Distler, Kryachko and Yudovskaya, 2008.
80. Melcher, Grum, Simon, Thalhammer and Stumpfl, 1997.
81. Appendix note to this page.
82. Schuiling and Krijgsman, 2006.
83. Hangx and Spiers, 2009.
84. Goldberg, Chen, O'Connor, Walters and Ziock.
85. House, House, Aziz and Schrag, 2007.
86. Doughton, 2006.
87. An ultrabasic rock consisting mostly of the minerals olivine and pyroxene.

88. Appendix note to this page.
89. McKee, 2005.
90. Appendix note to this page.
91. Appendix note to this page.
92. See Appendix note to this page.
93. Of course emissions will tend to grow, at least in the medium term, so actual capacity in terms of years of emissions is somewhat less. But only a fraction of these emissions will be captured and stored.
94. Metz, Davidson, de Coninck, Loos and Meyer, 2005.
95. Kapila, Chalmers and Leach, 2009.
96. Dahowski, Davidson, Dooley and Gentile, 2008. The IEA (International Energy Agency, 2008) has somewhat lower figures for aquifers.
97. Metz, Davidson, de Coninck, Loos and Meyer (2005) has the lower estimate; Dahowski, Davidson, Dooley and Gentile (2008) the higher.
98. Haszeldine,et al., 2005.
99. Shackley and Gough, 'Conclusions and Recommendations', 2006. Recent studies suggest 60–150 billion tonnes (Haszeldine S., 2009).
100. Kunzig and Broecker, 2008.
101. Appendix note to this page.
102. A break where rocks on either side have moved laterally or vertically relative to the other side.
103. A quadrillion is a thousand million million, or 10^{15}.
104. Analogous to 'full to the brim'.
105. Or that natural gas is continuing to migrate into the structure from actively-generating source rocks. This is occurring in some places, but is not very common, and happens only slowly, thus setting an upper bound for leakage rates.
106. Mondal, 2008; Stevens S. H., 2005.
107. Wilson and Monea, 2004.
108. Ballentine, Schoell, Coleman and Cain, 2001.
109. Wilson N., 2008.
110. Heinrich, Herzog and Reiner, 2003.
111. Jordan and Benson, 2009.
112. Wilson, Friedmann and Pollak, 2007.
113. Taku Ide, Friedmann and Herzog, 2006.
114. Nicot, 2009.
115. Wilson and Monea, 2004.
116. Cooper C., 2009.
117. Kulakofsky and Tahmourpour, 2009.
118. Taku Ide, Friedmann and Herzog, 2006.
119. Cooper C., 2009.
120. Heinrich, Herzog and Reiner, 2003.
121. Appendix note to this page.
122. Biello, 'Storing the Carbon in Fossil Fuels Where It Came from: Deep Underground', 2009.

123. Nelson, Evans, Sorensen, Steadman and Harju, 2005.
124. Wilson and Monea, 2004.
125. Franklin and Schwartz, 2009.
126. International Energy Agency, 2008.
127. Appendix note to this page. *Encyclopædia Britannica*, 2009.
128. 0.02%.
129. Appendix note to this page.
130. Nelson, Evans, Sorensen, Steadman and Harju, 2005.
131. Xue, Tanase, Saito, Nobuoka and Watanabe, 2005.
132. Dooley et al., 2006.
133. Giardini, Grynthal, Shedlock and Zhang, 1999.
134. Wilson, Friedmann and Pollak, 2007.
135. Colasurdo, 2000.
136. Heinrich, Herzog and Reiner, 'Environmental Assessment of Geologic Storage of CO_2', 2003.
137. Holloway, Bentham, & Kirk, 'Underground Storage of Carbon Dioxide', 2006.
138. Beaubien et al., 2008.
139. Heinrich, Herzog and Reiner, 'Environmental Assessment of Geologic Storage of CO_2', 2003.
140. Wilson, Friedmann and Pollak, 2007; Kharaka, Cole, Hovorka, Gunter, Knauss and Freifeld, 2006.
141. Hepple and Benson, 2004.
142. Archer et al., 2009; however, a noticeable fraction, probably 10–15%, still remains in the atmosphere after 10,000 years (Archer and Brovkin, Millennial Atmospheric Lifetime of Anthropogenic CO_2, 2006).
143. Haszeldine et al., 2005.
144. Some argue that it will take some 7,000 years for the atmosphere to recover from the very high emission scenarios, and therefore that carbon dioxide storage has to be effective for at least this long. But this misses the point that a world with extensive use of CCS will, *ipso facto*, not be a high emissions world.
145. Zeman and Keith, 2008.
146. Wilson and Monea, 2004.
147. As long as other surface disturbances, such as ploughing, can be avoided or corrected for.
148. Cooper C., 2009.
149. Of course, carbon dioxide injected from biomass or air capture would be the same as atmospheric carbon dioxide.
150. Metz, Davidson, de Coninck, Loos and Meyer, 2005.
151. Myers, Dickinson and Dickinson, 2009.
152. 2,500–7,500 tonnes.

4. BIO-SEQUESTRATION

1. Charles, 2007.
2. *The Economist*, 2009.
3. Trumper, Bertzky, Dickson, van der Heijden, Jenkins and Manning, 2009.
4. Again, stressing the point that the natural flux of carbon dioxide is very large, but there is no net addition of carbon dioxide to the atmosphere. In comparison, the small activities of humankind are enough to put the entire cycle out of balance. Schuur et al., 2008.
5. Much of the carbon in the biosphere and ocean is not held as carbon dioxide, but in various other chemical forms. It is more usual to talk about tonnes of carbon (1 tonne C = 3.67 tonnes CO_2) when discussing bio-sequestration, but I have quoted tonnes of CO_2-equivalent for easy comparison with the rest of the book.
6. Based on Stavins and Richards, 2005; proportion of carbon by biome from Trumper, Bertzky, Dickson, van der Heijden, Jenkins and Manning, 2009.
7. CO_2e is the notation for 'CO_2-equivalent', converting methane and other greenhouse gases into the mass of CO_2 that has the same global warming potential over a defined period of time, usually a century. Methane and nitrous oxide are much more potent greenhouse gases, tonne for tonne, than carbon dioxide, but are present in the atmosphere in much smaller quantities.
8. Mendelsohn and Dinar, 2009.
9. Though Greenpeace, to their credit, have fought strenuously to build a grassroots network in Brazil to combat Amazonian deforestation.
10. Lovelock, *The Vanishing Face of Gaia*, 2009.
11. Hamilton G., 2010.
12. Nevle and Bird, 2008. There are, of course, other theories for the causes of the Little Ice Age.
13. Fargione, Hill, Tilman, Polasky and Hawthorne, 2008.
14. Sielhorst, Molenaar and Offermans, 2008.
15. Shukman, 2009.
16. E.g. Friends of the Earth International, 2009.
17. Chakravorty, Hubert and Nøstbakken, 2009.
18. Institution of Mechnical Engineers, 2009.
19. Afforestation, Reforestation and [avoiding of] Deforestation, which sounds to me too much like ADD.
20. Mendelsohn and Dinar, 2009.
21. Kindermann et al., 2008.
22. Based on figures in Rametsteiner, Obersteiner, Kindermann and Sohngen (2008) and Kindermann et al. (2008).
23. Salleh, 2009.

24. 7%.
25. Canadell and Raupach, 2008.
26. Watkins, 2007.
27. Food and Agriculture Organization of the United Nations, 2007.
28. Hope, 'Valuing the climate change impacts of tropical deforestation', 2008.
29. 2%.
30. Stavins and Richards, 2005.
31. Trumper, Bertzky, Dickson, van der Heijden, Jenkins and Manning, 2009.
32. E.g. Malhi et al., 2009.
33. Harsch, Hulme, McGlone and Duncan, 2009.
34. Mendelsohn and Dinar, 2009.
35. Trumper, Bertzky, Dickson, van der Heijden, Jenkins, and Manning, 2009.
36. Richards and Stokes, 2004.
37. Intergovernmental Panel on Climate Change, 2000.
38. They also support vastly poorer and less complex ecosystems.
39. The Royal Society, 2009.
40. Sohngen and Brown, 2006.
41. Biopact, 2007.
42. Lovelock, *The Vanishing Face of Gaia*, 2009.
43. Bhagwat, 2009.
44. Appropedia, 2007.
45. Foley, 2009.
46. E.g. IRIN, 2004; Farah, 2002.
47. As well as 70% in Papua New Guinea, up to 60% in Brazil and 50% in Cameroon (Saunders and Nussbaum, 2008).
48. Hoare, 2008.
49. Meyfroidt and Lambin, 2009.
50. For an example of conflicts in Kenya arising from water use, deforestation/reforestation, agriculture and ecotourism, see Morgan, 2009.
51. 35–53% (Metzger and Benford, 2001).
52. Tweeten, Sohngen and Hopkins, 2000.
53. Bock B. et al., 2003.
54. Intergovernmental Panel on Climate Change, 2000.
55. About 40% each for Russia and the Americas, and 6% in South East Asia.
56. Wetlands International, 2006.
57. Krajick, 2005.
58. European Commission.
59. Metzger and Benford, 2001.
60. About 9,200 kWh.
61. Allowing for energy consumed in the capture process. Of course, if we also sequestered the carbon dioxide from burning natural gas, we would

have a carbon-negative energy system. Keith and Rhodes, 'Bury, burn or both: A two-for-one deal on biomass carbon and energy', 2002.
62. Goodall, 2008.
63. 'Black earth' in Portuguese.
64. *Sustainable Land Development Today*, 2009.
65. Lehmann, Gaunt and Rondon, 'Bio-char Sequestration in Terrestrial Ecosystems—a Review', 2006.
66. Pathirana, 2009.
67. Wardle, Nilsson and Zackrisson, 2008.
68. Woolf, 2008.
69. Austin, 2009.
70. Read and Lermit, 'Bio-energy with carbon storage (BECS): A sequential decision approach to the threat of abrupt climate change', 2005.
71. Woolf, 2008.
72. MacKay, 2009.
73. Markels and Barber, 2001.
74. Lutz, Caldeira, Dunbar and Behrenfeld, 2007.
75. Woods Hole Oceanographic Institution, 2003.
76. Royal Society, 2009.
77. Boyd, 2008.
78. E.g. Institution of Mechnical Engineers, 2009; Royal Society, 2009.
79. Boyd, 2008.
80. Weitzman, 2008.

5. SCALE, COSTS AND ECONOMICS

1. Millard, 2009.
2. Haszeldine S., 2009.
3. Appendix note to this page.
4. Enkvist, Nauclér and Rosander, 2007.
5. Dahowski, Davidson, Dooley and Gentile, 2008.
6. Appendix note to this page.
7. International Energy Agency, 2008.
8. Appendix note to this page.
9. Ibid.
10. Appendix note to this page.
11. MacKay, *Sustainable Energy—Without the Hot Air*, 2009.
12. World Bank, 2007.
13. Metz, Davidson, de Coninck, Loos and Meyer, 2005.
14. eRedux, 2005.
15. From 10.4 million bbl/day in 1950 to 75 million bbl/day in 2000 (BP, 2009).
16. Heston, Summers and Aten, 2006.
17. The Interstate Natural Gas Association of America estimates 46,000–100,000 km of new gas pipelines to be built by 2030 (Snow N., 2009).

18. World Economic Forum, Cambridge Energy Research Associates, 2009.
19. Appendix note to this page.
20. Global CCS Institute, 2009.
21. Metz, Davidson, de Coninck, Loos and Meyer, 2005.
22. See Appendix note to this page.
23. IHS-CERA power costs index, quoted in Al-Juaied and Whitmore, 2009.
24. See footnote, page 58
25. See, for example, Royal Academy of Engineering, May 2008.
26. See Appendix note to this page.
27. Total costs over the plant life, including capital and operating costs, amortising the capital costs at a rate allowing for the cost of capital.
28. International Energy Agency, 2008.
29. See Appendix note to this page.
30. Capturing only a small fraction of emissions should be cheaper than trying to capture 90–95% as envisioned for full CCS plants. However, economies of scale should also apply when building larger systems. In addition, $100 per tonne is the price that Shady Point is able to charge its customers; the true capture cost may be lower.
31. Al-Juaied and Whitmore, 2009.
32. See Appendix note to this page.
33. See Appendix note to this page.
34. Dooley et al., 2006.
35. See Appendix note to this page.
36. The corresponding chart showing electricity generation costs is in the Appendix note to this page.
37. A kilowatt hour is a common unit for energy, and the normal way in which electricity bills are expressed. It is the energy expended by a one kilowatt device (such as a small electric fire) running for one hour. In 2008, the average retail electricity price was approximately $0.9–10 per kWh in the USA, and about £0.11 per kWh in the UK (exclusive of tax).
38. Unlike renewables, which are often in remote locations and require grid upgrades and back-up due to intermittency.
39. About 4% more.
40. Bohm, Herzog, Parsons and Sekar, 2007.
41. International Energy Agency, 2008.
42. Ibid.
43. Merour, 2009.
44. Appendix not to this page.
45. *Middle East Economic Digest*, 2006.
46. Appendix note to this page.
47. US Department of Energy, 2009.
48. 12% for LNG and 13% for pollution scrubbers (Nauclér, Campbell and Ruijs, 2008).
49. International Energy Agency, 2008.

50. Learning factors from Al-Juaied and Whitmore, 2009; final costs in 2030 selected from International Energy Agency (2008), to be consistent with discussion earlier in chapter.
51. Riahi, Rubin, Taylor, Schrattenholzer and Hounshell, 2004.
52. On a discounted basis over project life.
53. International Energy Agency, 2008.
54. Appendix note to this page.
55. Global CCS Institute, 2009.
56. CA$40 million during the period 2000–2004 (Wilson and Monea, 2004); I have converted this to US$ assuming the 2002 exchange rate.
57. Klemeš, Cockerill, Bulatov, Shackley and Gough, 2006.
58. Appendix note to this page.
59. International Energy Agency, 2008.
60. Wilson and Monea, 2004.
61. International Energy Agency, 2006; (Metz, Davidson, de Coninck, Loos and Meyer, 2005.
62. Mostly based on Metz, Davidson, de Coninck, Loos and Meyer, 2005.
63. Ibid.
64. Stolaroff, Lowry and Keith, 'Using CaO- and MgO-rich industrial waste streams for carbon sequestration', 2005.
65. In Kunzig and Broecker, 2008; International Energy Agency, 2008.
66. Not the highest costs that could be imagined, but a reasonable value for one of the more expensive capture options.
67. International Energy Agency, 2008.
68. Dooley et al., 2006.
69. Metz, Davidson, de Coninck, Loos and Meyer, 2005.
70. Consistent with the IEA's view for CCS in the OECD around 2020 (International Energy Agency, 2008).
71. For details, see Appendix note to this page.
72. Kunzig and Broecker, 2008.
73. Lower estimate in Stolaroff, 'Capturing CO_2 from ambient air: a feasibility assessment' (2006); higher based on discussion in Keith, Ha-Duong and Stolaroff (2005).
74. Keith, Ha-Duong and Stolaroff, 2005.
75. By way of illustration only. Even allowing for some global warming, Canada tends to be too cold for viable biomass plantations such as sugarcane.
76. House, House, Aziz and Schrag, 2007.
77. Institution of Mechnical Engineers, 2009.
78. Most of this (perhaps 85%) is cheap, low-grade thermal energy; about 15% is carbon-neutral electricity. Keith, Ha-Duong and Stolaroff, 2005; Stolaroff, Keith and Lowry, 2008; Stolaroff, Keith and Lowry, 'Carbon Dioxide Capture from Atmospheric Air Using Sodium Hydroxide Spray', 2008; Herzog, 2003, although Stolaroff, 'Capturing CO_2 from ambient air: a feasibility assessment' (2006) argues that Herzog's system is insuf-

ficiently optimised; Baciocchi, Storti and Mazzotti (2006) correcting the CO_2 concentration from 500 ppm to 355 ppm; Zeman F., 'Energy and Material Balance of CO_2 Capture from Ambient Air', 2007; Zeman, 2006; Lackner, K., quoted in *The Economist*, 2009.

79. Based on a US population of 300 million, and other statistics as given in BP, 2008.
80. See Appendix note to this page.
81. Using price elasticity of -0.25 for business travel and -1.5 for short-haul leisure travel (Gillen, Morrison and Stewart, 2008).
82. Pielke, 2009.
83. Azar, Lindgren, Larson and Möllersten, 2006.
84. Assuming a 90% mitigation rate.
85. Appendix note to this page.
86. Rhodes and Keith, 2008.
87. Intergovernmental Panel on Climate Change, 2000; Metz, Davidson, de Coninck, Loos and Meyer, 2005.
88. Of course biofuels may have other advantages, such as energy security.
89. Richards and Stokes, 2004. Most of their sources indicate around $400/tonne, but they are all from 2001 and earlier, so I have allowed for about 30% inflation since then. Also they mostly do not include a failure rate.
90. Wetlands International, 2006.
91. Intergovernmental Panel on Climate Change, 2000.
92. Sathaye, Dale, Makundi and Chan, 2008.
93. Bock, Rhudy and Herzog, 2003.
94. Haszeldine et al., 2005.
95. Enkvist, Nauclér and Rosander, 2007, assuming $1.3:€1.
96. Kindermann et al., 2008.
97. Typical wood prices (which vary greatly by region) are about $75/m^3$, and 1 m^3 of wood releases about 0.9 tonnes of CO_2 (if burned).
98. Mendelsohn and Dinar, 2009.
99. Stavins and Richards, 2005.
100. New Zealand Trade and Enterprise.
101. Buchanan and Levine, 1999.
102. Hooijer, Silvius, Wösten and Page, 2006.
103. Wetlands International, 2006.
104. Schuur et al., 2008.
105. Appendix note to this page.
106. Appendix note to this page.
107. Read and Lermit, 'Bio-energy with carbon storage (BECS): A sequential decision approach to the threat of abrupt climate change', 2005.
108. Woolf, 2008.
109. Ibid.
110. Lehmann, 2007.
111. Appendix note to this page.

112. Markels and Barber, 2001.
113. To be precise, 1990–2100.
114. Appendix note to this page.
115. These prices have been driven down by the economic crisis, and should increase when and if solid economic growth resumes.
116. Appendix note to this page.
117. Jaccard, *Sustainable Fossil Fuels*, 2005)
118. Pacala and Socolow, 2004; Socolow R., 2005; Cascio, 'Stabilization Wedges', 2005.
119. Enkvist, Nauclér and Rosander, 2007.
120. Though the solar figures may be pessimistic. Suntech, the world's largest solar panel manufacturer, suggests 15–20 ¢ per kWh rather than the 40¢ given here. Nader, 2009.
121. Al-Juaied and Whitmore, 2009 and Global CCS Institute, 2009.
122. Energy Information Administration, 2009.
123. Metz, Davidson, de Coninck, Loos and Meyer, 2005.
124. International Energy Agency, 2008.
125. E.g. Rochon, 2008.
126. 'BLUE' scenario (International Energy Agency, 2008).
127. Metz, Davidson, de Coninck, Loos and Meyer, 2005.
128. Appendix note to this page.
129. Which they label 'BLUE'.
130. Appendix note to this page. For comparison, global GDP in 2008 was $54.6 trillion (at market exchange rates). Worldwide military spending is about 2% of GDP.
131. World Bank, 2006.
132. Azar, Lindgren, Larson and Möllersten, 2006.
133. Assuming a typical long-run coal price of $100/tonne and oil price of around $60 per barrel.
134. Hope, 'Valuing the climate change impacts of tropical deforestation', 2008.
135. Hope and Castilla-Rubio, 'A first cost-benefit analysis of action to reduce deforestation, 2008'.
136. Ibid.

6. POLICY

1. van der Schaaf, 2009,
2. Mulkern, 2009.
3. Think Carbon, 2009.
4. Wynn, Gardner and Maeda, 2009.
5. Upstream Online, 2009.
6. In a perfect world, of course. In practice, considerations of domestic and international politics, and scientific and economic uncertainties, mean that agreed systems, caps and tax levels are likely to be far from optimal. Hence

the need for flexible policy-making, and the avoidance of regulatory or technological 'lock-in'.

7. Adam, 2009.
8. Indymedia UK, 2008.
9. BBC News, 2009.
10. Wilcox, 2008.
11. Capitol Climate Action, 2009.
12. Hoskins and Buckle, 2009.
13. At least, at an industry level. At the level of individual firms, it may be attractive for those firms that can cut carbon dioxide intensity to do so, and sell credits to the laggards.
14. E.g. Jaccard, Rivers, Bataille, Murphy, Nyboer and Sadownik, May 2006.
15. Peltier, 2007.
16. International Energy Agency, 2008.
17. Aghion, Hemous and Veugelers, 2009. For economic modelling of this in the context of CCS, see Otto and Reilly, 2008.
18. Appendix note to this page.
19. Appendix note to this page.
20. The full bill is larger than this, but most is covered by carbon costs and electricity sales, and therefore is not a subsidy.
21. McGrath, 2009.
22. van Loon, 2008.
23. Watkins, 2007.
24. Strictly speaking, the period 1973–2003 (Pica, 2002).
25. Randerson, 2009.
26. Dooley et al., 2006.
27. European Technology Platform for Zero Emission Fossil Fuel Power Plants (ZEP), 2008.
28. Al-Juaied and Whitmore, 2009.
29. European Technology Platform for Zero Emission Fossil Fuel Power Plants (ZEP), 2008.
30. EUTurbines, 2007.
31. BBC News, 2009.
32. See, for instance, the furore over Tata Motors' proposed new car plant, e.g. *The Hindu*, 2006.
33. Mountford, 2009.
34. Randerson, 2009.
35. Watkins, 2007.
36. BBC News, 2009.
37. Ibid.
38. Schoenbrod and Stewart, 2009.
39. Franklin and Schwartz, 'For the Coal Industry, the Waxman-Markey Bill's CCS Provisions are a Mixed Bag', 2009.
40. Terracina, 2009.

41. Kahn, 2009.
42. FCCs, units used to convert the heavy fractions of the oil to light products, particularly petrol (gasoline).
43. European Technology Platform for Zero Emission Fossil Fuel Power Plants (ZEP), 2008.
44. Nauclér, Campbell and Ruijs, 2008.
45. Partly based on Nauclér, Campbell and Ruijs, 2008.
46. Buchwitz M., 2009.
47. van der Schaaf, 2009.
48. If nuclear is considered 'green', true in climate terms but possibly otherwise controversial.
49. Amongst others, Enkvist, Nauclér and Rosander, 2007.
50. Advertisement in the *Carbon Capture Journal*, May-June 2009, p. 26.
51. Gluyas, 2010.
52. Aldy and Pizer, 2009.
53. Roughly, by the fraction of actual participants raised to the power of -1.8. (Nordhaus W., 2009), i.e. a deal covering only 50% of emissions is 3.5 times as expensive as a comprehensive arrangement.
54. Ritz, 2009.
55. EurActiv, 2009.
56. Kerr, 2009.
57. EurActiv, 2009. Some legal opinions suggest that carbon tariffs might be possible under WTO law (Herman, 2009), but retaliation remains a likely outcome.
58. Hoare, 2008; Hamilton, Sjardin, Shapiro and Marcello, 2009.
59. Schlamadinger and Marland, 2000.
60. Although the Boxer-Kerry Senate bill does not guarantee that agricultural offsets will be accepted.
61. Monbiot, 2007.
62. Bradsher, 2006.
63. Or, even if a project is not 'economic' in the strict sense, it might be required by government regulation and would therefore not be 'additional'. For instance, many countries have restrictions on flaring unwanted gas from oilfields.
64. Bohm, Herzog, Parsons and Sekar, 2007.
65. E.g. Australian gas supplier Origin Energy (Upstream Online, 2009).
66. Shukman, 'Project to "grow carbon sinks"', 2009.
67. Luce, 2007.
68. Watkins, 2007.
69. Saunders and Nussbaum, 2008.
70. Rametsteiner, Obersteiner, Kindermann and Sohngen, 2008.
71. Fogarty, 2009.
72. AFP, 2009.
73. Hoare, 2008.
74. Foley, 2009.

75. Henderson, 2009.
76. Diamond, 2005.
77. Hoare, 2008.
78. Leng, 2007; Canadell and Raupach, 2008.
79. Chatham House, 2007.
80. Williams, 2009.
81. EurActiv, 2009.
82. *Massachusetts v EPA*, 2007.
83. The dispute over the name is indicative of wider problems bedevilling Gulf cooperation.
84. Metz, Davidson, de Coninck, Loos and Meyer, 2005.
85. Wilson and Keith, 'Geologic Carbon Storage: Understanding the Rules of the Underground', 2003.
86. Wilson, Friedmann and Pollak, 2007.
87. Riley, 2004.
88. Upstream Online, 2009.
89. Australian Government: Department of Energy, Resources and Tourism, 2009.
90. Oilbarrel, 2008.
91. Baugh.
92. Fitzgerald, Winchester and Johns, 2008.
93. Though unconventional gas may emerge as a major source of supply, as has happened in North America.
94. E.g. German economics minister Michael Glos said in January 2009 that, because of security of supply concerns, the role of gas in German power generation should not increase (Lohmann, 2009).
95. Tusiani and Shearer, 2007.
96. Upstream Online, 2009.
97. E.g. Upstream Online, 2009.
98. The various kinds of unconventional oil face other non-carbon environmental challenges, it is fair to point out, though these are not insoluble either.
99. Mills R. M., 2008; Odell, 2004.
100. Johansson, Azar, Lindgren and Persson, 2009.
101. Carlisle, 2009.
102. Baugh.
103. Murray, 2009.
104. Cookson, 2009.
105. Lahn, 2009.
106. Biopact, 2007.
107. Upstream Online, 2009.
108. BBC News, 2009.
109. ScottishPower, 2009.
110. HM Government, 2009.
111. Hernandez, 2007.

112. Méndez, 2008.
113. Coal Utilization Research Council.
114. Sommer, 2009.
115. Price andCaruso, 2009.
116. Reuters, 2009.
117. Environmental Expert, 2009.
118. FutureGen Alliance, 2009.
119. Gunter, Bachu, Palombi, Lakeman, Sawchuk and Bonner, 2009.
120. International Energy Agency, 2008.
121. Appendix note to this page.
122. Star Staff, 2009; Jones J., 2009.
123. Schmidt L., 2009.
124. Morales and van Loon, 2009.
125. Fung, 2009.
126. Kapila R. V.
127. Hyun-cheol, 2009.
128. Australian Government: Department of Resources, Energy and Tourism, 2008.
129. Macdonald-Smith, 2008.
130. Rigzone, 2009.
131. E.g. Tamayo, 2005.
132. As I describe in Mills R. M., 2008.
133. Merour, 2009.
134. Coal mining and oil sands, in particular, have other non-climate environmental challenges to overcome, particularly the destruction of land and habitats, and water use.
135. In this section, I mention a number of companies to illustrate those who may gain (or lose) from CCS. None of this, of course, should be construed as constituting investment advice.
136. Bhadeshia.
137. Leber, 2009.
138. Assuming the EOR operator could capture the difference between their costs and those of an average site.
139. E.g. Mills R. M., 2008; Goodall, 2008; Chaffin, 2009; Monbiot, 2007.
140. Shackley, Gough and McLachlan, 2006.
141. 'Not In My Back-Yard': local hostility to developments that would have wider economic and/or environmental benefits. Possibly NUMBY ('Not Under My Backyard') for carbon storage. In its more serious forms, it becomes BANANA—Build Absolutely Nothing Anywhere Near Anyone.
142. Chaffin, 'Public wary of carbon capture', 2009.
143. Rydberg, 2009.
144. Price and Caruso, 2009.
145. Fröhlingsdorf, Knauer and Schwägerl, 2009.
146. Comfort, 2010.

147. Baeza, Balagopal, Boudler, Buffet and Velken, 2008.
148. ZINC Research/Dufferin Research, 2009.
149. Sharp, Jaccard and Keith, 2009.
150. Bell, 2009.
151. Brennan, 2009.
152. See, for instance, comments to Brennan, 2009; ugent, 25 April 2009.
153. Flyvbjerg, Bruzelius and Rothengatter, 2003.
154. Bellona.
155. Vattenfall.
156. Cross, 2009.
157. Jaccard, *Sustainable Fossil Fuels*, 2005.
158. Rochon, 2008.
159. Mulkern, 2009.
160. Lucas, 2009.
161. An industry-funded group opposing carbon dioxide emission caps in the USA. Institute for Energy Research, 2009.
162. American free-market think tank. Mulkern, 2009.
163. WWF.
164. US environmental organisation. Sierra Club, 2008.
165. Mulkern, 2009.
166. Norwegian environmental organisation. Stangeland, 2007.
167. Anderson, 2009.
168. Green Alliance, 2006.
169. American Petroleum Institute, 2007.
170. Watkins, 2007.
171. Lovelock, 2009.
172. Lovelock, 'James Lovelock on Biochar: let the Earth remove CO_2 for us', 2009.
173. Pearce, 2009.
174. Rochon, 2008.
175. 'Environmental' because the original Kuznets curve deals with economic inequality, not pollution.
176. Tierney, 2009.
177. Kauppi, Ausubel, Fang, Mather, Sedjo and Waggoner, 2006.
178. Dutt, 2009.
179. The wealthy oil-producing states probably represent an exception—though Abu Dhabi feels rich enough to have devoted $25 billion to environmental projects.
180. Müller-Fürstenberger, Wagner and Müller, 2004.
181. Yandle, Bhattarai and Vijayaraghavan, 2004.
182. After Read and Lermit, 'Bio-energy with carbon storage (BECS): A sequential decision approach to the threat of abrupt climate change', 2005.
183. Parson, 2006.
184. Pielke, 2009.

185. Parson, 2006.
186. Though there is no empirical evidence for this at the moment. It is also possible that the prospect of large-scale geo-engineering might spur people to make more effort on conventional mitigation.
187. For example, the IPCC's B2 ASF scenario.
188. 'Hysteresis', in scientific jargon.
189. Hope and Castilla-Rubio, 'A first cost-benefit analysis of action to reduce deforestation', 2008.

7. RISKS

1. Chaffin, 'Public wary of carbon capture', 2009.
2. Biello, 2009.
3. Katz, 2007.
4. World Bank, 'Assessment of the World Bank/GEF Strategy for the Market Development of Concentrating Solar Thermal Power'.
5. World Economic Forum, Cambridge Energy Research Associates, 2009.
6. Rochon, 2008.
7. Cohen, *World Oil*, 2009.
8. *The Economist*, 2007.
9. Little, 2005.
10. Goldenberg, 2009.
11. Krauss and Mouawad, 2009.
12. Reuters, 2005.
13. Upstream Online, 2009.
14. E.g. McKillop, 2005.
15. I say 'appears' because, as I've argued earlier, inefficient use of labour destroys jobs elsewhere in the economy.
16. E.g. Darley, 2004; Heinberg, 2005.
17. E.g. Jaccard, Rivers, Bataille, Murphy, Nyboer and Sadownik, May 2006 Chow, Bovair, Pumphrey and Nichols, 2007.
18. The Royal Society, 2009.
19. Dewees, 2008.
20. Upstream Online, 2009.
21. Corporations are at least equally skilled in litigation, and usually better funded, but in this case most companies are more likely to be going to court to defend CCS rather than to prevent it.
22. Biomass and biofuels face perhaps similar problems; neither electricity nor oil companies are familiar or comfortable with agriculture.
23. I believe, by the way, that this factor is more significant than a lack of environmental commitment in explaining the limited progress made by companies such as Shell and BP in their wind and solar investments.
24. Other than BHP Billiton, which is really a mining company with a secondary petroleum business. The FutureGen web page does not even mention BHP Billiton's oil and gas activities.

25. Though it is conceivable that entities such as the BP-Rio Tinto joint venture Hydrogen Energy might be spun off as independent ventures at some point.
26. Krauss, 'Natural Gas Hits a Roadblock in New Energy Bill', 2009.
27. E.g. Monbiot, 'Woodchips with everything. It's the Atkins plan of the low-carbon world', 2009; Rughani, 2009.
28. E.g. Read (2009) attacks the idea that biochar is an 'easy way out'.
29. *The Economist*, 2009.
30. Evans A., 2009.
31. *The Economist*, 2009.
32. Watts, 2009.
33. Leng, 2007.
34. Quoted in Kunzig and Broecker, 2008.

8. CONCLUSIONS

1. Quoted in Chaffin, 2009.
2. Monbiot, 'Woodchips with everything. It's the Atkins plan of the low-carbon world', 2009.
3. Friends of the Earth Action.
4. Rochon, 2008.
5. Contrary to what some commentators suggest, air travel is not only a frivolous luxury used for wealthy people to reach their holiday homes in Tuscany or the Bahamas. Without aviation, how can environmental campaigners argue their case with a global audience (I have met them in Dubai, and doubt they cycled there), the IPCC meet to compile their reports, engineers get to Algeria to build the massive solar plants that will supply Europe, scientists travel to Congo to study deforestation, and people of different cultures build the mutual understanding to avoid the genocidal wars predicted by the 'neo-Luddites'?
6. Monbiot, *Heat: How to Stop the Planet Burning*, 2007.
7. Piling, 2009.
8. Randerson, 2009.
9. Cooper C., 2009.

APPENDIX: FURTHER DETAILS

1. Lichter, 2008.
2. Doran and Kendall Zimmerman, 2009.
3. The Royal Society, 2005.
4. Archer et al., 2009.
5. Barker, Koehler and Villena, 2002.
6. Enkvist, Nauclér and Rosander, 2007.
7. Nordhaus, 2007.

8. E.g. Sivertsson, 2004; Zittel and Schindler, March 2007; Aleklett, 2007; Rutledge, 2007.
9. *Mining Exploration News*, 2009.
10. Zittel and Schindler, March 2007.
11. Mayer.
12. Thornley, 2009.
13. Hao, Dong, Yang, Xu and Li, 2007.
14. From Aguilera, Eggert, Gustavo Lagos and Tilton, 2009.
15. Nakićenović, 2006.
16. Smil, 2008.
17. Gipe, 1995.
18. Narain, 2008.
19. Environmental Science Activities for the twenty-first Century.
20. Intergovernmental Panel on Climate Change, 2007.
21. Hartley, 2004.
22. Barnes, 2001.
23. Ibid.
24. Ibid.
25. Stolaroff, Lowry and Keith, 'Using CaO- and MgO-rich industrial waste streams for carbon sequestration', 2005.
26. Index Mundi.
27. Stolaroff, Lowry and Keith, 'Using CaO- and MgO-rich industrial waste streams for carbon sequestration', 2005)
28. Gieré and Stille, 2004.
29. Uibu, Uus and Kuusik, 2009.
30. Alumina Limited.
31. Index Mundi.
32. Jones, Joshi, Clark and McConchie, 2006.
33. Hitch, 2009.
34. MacKay, 2009.
35. Stuber, Forster, Rädel and Shine, 2006.
36. Geological Society, 2006.
37. US Geological Survey, 2009.
38. Metz, Davidson, de Coninck, Loos and Meyer, 2005.
39. UNFCCC.
40. E.g. Sarkarinejad, 2003.
41. Jassim, Suk and Waldhausrová, 2008.
42. BP, 2009.
43. Data from Dooley et al. (2006) with the exception of China (Dahowski, Davidson, Dooley and Gentile, 2008).
44. World Resources Institute, 2009.
45. Stevens, Pearce and Rigg, 2001.
46. Baines and Worden, 2004. Their estimates are drawn from sources published 1995, 1998 and 1999.
47. World Coal Institute, May 2005.

48. Metz, Davidson, de Coninck, Loos and Meyer, 2005.

49. Dooley et al., 2006.

50. Sonde.

51. International Energy Agency, 2008.

52. 240 Gt capacity given for USA only.

53. India only.

54. Stern, 2007.

55. International Energy Agency, 2008.

56. Nauclér, Campbell and Ruijs, 2008.

57. Primarily International Energy Agency, 2009.

58. International Energy Agency, 2008.

59. Al-Juaied and Whitmore, 2009.

60. International Energy Agency, 2008.

61. The 'ACT' scenario.

62. The 'BLUE' scenario.

63. Strictly speaking, the A1B scenario.

64. Strictly speaking, the B1 scenario.

65. Akimoto and Tomoda, 2005.

66. Stangeland, 'Why CO_2 Capture and Storage (CCS) is an Important Strategy to Reduce Global CO_2 Emissions, 2007'.

67. European Technology Platform for Zero Emission Fossil Fuel Power Plants (ZEP), 2008.

68. UNData, 2008.

69. BP, 2009.

70. Metz, Davidson, de Coninck, Loos and Meyer, 2005; International Energy Agency, 2008.

71. Al-Juaied and Whitmore, 2009.

72. Ibid.

73. Dooley et al., 2006.

74. Hildebrand and Herzog, 2009.

75. Metz, Davidson, de Coninck, Loos and Meyer, 2005.

76. Various sources, including International Energy Agency, 2008; Klemeš, Cockerill, Bulatov, Shackley and Gough, 2006; Metz, Davidson, de Coninck, Loos and Meyer, 2005. Refinery capture costs estimated as $57/tonne captured, assuming 10% efficiency loss to convert from tonnes captured to tonnes avoided, $4/MMBtu cost of gas, thirty-year plant life and 12% discount rate.

77. Al-Juaied and Whitmore, 2009.

78. United Nations Statistics Division, 2008.

79. BP, 2009.

80. Al-Juaied and Whitmore, 2009.

81. Simmonds, Hurst, Wilkinson, Reddy and Khambaty, 2003.

82. Global CCS Institute, 2009.

83. Assuming, from Al-Juaied and Whitmore (2009), that a non-capture IGCC of 538 MW (75% capacity factor) emits 822 kg CO_2/MWh, and

that a capture IGCC of 493 MW (75% capacity factor) emits 97 kg CO_2/MWh.

84. Based mainly on Dahowski, Davidson, Dooley and Gentile, 2008.
85. E.g. Johnston, 1994.
86. Actually, to give a 10% internal rate of return on invested capital, a likely minimum that an oil company would accept.
87. Metz, Davidson, de Coninck, Loos and Meyer, 2005.
88. House, House, Aziz and Schrag, 2007.
89. IPCC in Rhodes and Keith, 2008.
90. BP, 2008.
91. Netherlands Environmental Assessment Agency, 2009.
92. Livingood, Winicaties and Stein, 1999.
93. Woolf, 2008.
94. Lehmann, 2007.
95. GreenFacts, 2009.
96. Lehmann, 2007.
97. Tweeten, Sohngen and Hopkins, 2000.
98. Lal, 2004.
99. Ammonia from Greenhouse Gas Protocol Initiative, 2009; cement from Mahasenan, Dahowski and Davidson, 2003; aluminium from Allwood and Cullen, 2009; steel from Carbon Trust, 2009.
100. Oil price is Dated Brent from BP, 2009. Coal price is US Central Appalachian coal spot price from BP, 2009. Gas price is US Henry Hub from BP, 2009. Gasoline is US retail price from Energy Information Administration, 2009. Electricity is the average of retail prices to US consumers from Energy Information Administration, 2009. Aluminium and steel prices are average 2000–2009 from Mongabay, 2009. Ammonia price is approximately estimated from Fertilizer Institute, 2009. Cement price is for 2008, from climatelab.
101. Handley, 2008.
102. PointCarbon, 2009.
103. Paltsev et al., 2007.
104. Typical cost from Dooley et al., 2006.
105. Al-Juaied and Whitmore, 2009.
106. Smith, 2009.
107. Stangeland, 'A Model for the CO_2 Capture Potential', 2007.
108. International Monetary Fund, 2009.
109. Natural Resources Canada, 2009.

GLOSSARY

1. BP, 2009.

BIBLIOGRAPHY

Adam, D. (August 13, 2009). 'Coal stations will be "lightning rod" for global dissent, warns watchdog's head'. *The Guardian.*

Agence France-Presse (October 13, 2009). 'Brazil's Lula vows to slow rate of Amazon deforestation'. AFP.

Agrawala, S., Raksakulthai, V., van Aalst, M., Larsen, P., Smith, J. and Reynolds, J. (2003). 'Development and Climate Change in Nepal: Focus on Water Resources and Hydropower'. OECD, Paris.

Aguilera, R., Eggert, R., Gustavo Lagos, C. C. and Tilton, J. E. (2009). 'Depletion and the Future Availability of Petroleum Resources'. *The Energy Journal*, 30 (1).

Ahlbrandt, T. S., Charpentier, R. R., Klett, T. R., Schmoker, J. W. and Schenk, C. J. (2000). *US Geological Survey World Petroleum Assessment 2000.* United States Geological Survey.

Akimoto, K. and Tomoda, T. (2005). 'Cost and Technology Role for Different Levels of CO2 Concentration Stabilisation'. *International Scientific Symposium: Avoiding Dangerous Climate Change.* Exeter, UK.

Aldy, J. and Pizer, W. (2009). 'Issues in Designing U.S. Climate Change Policy'. *The Energy Journal*, 30 (3), 179–209.

Aleklett, K. (2007). 'Reserve Driven Forecasts for Oil, Gas and Coal and Limits in Carbon Dioxide Emissions'. Joint Transport Research Centre, Discussion Paper No. 2007–18, 20.

Al-Juaied, M. and Whitmore, A. (2009). 'Realistic Costs of Carbon Capture'. John F. Kennedy School of Government.

Allen, M. R. (2007). 'Atmosphere: A Reappraisal of Climate Sensitivity'. *Science*, 318 (5850), 582–3.

Allen, M. R., Frame, D. J., Huntingford, C., Jones, C. D., Lowe, J. A., Meinshausen, M. et al. (2009). 'Warming caused by cumulative carbon emissions towards the trillionth tonne'. *Nature*, 458, 1163–6.

Allen, M., Frame, D., Frieler, K., Hare, W., Huntingford, C., Jones, C. et al. (2009). 'The exit strategy'. *Nature reports*, 3, 56–8.

BIBLIOGRAPHY

Allison, L. (May 21, 2007). 'An Arizona Tellurium Rush?' Retrieved 9 May 2009 from Arizona Geology: http://arizonageology.blogspot.com/2007/05/arizona-tellurium-rush.html

Allwood, J. M. and Cullen, J. M. (2009). 'Steel, aluminium and carbon: alternative strategies for meeting the 2050 carbon emission targets'. Low Carbon Materials Processing, University of Cambridge.

Alumina Limited (n.d.). 'Global Operations'. Retrieved 30 May 2009 from Alumina Limited: http://www.aluminalimited.com/index.php?s=awac_biz&ss= global&p=global_op#Kwinana

American Petroleum Institute (2007). 'The promise of energy with lower CO_2 emissions'. Retrieved 7 September 2009 from American Petroleum Institute: http://www.api.org/ehs/climate/new/upload/CCS_Brochure_final.pdf

Anderson, A. S. (May 14, 2009). 'Statement by A. Scott Anderson, Environmental Defense Fund, Regarding S.1013, the Department of Energy Carbon Capture and Sequestration Program Amendments Act of 2009'. Retrieved 8 September 2009 from Environmental Defense Fund: http://www.edf.org/documents/9741_CCS-Anderson-Senate-Testimony-2009-May-12.pdf

Ansolabehere, S., Beer, J., Deutch, J., Ellerman, A. D., Friedmann, S. J., Herzog, H. et al. (2007). *The Future of Coal*. Massachusetts Institute of Technology.

Appropedia (May 19, 2007). 'Kenya Ceramic Jinko'. Retrieved 25 August 2009 from Appropedia: http://www.appropedia.org/Kenya_Ceramic_Jinko

Archer, D., Eby, M., Brovkin, V., Ridgwell, A., Cao, L., Mikolajewicz, U. et al. (2009). 'Atmospheric Lifetime of Fossil Fuel Carbon Dioxide'. *Annual Review of Earth and Planetary Sciences*, 37, 117–34.

Arcoumanis, C. and Kamimoto, T. (2008). *Flow and Combustion in Automotive Engines*. Springer.

Aresta, M. (2003). *Carbon Dioxide Recovery and Utilization*. Springer.

Argonne National Laboratory. 'Well to Wheel Analysis: Energy use and GHG emissions of selected fuel/powertrain combinations'. US Department of Energy.

Arrow, K. J. (2007). 'The Case for Mitigating Greenhouse Gas Emissions'. Retrieved 24 April 2009 from Project Syndicate: http://www.project-syndicate.org/commentary/arrow1

Aune, F. R., Rosendahl, K. E. and Sagen, E. L. (2009). 'Globalisation of Natural Gas Markets—Effects on Prices and Trade Patterns'. *Energy Journal* (Special Issue).

Ausebel, J. H. (2002). *Climate Change: Some Ways to Lessen Worries*. New Delhi, India: Liberty Institute.

Austin, A. (May 12, 2009). 'Dynamotive, BlueLeaf release biochar test results'. *Biomass Magazine*.

Australian Government: Department of Energy, Resources and Tourism (April 6, 2009). 'Guidance Notes for Applicants'. Retrieved 15 August 2009 from Australian Government: Department of Energy, Resources and Tourism: http://www.ret.gov.au/resources/carbon_dioxide_capture_and_geological_

storage/carbon_capture_and_storage_acreage_release/guidance_notes_for_
applicants/Pages/guidance_notes_for_applicants.aspx

Australian Government: Department of Resources, Energy and Tourism (December 19, 2008). 'National Low Emissions Coal Initiative (NLECI)'. Retrieved 15 August 2009 from Australian Government: Department of Resources, Energy and Tourism: http://www.ret.gov.au/resources/resources_programs/nleci/Pages/NationalLowEmissionsCoalInitiative.aspx

Azar, C., Lindgren, K., Larson, E. and Möllersten, K. (2006). 'Carbon capture and storage from fossil fuels and biomass—Costs and potential role in stabilizing the atmosphere'. *Climatic Change*, 74, 47–79.

AZO Materials (April 6, 2009). 'Global Cement Production to Approach 4 Billion Tonnes by 2012'. Retrieved 11 August 2009 from AZO Materials: http://www.azom.com/news.asp?NewsID=16364

Baciocchi, R., Storti, G. and Mazzotti, J. (2006). 'Process design and energy requirements for the capture of carbon dioxide from air'. *Chemical Engineering Processing*, 45, 1047–58.

Baeza, R., Balagopal, B., Boudler, E., Buffet, M. and Velken, I. (2008). 'Carbon Capture and Storage: A Solution to the Problem of Carbon Emissions'. Boston Consulting Group.

Baines, S. J. and Worden, R. H. (2004). *Geological Storage of Carbon Dioxide*. Special Publication no. 233. London, UK: Geological Society.

Ballentine, C. J., Schoell, M., Coleman, D. and Cain, B. A. (2001). '300-Myr-old magmatic CO_2 in natural gas reservoirs of the west Texas Permian basin'. *Nature*, 409 (6818), 327–31.

Bamber, J. L., Riva, R. E., Vermeersen, B. L. and LeBrocq, A. M. (2009). 'Reassessment of the Potential Sea-Level Rise from a Collapse of the West Antarctic Ice-Sheet'. *Science*, 324 (5929), 901–3.

Banerjee, R., Phan, A., Wang, B., Knobler, C., Furukawa, H., O'Keeffe, M. et al. (2008). 'Imidazolate Frameworks and Application to CO_2 Capture'. *Science*, 319, 939–43.

Barker, T., Koehler, J. and Villena, M. (2002). 'The costs of greenhouse gas abatement: a meta-analysis of post-SRES mitigation scenarios'. *Environmental Economics and Policy Studies*, 5, 135–66.

Barnes, P. (2001). *Structure and Performance of Cements* (2nd ed.). Taylor & Francis.

Baugh, L. L. (n.d.). 'The Promise and the Challenge of Carbon Capture and Geologic Sequestration'. Fulbright & Jaworski LLP Global Energy Law Brief, p. 5.

Baxter, L. (July-August, 2009). 'Cryogenic carbon capture technology'. *Carbon Capture Journal*, pp. 18–21.

BBC News (August 13, 2009). 'Australia emissions plan rejected'. Retrieved 23 August 2009 from BBC News: http://news.bbc.co.uk/2/hi/asia-pacific/8198690.stm

BBC News (October 16, 2009). 'Carbon capture plant backed by EU'. Retrieved 18 October 2009 from BBC News: http://news.bbc.co.uk/2/hi/uk_news/england/south_yorkshire/8311286.stm

BIBLIOGRAPHY

BBC News (April 23, 2009). '"Clean" coal plants get go-ahead'. Retrieved 13 August 2009 from BBC News: http://news.bbc.co.uk/2/hi/uk_news/politics/8014295.stm

BBC News (October 8, 2009). 'New Kingsnorth coal plant delayed'. Retrieved 9 October 2009 from BBC News: http://news.bbc.co.uk/2/hi/uk_news/8296076.stm

BBC News (August 20, 2009). 'Russia tackles Siberia oil slick'. Retrieved 23 August 2009 from BBC News: http://news.bbc.co.uk/2/hi/europe/8209663.stm

BBC News (May 21, 2009). 'Wind farm "kills Taiwanese goats"'. Retrieved 21 May 2009 from BBC News: http://news.bbc.co.uk/2/hi/asia-pacific/8060969.stm

Beaubien, S., Ciotoli, G., Coombs, P., Dictor, M., Krüger, M., Lombardi, S. et al. (2008). 'The impact of a naturally occurring CO_2 gas vent on the shallow ecosystem and soil chemistry of a Mediterranean pasture (Latera, Italy)'. *International Journal of Greenhouse Gas Control*, 2 (3), 373–87.

Bell, D. (October 2, 2009). 'Poll finds support for Michigan coal-fired plants'. *Michigan Business*.

Bellona (n.d.). 'CO_2-håndtering'. Retrieved 29 October 2009 from Bellona: http://www.bellona.no/subjects/1138831369.22

Bhadeshia, H. K. (n.d.). 'Lecture 12: High Temperature Alloys' . Retrieved 9 August 2009 from http://www.msm.cam.ac.uk/phase-trans/abstracts/L12.pdf

Bhagwat, S. (September 24, 2009). 'Could agroforestry solve the biodiversity crisis and address poverty?' (J. Hance, Interviewer)

Biello, D. (June 22, 2009). 'A Scientific Look at "Clean" Coal'. *Scientific American*.

Biello, D. (April 10, 2009). 'How Fast Can Carbon Capture and Storage Fix Climate Change?' *Scientific American*.

Biello, D. (April 8, 2009). 'Storing the Carbon in Fossil Fuels Where It Came from: Deep Underground'. *Scientific American*.

Biopact (October 25, 2007). 'Carbon-negative bioenergy is here: GreatPoint Energy to build biomass gasification pilot plant with carbon capture and storage'. Retrieved 25 May 2009 from Biopact: http://news.mongabay.com/bioenergy/2007/10/carbon-negative-bioenergy-is-here.html

Biopact (October 24, 2007). 'Carbon-negative bioenergy recognized as Norwegian CO2 actors join forces to develop carbon capture technologies' . Retrieved 2 September 2009 from Biopact: http://news.mongabay.com/bioenergy/2007/10/carbon-negative-bioenergy-recognized-as.html

Biopact (September 17, 2007). 'Scientists develop low-lignin eucalyptus trees that store more CO2, provide more cellulose for biofuels'. Retrieved 2 September 2009 from Biopact: http://news.mongabay.com/bioenergy/2007/09/scientists-develop-low-lignin.html

Black, R. (June 19, 2009). 'Climate "meltdown", yet fusion lags'. Retrieved 20 June 2009 from BBC News: http://www.bbc.co.uk/blogs/thereporters/richardblack/

Black, R. (August 27, 2009). 'Climate protection "to cost more"'. Retrieved 28 August 2009 from BBC News: http://news.bbc.co.uk/2/hi/science/nature/8224823.stm

Black, R. (August 25, 2009). 'Forest fires—a continuing Greek tragedy?' Retrieved 25 August 2009 from BBC News: http://news.bbc.co.uk/2/hi/science/nature/8220491.stm

Black, R. (April 16, 2009). 'West Africa faces "megadroughts"'. Retrieved 17 April 2009 from BBC News: http://news.bbc.co.uk/2/hi/science/nature/8003060.stm

Blue Flint Ethanol (2009). 'Blue Flint Ethanol'. Retrieved 23 August 2009,from http://www.blueflintethanol.com/index.asp

Bock, B. R., Rhudy, R. G. and Herzog, H. J. (2003). 'CO$_2$ Storage and Sink Enhancement: Developing Comparable Economics'. *Second Annual Carbon Sequestration Conference* (p. 14).

Bock, B., Rhudy, R., Herzog, H., Klett, M., Davison, J., De La Torre Ugarte, D. G. et al. (2003). 'Economic Evaluation of CO$_2$ Storage and Sink Enhancement Options' . TVA Public Power Institute.

Bohm, M. C., Herzog, H. J., Parsons, J. E. and Sekar, R. C. (2007). 'Capture-ready coal plants—Options, technologies and economics'. *International Journal of Greenhouse Gas Control*, 1, 113–20.

Bond, G. M., Medina, M.-G., Stringer, J. and Simsek-Ege, F. A. 'CO$_2$ Capture from Coal-Fired Utility Generation Plant Exhausts, and Sequestration by a Biomimetic Route Based on Enzymatic Catalysis—Current Status'. National Energy Technology Laboratory.

Borenstein, S. (June 10, 2009). 'Not so windy: Research suggests winds dying down'. Associated Press.

Boyd, P. W. (2008). 'Ranking geo-engineering schemes'. *Nature Geoscience*, 1, 722–4.

BP (2008). *BP Statistical Review of World Energy.*

BP (2009). *BP Statistical Review of World Energy.*

Bradsher, K. (December 21, 2006). 'Outsize Profits, and Questions, in Effort to Cut Warming Gases'. *New York Times.*

Brehm, D. (September 2, 2009). 'Methane gas likely spewing into the oceans through vents in sea floor'. Retrieved 13 September 2009 from MIT News: http://web.mit.edu/newsoffice/2009/methane-0902.html

Brennan, R. J. (April 14, 2009). 'Tiny Saskatchewan town turns carbon trap into cash'. *Toronto Star.*

Bressand, F., Farrell, D., Haas, P., Morin, F., Nyquist, S., Remes, J. et al. (May, 2007,). 'Curbing Global Energy Demand Growth: the Energy Productivity Opportunity'. *McKinsey Quarterly.*

Breunese, J. N. (2006). 'The Netherlands: A Case of Optimisation of Recovery and Opportunities for Re-Use of Natural Gas Assets'. *23rd World Gas Conference* (p. 15). Amsterdam, Netherlands.

Breunese, J. and van Leverink, D. (2008). 'Exploration Country Focus: the Netherlands'. American Association of Petroleum Geologists Newsletter, 4.

Broecker, W. S. (June 18, 2008). 'Deep divisions'. *The Guardian*.

Broomby, R. (July 8, 2009). 'Nuclear dawn delayed in Finland'. Retrieved 8 July 2009 from BBC News: http://news.bbc.co.uk/2/hi/europe/8138869. stm

Brown, J. (2009). Personal communication.

Bryce, R. (July 19, 2005). 'Corn Dog: the ethanol subsidy is worse than you can imagine'. Retrieved 11 May 2009 from Slate: http://www.slate.com/id/2122961/

Bryngelsson, M. and Westermark, M. (2009). 'CO_2 capture pilot test at a pressurized coal fired CHP plant'. *Energy Procedia*, 1 (1), 1403–10.

Buchanan, A. H. and Levine, S. B. (1999). 'Wood-based building materials and atmospheric carbon emissions'. *Environmental Science & Policy*, 2 (6), 427–37.

Buchwitz, M. (August 10, 2009). 'IUP/IFE SCIAMACHY WFM-DOAS'. Retrieved 22 August 2009 from University of Bremen: http://www.iup.uni-bremen.de/sciamachy/NIR_NADIR_WFM_DOAS/

Buchwitz, M., Schneissing, O., Burrows, J. P., Bovensmann, H., Reuter, M. and Notholt, J. (2007). 'First direct observation of the atmospheric CO_2 year-to-year increase from space'. *Atmospheric Chemistry and Physics*, 7, 4249–56.

Bullis, K. (September 4, 2009). 'Mixing Solar with Coal to Cut Costs'. Retrieved 13 September 2009 from *Technology Review*: http://www.technologyreview.com/energy/23349/?a=f

Bundesanstalt für Geowissenschaften und Rohstoffe (BGR) (2005). *Resources and Availability of Energy Resources*.

Burns, J. (August 18, 2009). 'Methane seeps from Arctic sea bed'. Retrieved 18 August 2009 from BBC News: http://news.bbc.co.uk/2/hi/science/nature/8205864.stm

Campbell, C. J. (October, 2005). *Oil Crisis*. Multi-Science Publishing.

Campbell, C. J. (August 15, 2006). 'ProductionDepletion2005.xls' . Retrieved 28 March 2009, from The Coming Global Oil Crisis: http://www.oilcrisis.com/campbell/ProductionDepletion2005.xls

Canadell, J. G. and Raupach, M. R. (2008). 'Managing Forests for Climate Change Mitigation'. *Nature*, 320, 1456–7.

Capitol Climate Action (March 3, 2009). 'Mass civil disobedience at the coal-fired Capitol Power Plant, Washington, DC'. Retrieved 15 August 2009 from Capitol Climate Action: http://www.capitolclimateaction.org/

Carbon Capture Journal (May–June, 2009). 'Nuon starts construction of CO_2 capture test at Buggenum', p. 20.

Carbon Trust (2009). 'Life-cycle energy and emissions of marine energy devices'. Retrieved 5 September 2009 from Carbon Trust: http://www.carbontrust.co.uk/technology/technologyaccelerator/life-cycle_energy_and_emissions.htm

Carlisle, T. (August 17, 2009). 'Gulf exporters will need to make a clean break'. *The National*.

Cascio, J. (December 14, 2005). 'Stabilization Wedges'. Retrieved 26 July 2009 from WorldChanging: http://www.worldchanging.com/archives/003861. html

Cascio, J. (May/June, 2009). 'The Next Big Thing: Resilience'. *Foreign Policy*.

Central Intelligence Agency (April 23, 2009). 'Country Comparison—Military Expenditure'. Retrieved 13 May 2009 from *The World Factbook*: https://www.cia.gov/library/publications/the-world-factbook/rankorder/2034rank. html

Chaffin, J. (July 29, 2009). 'Public wary of carbon capture'. *Financial Times*.

Chaffin, J. (July 28, 2009). 'The carbon-capture challenge'. *Financial Times*.

Chakravorty, U., Hubert, M.-H. and Nøstbakken, L. (2009). 'Fuel versus Food'. *The Annual Review of Resource Economics*, 1 (23), 1–19.

Challenor, P., Hankin, R. and Marsh, B. (2005). 'The Probability of Rapid Climate Change'. *Avoiding Dangerous Climate Change*, p. 23. Exeter.

Chand, S. (September 10, 2009). 'UK climate scepticism more common'. Retrieved 12 September 2009 from BBC News: http://news.bbc.co.uk/2/hi/science/nature/8249668.stm

Charles, P. (October 25, 2007). Speech to a WWF Gala Dinner. Hampton Court Palace, UK.

Chatham House (2007). 'Workshop on reducing emissions from tropical deforestation'. London, UK.

Chow, E. C., Bovair, J. L., Pumphrey, D. L. and Nichols, M. W. (2007). Topic Paper #30: *Historical Perspective on Energy Crises and U.S. Policy Responses*. National Petroleum Council.

Christiansen, T. (n.d.). 'Sargas Pressurized Power plant with integrated CO_2-capture'. Retrieved 16 May 2009 from Sargas AS: http://www.sargas.no/Assets/Text%20files/sargas_websiteversion_summary_report.pdf

Cleanteach Group LLC (May 15, 2007). 'Alcoa develops carbon capture for aluminum plants'. Retrieved 23 May 2009 from Cleantech Group LLC: http://cleantech.com/news/1165/alcoa-develops-carbon-capture-for-alum

Clement, A. C., Burgman, R. and Norris, J. R. (2009). 'Observational and Model Evidence for Positive Low-Level Cloud Feedback'. *Science*, 325 (5939), 460–64.

climatelab (n.d.). 'Cement'. Retrieved 2 September 2009 from climatelab: http://climatelab.org/Cement

CNG Services Ltd (2009). 'GTL v LNG/CNG for transport applications' .

CO_2SINK Project Team (September, 2005). 'CO_2SINK—The First Year'. Retrieved 13 June 2009 from IEA Greenhouse R&D Programme: http://www.ieagreen.org.uk/sept79.htm#15

Coal Utilization Research Council (n.d.). 'DOE Large-scale CO_2 Injection Projects' . Retrieved 14 August 2009 from Coal Utilization Research Council: http://www.coal.org/userfiles/File/DOE_Large-Scale_sequestration_projects. pdf

Cohen, D. M. (July, 2009). *World Oil*, p. 23.

Cohen, D. M. (May, 2009). 'Hydate production and CO_2 injection: Two birds with one stone'. *World Oil*, p. 21.

Colasurdo, C. (2000). 'Mammoth's Perilous Magma'. *California Wild*.

Conn, C., Baum, G., Mudd, C. and Gunnulfsen, J. 'Potential for Geologic Storage of CO_2 in Western Maryland—Phase I Studies'. Maryland Department of Natural Resources.

Cookson, C. (September 9, 2009). 'Britain urged to exploit potential of CO_2 capture'. *Financial Times*.

Cooper, C. (2009). 'A Technical Basis for Carbon Dioxide Storage' . CO_2 Capture Project.

Cooper, J. F. and Berner, K. (2005). 'The Carbon/Air Fuel Cell: Conversion of Coal-Derived Carbons'. *The Carbon Fuel Cell Seminar* (p. 15). Palm Springs, California, USA.

Council of the European Union (11 March, 2005). 'Climate Change: Medium and longer term emission reduction strategies, including targets'. Brussels.

Cox, P., Jones, C. and Huntingford, C. (2005). 'Conditions for Positive Feedbacks from the Land Carbon Cycle'. *Avoiding Dangerous Climate Change* (p. 35). Exeter.

Cross, A. (August 30, 2009). 'Winning over the "Nimby blockade"'. Retrieved 13 September 2009 from BBC News: http://news.bbc.co.uk/2/hi/science/nature/8223048.stm

Dahowski, R. T., Davidson, C. L., Dooley, J. J. and Gentile, R. H. (2008). *Regional Opportunities for Carbon Dioxide Capture and Storage in China*. New Orleans, USA: Battelle.

Dansgaard, W., White, J. W. and Johnsen, S. J. (1989). 'The abrupt termination of the Younger Dryas climate event'. *Nature*, 339, 532–4.

Darley, J. (2004). *High Noon for Natural Gas*. Vermont, USA: Chelsea Green Publishing.

Dasgupta, P. (November 11, 2006). 'Comments on the Stern Review's Economics of Climate Change' . Retrieved 18 April 2009 from http://www.econ.cam.ac.uk/faculty/dasgupta/STERN.pdf

Day, D. (2005). 'Bioenergy and Land Stewardship in China'. *Conserve or Invest? What We Earn from Carbon Utilization*, (p. 63). University of Georgia Bioconversion Center, Presentation to National Association of Conservation Districts.

Day, P. (August 15, 2009). 'Prosperity promise of Bolivia's salt flats'. Retrieved 16 August 2009 from BBC News: http://news.bbc.co.uk/2/hi/programmes/from_our_own_correspondent/8201058.stm

Dewees, D. N. (2008). 'Pollution and the Price of Power'. *The Energy Journal*, 28 (2).

Diamond, J. (2005). *Collapse: How Societies Choose to Fail or Survive*. London UK: Penguin.

Dickens, G. R., Castillo, M. M. and Walker, J. G. (2001). 'A blast of gas in the latest Paleocene: Simulating first-order effects of massive dissociation of oceanic methane hydrate'. *Geology*, 25, 259–62.

BIBLIOGRAPHY

Dikeman, N. (April 22, 2009). 'The REAL story on Moore's Law and solar'. Retrieved 11 May 2009 from CleanTech Group LLC: http://cleantech.com/news/4395/moores-law-solar-neal-dikeman

DirectCarbon (2008). 'A new kind of fuel cell'. Retrieved 25 May 2009 from directcarbon: http://www.directcarbon.com/technology.html

Distler, V. V., Kryachko, V. V. and Yudovskaya, M. A. (2008). 'Ore petrology of chromite-PGE mineralization in the Kempirsai ophiolite complex'. *Mineralogy and Petrology*, 92 (1–2), 31–58.

Dodson, S. (May 11, 2006). 'A cracking alternative to cement'. *The Guardian*.

Dooley, J. J., Dahowski, R. T., Davidson, C. L., Wise, M. A., Gupta, N., Kim, S. H. et al. (2006). 'Carbon Dioxide Capture and Geologic Storage'. Global Energy Technology Strategy Program.

Doran, P. T. and Kendall Zimmerman, M. (2009). 'Examining the Scientific Consensus on Climate Change'. *Climate Change*, 90 (3), 22–3.

Doughton, S. (November 24, 2006). 'Can we lock greenhouse gases away in rocks?' *Seattle Times*.

Dunnahoe, T. (June 1, 2009). 'Dirty talk should not be allowed'. *E&P Magazine*.

Dutt, K. (2009). 'Governance, institutions and the environment-income relationship: a cross-country study'. *Environmental Development and Sustainability*, 705–23.

Economist, The (July 30, 2007). 'Finally'.

Economist, The (March 5, 2009). 'Scrubbing the skies'.

Economist, The (June 11, 2009). 'The future of the forest'.

Edgar (September 1, 2009). 'Background information: EDGAR v4.0 greenhouse gas emissions dataset'. Retrieved 1 September 2009 from European Commission: http://edgar.jrc.ec.europa.eu/background.php

Edmonds, J. (2005). 'Two-Degrees of Climate Change'. *Avoiding Dangerous Climate Change*, (p. 27). Exeter.

Edner, D. (August 21, 2009). 'Natural gas price sinks to lowest in 7 years'. Vancouver: *Globe and Mail*.

Encyclopædia Britannica (August 1, 2009). 'Sedimentary rock'. Retrieved 1 August 2009 from Encyclopædia Britannica Online: http://www.britannica.com/EBchecked/topic/532232/sedimentary-rock

Energy Information Administration (2009). *Annual Energy Outlook 2009* (revised). Washington DC.

Energy Information Administration (July 10, 2009). 'Average Retail Price of Electricity to Ultimate Consumers: Total by End-Use Sector'. Retrieved 25 July 2009 from Energy Information Administration: http://www.eia.doe.gov/cneaf/electricity/epm/table5_3.html

Energy Information Administration. (July, 2009). 'Short-Term Energy Outlook—Real Petroleum Prices'. Retrieved 25 July 2009 from Energy Information Administration: http://www.eia.doe.gov/emeu/steo/pub/fsheets/real_prices.html

Energy Information Administration (2008). Table 6.2. 'World Total Net Electricity Consumption, 1980–2006'.

Energy Innovation Financial Network (n.d.) 'Project aims to extract dam methane'. Retrieved 11 May 2009 from Energy Innovation Financial Network: http://www.eifn.ipacv.ro/newsletters/uploads/newsletter20_Project_aims_to_extract_dam_methane.pdf

Enkvist, P.-A., Nauclér, T. and Rosander, J. (2007). 'A cost curve for greenhouse gas reduction'. *McKinsey Quarterly*, 1.

Environmental Expert (October 5, 2009). 'Secretary Chu announces first awards from US$1.4bn for industrial carbon capture and storage projects'. Retrieved 6 October 2009 from Environmental Expert: http://air.environmental-expert.com/resultEachPressRelease.aspx?cid=29287&codi=71863&lr=1&idCategory=1

Environmental Science Activities for the 21st Century (n.d.). 'Trees and Carbon'. Retrieved 9 May 2009 from Environmental Science Activities for the 21st Century: http://esa21.kennesaw.edu/activities/trees-carbon/trees-carbon.pdf

eRedux (2005). 'Florida Energy Consumption Information'. Retrieved 3 July 2009 from eRedux: http://www.eredux.com/states/state_detail.php?id=1114

EurActiv (July 28, 2009). 'Carbon tariffs falling out of favour as trade war looms'. Retrieved 15 August 2009 from EurActiv: http://www.euractiv.com/en/climate-change/carbon-tariffs-falling-favour-trade-war-looms/article-184449

EurActiv (May 26, 2009). 'Commission faces revolt over "carbon leakage" plans'. Retrieved 15 August 2009 from EurActiv: http://www.euractiv.com/en/climate-change/commission-faces-revolt-carbon-leakage-plans/article-182634

European Commission (n.d.). 'Planning Forests for Climate Conservation'. Retrieved 28 August 2009 from Wonders of Life: http://ec.europa.eu/research/quality-of-life/wonderslife/project08_en.html

European Technology Platform for Zero Emission Fossil Fuel Power Plants (ZEP) (2008). *EU Demonstration Programme for CO_2 Capture and Storage (CCS)*.

EUTurbines (2007). *EUTurbines Position on Zero Emission Fossil Fuel Power Plants (ZEP)/Carbon Capture Storage (CCS)*.

Evans, A. (2009). *The Feeding of the Nine Billion: Global Food Security for the 21st Century*. London, UK: Chatham House.

Evans, R. K. (March 2008). *An Abundance of Lithium*.

Faaij, A. (2007). *Biomass and biofuels: A background report for the Energy Council of the Netherlands*. Universiteit Utrecht, Utrecht, Netherlands.

Farah, D. (June 4, 2002). 'Liberian Leader Again Finds Means to Hang On'. *Washington Post*.

Fargione, J., Hill, J., Tilman, D., Polasky, S. and Hawthorne, P. (2008). 'Land Clearing and the Biofuel Carbon Debt'. *Science*, 319 (5867), 1235–8.

Farrell, A. (2005). 'How Clean Is Clean, At What Cost, And When?' *Clean Coal Technology Status and Potential Issues* (p. 52). Sacramento, California, USA.

Faust, M. (September 26, 2007). Testimony before Sacramento Metropolitan Air Quality Management District (SMA QMD) RE: Wood Burning Rule 421.

Feng, M. (May–June, 2009). 'Recent development on solid sorbents for CO_2 capture'. *Carbon Capture Journal*, pp. 21–4.

Fertilizer Institute, The (2009). 'The Fertilizer Institute'. Retrieved 26 July 2009 from The Fertilizer Institute: http://www.tfi.org/

Fitzgerald, J., Winchester, R. and Johns, J. (2008). *Carbon Capture and Storage country readiness index*. Ernst and Young.

Flyvbjerg, B., Bruzelius, N. and Rothengatter, W. (2003). *Megaprojects and Risk*. Cambridge, UK: Cambridge University Press.

Fogarty, D. (September 29, 2009). 'Indonesia CO_2 pledge to help climate talks: greens'. Reuters.

Foley, C. (June, 2009). 'The End of the Amazon?' *Foreign Policy*.

Food and Agriculture Organization of the United Nations (2007). *State of the World's Forests 2007*. Rome: Electronic Publishing Policy and Support Branch, Communication Division, Food and Agriculture Organization.

Franklin, C. and Schwartz, M. (April 24, 2009). 'For the Coal Industry, the Waxman-Markey Bill's CCS Provisions are a Mixed Bag'. Retrieved 11 September 2009 from ClimateIntel: http://climateintel.com/2009/04/24/for-the-coal-industry-the-waxman-markey-bill%E2%80%99s-ccs-provisions-are-a-mixed-bag/

Franklin, C. and Schwartz, M. (July 30, 2009). 'Regulation of Hydrofracturing: What Effect will it Have on CCS?' Retrieved 13 August 2009 from ClimateIntel: http://climateintel.com/2009/07/30/regulation-of-hydrofracturing-what-affect-will-it-have-on-ccs/

Friends of the Earth Action (n.d.). 'Friends of the Earth Action's Campaigns'. Retrieved 13 September 2009 from Friends of the Earth Action: http://action.foe.org/t/4027/content.jsp?content_KEY=3351

Friends of the Earth International (2009). *who benefits from gm crops?*

Fröhlingsdorf, M., Knauer, S. and Schwägerl, C. (May 26, 2009). 'German Carbon Sequestration Plans Stall'. Retrieved 13 September 2009, from Spiegel Online: http://www.spiegel.de/international/germany/0,1518,632620,00.html

Fry, T. (May 1, 2009). 'The changing policy landscape in a post-moratorium world'. *Offshore Magazine*.

Fung, P. (2009). *China's Energy Sector: A clearer view*. KPMG.

FutureGen Alliance (July 14, 2009). 'Department of Energy releases record of decision on FutureGen'. Retrieved 14 August 2009 from FutureGen Alliance: http://www.futuregenalliance.org/news/releases/pr_07–14–09.pdf

Garnaut, R. (2008). *The Garnaut Climate Change Review*. Cambridge, UK: Cambridge University Press.

Geological Society (2006). *Brick and other ceramic products*. London, UK.

Gerlach, T. M. (1991). 'Present-day CO_2 emissions from volcanoes'. *Transactions of the American Geophysical Union (EOS)*, 72, 249, 254–5.

Giardini, D., Grynthal, G., Shedlock, K. and Zhang, P. (1999). 'Global Seismic Hazard Map'. Retrieved 3 August 2009 from http://geology.about.com/gi/dynamic/offsite.htm?site=http://www.seismo.ethz.ch/gshap/

Gibbins, J. (2008). 'Making CCS Work: Economics and Critical Issues: Sequencing the Deployment'. The Senate Energy and Natural Resources Committee (p. 17). Washington DC; Imperial College, London.

Gielen, D. and Unander, F. (March 2005). *Alternative Fuels: An Energy Technology Perspective*. Paris: International Energy Agency.

Gieré, R. and Stille, P. (2004). *Energy, Waste and the Environment*. The Geological Society.

Gilfillan, S. M., Lollar, B. S., Holland, G., Blagburn, D., Stevens, S., Schoell, M. et al. (2009). 'Solubility trapping in formation water as dominant CO_2 sink in natural gas fields'. *Nature*, 458, 614–18.

Gillen, D. W., Morrison, W. G. and Stewart, C. (2008). 'Air Travel Demand Elasticities: Concepts, Issues and Measurement'. Waterloo, Canada: Department of Finance Canada.

Gillingham, K., Newell, R. G. and Palmer, K. (2009). 'Energy Efficiency Economics and Policy'. *Annual Review of Resource Economics* (1), 14.1–14.23.

Gipe, P. (1995). *Wind Energy Comes of Age*. NY: Wiley.

Giroudiere, F., Ambrosino, J. L., Fischer, B., Pavone, D., Sanz-Garcia, E., Le Gall, A. et al. (2009). 'Hygensys: A New Process for Power Production with Pre-Combustion CO_2' in L. I. Eide, *Carbon Dioxide Capture for Storage in Deep Geologic Formations—Results from the CO_2 Capture Project: Volume Three Advances in CO_2 Capture and Storage* (pp. 221–36). Berks, UK: CPL Press.

Glanz, J. (June 23, 2009). 'Deep in Bedrock, Clean Energy and Quake Fears'. *New York Times*, p. A1.

Global CCS Institute (2009). *Strategic Analysis of the Global Status of Carbon Capture and Storage*. Canberra, Australia: Global CCS Institute.

Gold, R. (December 26, 2008). 'Exxon could benefit from emissions work'. *Wall Street Journal*.

Goldberg, P., Chen, Z.-Y., O'Connor, W., Walters, R. and Ziock, H. 'CO_2 Mineral Sequestration Studies in US'. National Energy Technology Laboratory.

Goldenberg, S. (February 26, 2009). 'Coen brothers target US coal industry'. *The Guardian*.

Goodall, C. (2008). *Ten Technologies to Save the Planet*. London, UK: Profile Books.

Gore, A. (2006). *An Inconvenient Truth: the Planetary Emergency of Global Warming and What We Can Do About It*. London, UK: Bloomsbury.

GreatPoint Energy (2009). 'GreatPoint Energy'. Retrieved 25 May 2009 from GreatPoint Energy: http://www.greatpointenergy.com/

Green Alliance (May, 2006). 'Green Alliance response to HM Treasury consultation Carbon capture and storage: A consultation on barriers to commercial implementation' . Retrieved 8 September 2009 from HM Treasury: http://www.hm-treasury.gov.uk/d/carbon_40_green_alliance.pdf

Green Car Congress (July 20, 2007). 'Welsh "GreenBox": Carbon Capture and Algae-to-Biodiesel Scheme'. Retrieved 31 May 2009 from Green Car Congress: http://www.greencarcongress.com/2007/07/welsh-greenbox-.html

Green, M. (n.d.). 'Underground Coal Gasification—A clean indigenous energy option' . Retrieved 24 August 2009 from UCG Engineering Ltd: http://www.coal-ucg.com/publishedarticleonucg.html

GreenBang.(May 29, 2009). 'Ecobox developer wins monthly innovation award'. Retrieved 1 June 2009 from GreenBang: http://www.greenbang.com/ecobox-developer-wins-monthly-innovation-award/

GreenFacts (August 28, 2009). 'Scientific Facts on Forests'. Retrieved 29 August 2009 from GreenFacts: http://www.greenfacts.org/en/forests/l-2/6-forest-wood-timber-food.htm

Greenhouse Gas Protocol Initiative (2009). 'Ammonia'. Retrieved 26 July 2009 from the Greenhouse Gas Protocol Initiative: http://www.ghgprotocol.org/calculation-tools/ammonia

Gribbin, J. and Gribbin, M. (2009). *He Knew He Was Right: The Irrepressible Life of James Lovelock and Gaia*. London, UK: Penguin.

Grupo de Combustión y Gasificación (n.d.). 'Research Activities'. Retrieved 16 May 2009 from Grupo de Combustión y Gasificación: http://www.icb.csic.es/index.php?id=144&L=1

Gunter, W. D., Bachu, S., Palombi, D., Lakeman, B., Sawchuk, W. and Bonner, D. (2009). 'Heartland Area Redwater reef saline aquifer CO_2 storage project'. *Energy Procedia*, 1 (1), 3943–50.

Gustafson, T. (2006). *Crisis Amid Plenty: The Politics of Soviet Energy under Brezhnev and Gorbachev*. Princeton, US: Princeton University Press.

Hamilton, K., Sjardin, M., Shapiro, A. and Marcello, T. (2009). 'Fortifying the Foundation: State of the Voluntary Carbon Markets 2009' . Ecosystem Marketplace/New Carbon Finance.

Hamilton, T. (February 22, 2007). 'Capturing Carbon with Enzymes'. Retrieved 6 August 2009 from *Technology Review*: http://www.technology-review.com/energy/18217/page2/

Hamilton, T. (May 4, 2009). 'Turning Natural Gas Green'. Retrieved 22 May 2009 from *Technology Review*: http://www.technologyreview.com/energy/22580/

Handley, J. (18 October, 2008). 'A Question of Balance: Finding the Optimal Carbon Tax Rate'. Retrieved 10 July 2009 from Carbon Tax Center: http://www.carbontax.org/blogarchives/2008/10/18/a-question- of-balance-finding-the-optimal-carbon-tax-rate/

Hangx, S. J. and Spiers, C. J. (2009). 'Coastal spreading of olivine to control atmospheric CO_2 concentrations: A critical analysis of viaibility'. *International Journal of Greenhouse Gas Control* .

Hao, X., Dong, G., Yang, Y., Xu, Y. and Li, Y. (2007). 'Coal to Liquid (CtL): Commercialization Prospects in China'. *Chemical Engineering Technology*, 30 (9), 1157–65.

Hare, B. (June 18, 2008). 'CO_2 disposal in the ocean is a dangerous distraction'. *The Guardian*.

Harsch, M. A., Hulme, P. E., McGlone, M. S. and Duncan, R. P. (2009). 'Are treelines advancing? A global meta-analysis of treeline response to climate warming'. *Ecology Letters*.

Hartley, A. (October 19, 2004). 'Carbon Trust/GTS Study Proves the Benefits of Using Recycled Glass'. Retrieved 19 May 2009 from GTS: http://www.glass-ts.com/News/PressArchive/PressReleases6.html

Hassanzadeh, H., Pooladi-Darvish, M. and Keith, D. W. (2009). 'Accelerating CO_2 Dissolution in Saline Aquifers for Geological Storage—Mechanistic and Sensitivity Studies'. *Energy & Fuels*, 23, 3328–36.

Haszeldine, R. S., Quinn, O., England, G., Wilkinson, M., Shipton, Z., Evans, J. P. et al. (2005). 'Natural Geochemical Analogues for Carbon Dioxide Storage in Deep Geological Porous Reservoirs, a United Kingdom Perspective'. *Oil and Gas Science and Technology*, 60 (1), 33–49.

Haszeldine, S. (2009). 'Carbon capture and storage: the UK's fourth energy pillar, or broken bridge?' *British Science Festival* (p. 8). Guildford, UK.

Hawksley, H. (April 12, 2009). 'City air pollution "shortens life"'. Retrieved 10 May 2009 from BBC News: http://news.bbc.co.uk/2/hi/health/7946838.stm

Heathcote, M. and Fryer, C. (2008). 'Alternatives to Petroleum Feedstocks: a Leading Role for Asia'. Tecnon OrbiChem Marketing Seminar at APIC 2008 (p. 23). Singapore.

Heinberg, R. (2005). *The Party's Over*. London, UK: New Society Publishers.

Heinrich, J. J., Herzog, H. J. and Reiner, D. M. (2003). 'Environmental Assessment of Geologic Storage of CO_2'. *Second National Conference on Carbon Sequestration* (p. 8). Washington, DC, USA.

Henderson, P. (September 30, 2009). 'Biggest California forest owner enters carbon trade'. Reuters.

Hepple, R. P. and Benson, S. M. (2004). 'Implications of Surface Seepage on the Effectiveness of Geologic Storage of Carbon Dioxide as a Climate Change Mitigation Strategy'. *Environmental Geology*.

Hernandez, M. (2007). 'Driving Gasification Forward'. *Gasification Technologies Conference 2007* (p. 10). San Francisco, USA.

Herring, H. (April, 1998). *Does Energy Efficiency Save Energy: The Implications of Accepting the Khazzoom-Brookes Postulate*. Milton Keynes, UK: The Open University.

Herzog, H. (2009). 'A Research Program for Promising Retrofit Technologies'. MIT Symposium on Retro-fitting of Coal-Fired Power Plants for Carbon Capture.

Herzog, H. (2003). *Assessing the Feasibility of Capturing CO_2 from the Air*. Cambridge, Massachusetts, USA: Massachusetts Institute of Technology, Laboratory for Energy and the Environment.

Heston, A., Summers, R. and Aten, B. (2006). Penn World Tables. Retrieved 20 June 2009 from: http://pwt.econ.upenn.edu/php_site/pwt_index.php

Hetland, J. and Christensen, T. (2008). 'Assessment of a fully integrated SAR-GAS process operating on coal with near zero emissions'. *Applied Thermal Engineering*, 28, 2030–38.

Heydari, E. (2004). 'Anahita, Deev Jahi and Amordaad: Three Oceans of the Permian-Triassic Boundary'. Geological Society of America Annual Meeting, 36, p. 16. Denver, Colorado, USA.

Hildebrand, A. N. and Herzog, H. J. (2009). 'Optimization of Carbon Capture Percentage for Technical and Economic Impact of Near-Term CCS Implementation at Coal-Fired Power Plants'. *Energy Procedia* I, 4135–42.

Hindu, The (December 28, 2006). 'Competitors fuelling Singur controversy: Ratan Tata'.

Hitch, M. (July–August, 2009). 'Revaluing mine waste rock for carbon capture and storage'. *Carbon Capture Journal*, pp. 24–6.

HM Government (2009). *The UK Low Carbon Transition Plan*. Norwich, UK: TSO.

Hoare, A. (2008). *The Search for Innovative Options for the Forests of the Democratic Republic of Congo*. London, UK: Chatham House.

Holloway, S. *Geological sequestration of carbon dioxide—concepts and potential impacts*. Nottingham, UK: British Geological Survey.

Holloway, S., Bentham, M. and Kirk, K. (2006). 'Underground Storage of Carbon Dioxide' in S. Shackley and C. Gough, *Carbon Capture and Its Storage* (p. 313). Surrey, UK: Ashgate Publishing.

Hooijer, A., Silvius, M., Wösten, H. and Page, S. (2006). 'PEAT-CO$_2$—Assessment of CO$_2$ emissions from drained peatlands in SE Asia' . Delft Hydraulics.

Hope, C. (2008). 'Valuing the climate change impacts of tropical deforestation' . Cambridge, UK: Judge Business School.

Hope, C. and Castilla-Rubio, J. C. (2008). 'A first cost-benefit analysis of action to reduce deforestation' . Cambridge, UK: Judge Business School.

Hoskins, B. and Buckle, S. (2009). 'Response to *Towards Carbon Capture and Storage* from the Grantham Institute for Climate Change, Imperial College London' . Retrieved 28 August 2009 from Imperial College London: https://www8.imperial.ac.uk/content/dav/ad/workspaces/climatechange/pdfs/towardscarboncapture.pdf

House, K. Z., House, C. H., Aziz, M. J. and Schrag, D. P. (2007). 'Electrochemical Acceleration of Chemical Weathering as an Energetically Feasible Approach to Mitigating Anthropogenic Climate Change'. *Environmental Science and Technology*, 41, 8464–70.

Howell, K. (August 19, 2009). 'Company Taps Jet-Engine Technology in Bid to Cut Carbon-Capture Costs'. *New York Times*.

HTC Purenergy (May–June, 2009). 'HTC Purenergy—focused on environment, economy and energy security'. *Carbon Capture Journal*, pp. 4–5.

Hurley, J. P., Kadrmas, N. J. and Robson, F. (2005). 'Testing of a Very High-Temperature Heat Exchanger for IFCC Power Systems'. 19th Annual Conference on Fossil Energy Materials. Knoxville, Tennessee, USA.

Hyndman, R. D. and Dallimore, S. R. (2001). 'Gas Hydrates: Natural gas hydrate studies in Canada'. *Recorder, Canadian Society of Exploration Geophysicists*, 26, 11–20.

Hyun-cheol, K. (October 15, 2009). 'No Emission Power Plant Due in 2020'. *Korea Times* .

Imperial College London (n.d.). 'Research: Carbon Capture and Storage'. Retrieved 16 May 2009 from Imperial College London: http://www3.imperial.ac.uk/carboncaptureandstorage/carboncapture/calciumlooping

Index Mundi (n.d.). 'Alumina: World Production, by Country'. Retrieved 30 May 2009 from Index Mundi: http://www.indexmundi.com/en/commodities/minerals/bauxite_and_alumina/bauxite_and_alumina_table12.html

Index Mundi (n.d.). 'Iron and Steel Slag Sold or Used in the United States'. Retrieved 30 May 2009 from Index Mundi: http://www.indexmundi.com/en/commodities/minerals/iron_and_steel_slag/iron_and_steel_slag_t1.html

Index Mundi (n.d.). 'Raw Steel: World Production, by Country'. Retrieved 30 May 2009 from Index Mundi: http://www.indexmundi.com/en/commodities/minerals/iron_and_steel/iron_and_steel_t10.html

Indymedia UK (November 24, 2008). 'UK Coal Industry Braced for 48 hrs of Protests'. Retrieved 15 August 2009 from Indymedia UK: http://www.indymedia.org.uk/en/2008/11/413424.html

Institute for Energy Research (August 5, 2009). 'Abracadabra Energy Policy: Are the Generating Alternatives to Coal-Fired Electricity Ready for Waxman-Markey Targets?' Retrieved 29 August 2009 from Institute for Energy Research: http://www.instituteforenergyresearch.org/2009/08/05/abracadabra-energy-policy-are-the-generating-alternatives-to-coal-fired-electricity-ready-for-waxman-markey-targets/

Institution of Mechnical Engineers (2009). *Geo-Engineering: Giving Us the Time to Act?* London, UK.

Intergovernmental Panel on Climate Change (2007). *Climate Change 2007—Mitigation of Climate Change: Working Group III contribution to the Fourth Assessment Report of the IPCC.* Cambridge, UK: Cambridge University Press.

Intergovernmental Panel on Climate Change (2007). *Climate Change 2007: Synthesis Report.* Valencia, Spain.

Intergovernmental Panel on Climate Change (2000). *IPCC Special Report: Land Use, Land-Use Change, and Forestry.* Cambridge, UK: Cambridge University Press.

International Climate Change Taskforce (January, 2005). *Meeting the Climate Challenge.* London, UK: The Institute for Public Policy Research.

International Energy Agency (2008). *CO_2 Capture and Storage: a key carbon abatement option.* Paris, France: OECD/IEA.

International Energy Agency (2009). 'Statistics and Balances'. Retrieved 15 September 2009 from International Energy Agency: http://www.iea.org/Textbase/stats/index.asp

International Energy Agency (2006). *World Energy Outlook 2006*. Paris, France: OECD/IEA.

International Monetary Fund (2009). *World Economic Outlook*. Washington, DC, USA.

IRIN (December 23, 2004). 'COTE D'IVOIRE: Civil war allows rampant illegal logging'. Retrieved 12 September 2009 from IRIN: http://www.irinnews.org/report.aspx?reportid=52512

Izundu, U. (June 22, 2009). 'Groningen: Unconventional gas resources key to European supply'. *Oil and Gas Journal*.

Jaccard, M. (2005). *Sustainable Fossil Fuels*. Cambridge, UK: Cambridge University Press.

Jaccard, M., Rivers, N., Bataille, C., Murphy, R., Nyboer, J. and Sadownik, B. (May, 2006). *Burning Our Money to Warm the Planet: Canada's Ineffective Efforts to Reduce Greenhouse Gas Emissions*. C.D. Howe Institute, Toronto.

Jassim, S. Z., Suk, M. and Waldhausrová, J. (2008). 'Magmatism and metamorphism in the Zagros Suture' in S. Z. Jassim and J. C. Goff, *Geology of Iraq* (p. 341). The Geological Society.

Jha, A. (December 31, 2008). 'Revealed: The cement that eats carbon dioxide'. *The Guardian*.

Joerss, M., Woetzel, J. R. and Zhang, H. (May, 2009). 'China's green opportunity'. *McKinsey Quarterly*.

Johansson, D. J., Azar, C., Lindgren, K. and Persson, T. A. (2009). 'OPEC Strategies and Oil Rent in a Climate Conscious World'. *The Energy Journal*, 30 (3), 23–30.

Johnson, K. (August 16, 2009). 'A New Test for Business and Biofuel'. *New York Times* .

Johnson, T. (2005). 'Prospects for Brown Coal IDGCC'. Coal 21 1st Annual Conference. Sydney, Australia.

Johnson, T. and Pleasance, G. (October, 1996). 'Clean Coal Technology for Brown Coal Power Generation'. *Australian Coal Review*, pp. 38–41.

Johnston, D. (1994). *International Petroleum Fiscal Systems and Production Sharing Contracts*. Tulsa, OK, USA: Pennwell Books.

Jolly, J. (August 28, 2009). 'Oxfam warning over Nepal climate'. Retrieved 28 August 2009 from BBC News: http://news.bbc.co.uk/2/hi/south_asia/8225901.stm

Jones, C., Cox, P. and Huntingford, C. (2005). 'Impact of climate-carbon cycle feedbacks on emissions scenarios to achieve stabilisation'. *Avoiding Dangerous Climate Change* (p. 18). Exeter.

Jones, G., Joshi, G., Clark, M. and McConchie, D. (September 5, 2006). 'Carbon Capture and the Aluminium Industry: Preliminary Studies'. *Environmental Chemistry*, 3 (4), pp. 297–303.

Jordan, P. D. and Benson, S. M. (2009). 'Well blowout rates and consequences in California Oil and Gas District 4 from 1991 to 2005: implications for geological storage of carbon dioxide'. *Environmental Geology*, 57 (5), 1103–23.

Kahn, M. E. (May/June, 2009). 'Think Again: The Green Economy'. *Foreign Policy*.

Kalkbekken, S. and Rive, N. (2005). 'Why delaying climate action is a gamble'. *Avoiding Dangerous Climate Change* (p. 11). Exeter.

Kapila, R. V., Chalmers, H. and Leach, M. (2009). *Investigating the prospects for Carbon Capture and Storage technology in India*. Christian Aid.

Katz, A. (September 4, 2007). 'Nuclear Bid to Rival Coal Chilled by Flaws, Delay in Finland'. Bloomberg.

Kauppi, P. E., Ausubel, J. H., Fang, J., Mather, A. S., Sedjo, R. A. and Waggoner, P. E. (2006). 'Returning forests analyzed with the forest identity'. *Proceedings of the National Academy of Sciences*, 103 (46), 17574–9.

Keesom, W., Unnasch, S. and Moretta, J. (2009). *Life Cycle Assessment Comparison of North American and Imported Crudes*. Chicago, Illinois, USA: Jacobs Consultancy.

Keith, D. W. and Rhodes, J. S. (2002). 'Bury, burn or both: A two-for-one deal on biomass carbon and energy'. *Climatic Change*, 54, 375–7.

Keith, D. W., Ha-Duong, M. and Stolaroff, J. K. (2005). 'Climate Strategy with CO_2 Capture from the Air'. *Climatic Change*.

Kelemen, P. B. and Matter, J. (2008). 'In situ carbonation of peridotite for CO_2 storage'. *Proceedings of the National Academy of Sciences*, 105 (45), 17295–300.

Kerr, C. (August 18, 2009). 'Rudd warns of tariffs hit to economy if emissions trading scheme not passed'. *The Australian*.

Khadka, N. S. (July 31, 2009). 'Uncertainties surround future monsoons'. Retrieved 1 August 2009 from BBC News: http://news.bbc.co.uk/2/hi/science/nature/8178463.stm

Kharaka, Y. K., Cole, D. R., Hovorka, S. D., Gunter, W. D., Knauss, K. G. and Freifeld, B. M. (2006). 'Gas-water-rock interactions in Frio Formation following CO_2 injection: Implications for the storage of greenhouse gases in sedimentary basins'. *Geology*, 34 (7), 577–80.

Kheshgi, H. S. and Prince, R. C. (2005). 'Sequestration of fermentation CO_2 from ethanol production'. *Energy*, 30 (10), 1865–71.

Kindermann, G., Obersteiner, M., Sohngen, B., Sathaye, J., Andrasko, K., Rametsteiner, E. et al. (2008). 'Global cost estimates of reducing carbon emissions through avoided deforestation'. *Proceedings of the National Academy of Sciences*, 105 (30), 10302–7.

Klemeš, J., Cockerill, T., Bulatov, I., Shackley, S. and Gough, C. (2006). 'Engineering Feasibility of Carbon Dioxide Capture and Storage' in S. Shackley and C. Gough, *Carbon Capture and its Storage: An Integrated Assessment* (pp. 43–85). Surrey, UK: Ashgate Publishing.

Kohberg, T. (2009). *CCS—Carbon Capture and Storage*. Saarbrücken, Germany: Verlag Dr Müller.

Krajick, K. (May, 2005). 'Fire in the Hole'. *Smithsonian Magazine.*

Krasnodebski, A., Ziebinski, B. and Zys, B. (2009). 'Power at a time of change'. *International Financial Law Review* (Supplement).

Krauss, C. (September 6, 2009). 'Natural Gas Hits a Roadblock in New Energy Bill'. *New York Times.*

Krauss, C. and Mouawad, J. (August 18, 2009). 'Oil Industry Backs Protests of Emissions Bill'. *New York Times.*

Krevor, S. C., Graves, C. R., Van Gosen, B. S. and McCafferty, A. E. (2009). *Mapping the Mineral Resource Base for Mineral Carbon Dioxide Sequestration in the Conterminous United States.* United States Geological Survey.

Kulakofsky, D. and Tahmourpour, F. (June, 2009). 'Cementing solutions help protect the environment'. *E&P Magazine*, 41–2.

Kulish, N. and Glanz, J. (September 10, 2009). 'German Geothermal Project Leads to Second Thoughts After the Earth Rumbles'. *New York Times.*

Kunzig, R. and Broecker, W. (2008). *Fixing Climate: The Story of Climate Change and How to Stop Global Warming.* London, UK: Profile Books.

Lackner, K. S. (2006). 'Carbon Sequestration' in R. Mabro, *Oil in the 21st Century* (p. 351). Oxford, UK: Oxford University Press.

Lackner, K. S., Grimes, P. and Ziock, H.-J., *Capturing Carbon Dioxide from Air.* National Energy Technology Laboratory, US DOE.

Lahn, G. (2009). 'The Dilemma of Oil Depletion'. 1st Yemen Energy Forum: Energy Security and Development in Yemen, (p. 29). Sana'a, Yemen.

Lal, R. (2004). 'Soil carbon sequestration to mitigate climate change'. *Geoderma*, 123 (1–2), 1–22.

Landais, E. (October 22, 2009). 'Oil will stay as future energy source'. *Gulf News* .

Latimer, C., Kim, J., Tahara-Stubbs, M. and Wang, Y. (May 29, 2009). 'China's rare earth monopoly threatens global suppliers, rival producers claim'. *Financial Times.*

Le Thiez, P., Torp, T. A., Feron, P., Zweigel, P. and Lindeberg, E. (2005). '"CASTOR"—CO_2 from Capture to Storage—An innovative European Integrated Project'. *Fourth Annual Conference on Carbon Capture and Sequestration* (p. 21). Alexandria, Virginia, USA.

Leber, J. (August 26, 2009). 'Is There Some Light at the End of Coal's Long, Dark Tunnel?' *New York Times.*

Lehmann, J. (2007). 'A handful of carbon'. *Nature*, 447, 143–4.

Lehmann, J., Gaunt, J. and Rondon, M. (2006). 'Bio-char Sequestration in Terrestrial Ecosystems—a Review'. *Mitigation and Adaptation Strategies for Global Change*, 11, 403–27.

Leng, T. B. (May 9, 2007). 'China's reforestation programme hurts farmers' livelihood'. Retrieved 13 September 2009 from Channel NewsAsia: http://www.channelnewsasia.com/stories/eastasia/view/275056/1/.html

Lenton, T. M., Held, H., Kriegler, E., Hall, J. W., Lucht, W., Rahmstorf, S. et al. (2008). 'Tipping elements in the Earth's climate system'. *Proceedings of the National Academy of Sciences*, 105 (6), 1786–93.

Leonenko, Y. and Keith, D. W. (2008). 'Reservoir Engineering To Accelerate the Dissolution of CO_2 Stored in Aquifers'. *Environmental Science & Technology*, 42, 2742–7.

Lever, R. (March 18, 2007). 'Coal's future clouded by global warming debate'. AFP.

Lewandowski, S. R. (July 17, 2006). Intermountain Rural Electric Association memo. Retrieved 11 September 2009 from Desmogblog: http://www.desmogblog.com/files/IREA-memo.pdf

Lex (May 24, 2009). 'Polysilicon'. *Financial Times*.

Lichter, S. R. (April 24, 2008). 'Climate Scientists Agree on Warming, Disagree on Dangers and Don't Trust the Media's Coverage of Climate Change'. Retrieved 12 April 2009 from STATS: http://stats.org/stories/2008/global_warming_survey_apr23_08.html

Lima, I. B., Ramos, F. M., Bambace, L. A. and Rosa, R. R. (2008). 'Methane Emissions from Large Dams as Renewable Energy Resources: A Developing Nation Perspective'. *Mitigation and Adaptation Strategies for Climate Change*, 13 (2), 193–206.

Linsenmaier, J. (2009). 'Carbon Capture and Storage (CCS): A Challenge for the Future'. Vibrant Gujarat Summit (p. 27).

Little, A. (August 4, 2005). 'New Asia-Pacific climate pact is long on PR, short on substance'. Retrieved 13 September 2009 from Grist: http://www.grist.org/article/little-pact/

Livingood, M., Winicaties, J. and Stein, J. (1999). *Centralia Mine Fire Analysis: Presence of Sulfur-bearing Mineral Deposits at Thermal Vents.*

Lockwood, M. (2008). 'Recent changes in solar outputs and the global mean surface temperature. III. Analysis of contributions to global mean air surface temperature rise'. *Proceedings of the Royal Society.*

Lockwood, M. and Fröhlich, C. (2007). 'Recent oppositely directed trends in solar climate forcings and the global mean surface air temperature'. *Proceedings of the Royal Society.*

Lockwood, M. and Fröhlich, C. (2008). 'Recent oppositely directed trends in solar climate forcings and the global mean surface air temperature. II. Different reconstructions of the total solar irradiance variation and dependence on response time scale'. *Proceedings of the Royal Society.*

Lohmann, H. (2009). *The German Gas Market post 2005.* Oxford, UK: Oxford Institute for Energy Studies.

Lomborg, B. (2007). *Cool It: The Skeptical Environmentalist's Guide to Global Warming.* New York, USA: Vintage Books, Random House.

Lomborg, B. (November 2, 2006). 'Stern Review: the dodgy numbers behind the latest warming scare'. *Wall Street Journal.*

Lomborg, B. (2001). *The Skeptical Environmentalist.* Cambridge, UK: Cambridge University Press.

BIBLIOGRAPHY

Lovelock, J. (March 24, 2009). 'James Lovelock on Biochar: let the Earth remove CO_2 for us'. *The Guardian*.

Lovelock, J. (2009). *The Vanishing Face of Gaia*. London, UK: Penguin.

Lovins, A. B., Datta, E. K., Bustnes, O.-E., Koomey, J. G. and Glasgow, N. J. (2004). *Winning the Oil Endgame*. Rocky Mountain Institute.

Lubaś, J., Warchoł, M., Krępulec, P. and Wolnowski, T. (2008). 'Greenhouse gas sequestration in aquifers saturated by natural gases'. *Gospodarka Surowcami Mineralnymi*, 299–308.

Lucas, C. (April 23, 2009). 'No public funding for carbon capture, says Green Party'. Retrieved 8 September 2009 from Green Party: http://www.greenparty.org.uk/news/23–04–2009-clean-coal-distraction.html

Luce, E. (2007). *In Spite of the Gods: The Strange Rise of Modern India*. London, UK: Abacus, LittleBrown.

Lutz, M., Caldeira, K., Dunbar, R. and Behrenfeld, M. (2007). 'Seasonal rhythms of net primary production and particulate organic carbon flux describe biological pump efficiency in the global ocean'. *Journal of Geophysical Research, 112*.

Lynch, T. (2005). 'Clean Coal and Co-Production Potential'. Clean Coal for Transportation Fuels Workshop, (p. 18). West Lafayette, Indiana, USA.

Macdonald-Smith, A. (December 2, 2008). 'Shell, Anglo to Delay A$5 Billion Clean Fuels Project'. Bloomberg.

MacKay, D. J. (2009). *Sustainable Energy—Without the Hot Air*. Cambridge, UK: UIT Cambridge Limited.

Maddison, A. (March, 2009). 'Historical Statistics of the World Economy: 1–2006 AD'. Retrieved 16 August, 2009 from Groningen Growth and Development Centre: http://www.ggdc.net/maddison/Historical_Statistics/horizontal-file_03–2009.xls

Mahasenan, N., Dahowski, R. T. and Davidson, C. L. (2003). *The Role of Carbon Dioxide Capture and Storage in Reducing Emissions from Cement Plants in North America*. Richland, Washington, USA: Pacific Northwest National Laboratory.

Maksoud, J. (May 6, 2009). 'Energy-efficiency goals could reshape Europe's energy landscape'. *E&P Magazine*.

Malhi, Y., Aragão, L. E., Galbraith, D., Huntingford, C., Fisher, R., Zelazowski, P. et al. (2009). 'Exploring the likelihood and mechanism of a climate-change-induced dieback of the Amazon rainforest'. *Proceedings of the National Academy of Sciences*.

Mantripragada, H. C. and Rubina, E. S. (2009). 'CO_2 reduction potential of coal-to-liquids (CTL) plants'. *Energy Procedia* 1, 4331–8.

Marchetti, C. (1977). 'On Geoengineering and the CO_2 problem'. *Climatic Change*, 1, 59–68.

Markels, M. and Barber, R. T. (2001). 'Sequestration of CO_2 by Ocean Fertilization'. *National Energy Technology Laboratory Conference on Carbon Sequestration* (p. 9).

Massachussetts v EPA, 127 S. (Ct.1438 2007).

Mayer, A. J. (n.d.). 'Rethinking U.S. Coal Reserves and Resources'.

McCracken, R. (2Q 2009). 'The Unbearable Lightness of Wind'. *International Association for Energy Economics Forum*, 7–10.

McGrail, B. P., Schaef, H. T. and Spane, F. A. (2008). 'New Findings Regarding Carbon Dioxide Sequestration in Flood Basalts' in M. Goel, B. Kumar and S. N. Charan, *Carbon Capture and Storage: R&D Technologies for Sustainable Energy Future* (p. 224). New Delhi, India: Narosa Publishing House.

McGrath, M. (June 17, 2009). 'Fusion falters under soaring costs'. Retrieved 21 August 2009 from BBC News: http://news.bbc.co.uk/2/hi/science/nature/8103557.stm

McKee, B. N. (2005). 'Discussion Paper from the Task Force for Reviewing and Identifying Standards with Regards to CO_2 Storage Capacity Measurement'. Carbon Sequestration Leadership Forum.

McKillop, A. (2005). *The Final Energy Crisis*. London, UK: Pluto Press.

McKinsey. (June, 2009). 'Exploring global energy demand'. Retrieved 5 June 2009, from *McKinsey Quarterly*: http://www.mckinseyquarterly.com/Economic_Studies/Productivity_Performance/Exploring_global_energy_demand_2369

Meinshausen, M. (2005). 'On the risk of overshooting 2°C'. *Avoiding Dangerous Climate Change* (p. 32). Exeter.

Meinshausen, M., Meinshausen, N., Hare, W., Raper, S. C., Frieler, K., Knutti, R. et al. (2009). 'Greenhouse gas emission targets for limiting global warming to 2°C'. *Nature*, 458, 1158–62.

Melcher, F., Grum, W., Simon, G., Thalhammer, T. V. and Stumpfl, E. F. (1997). 'Petrogenesis of the Ophiolitic Giant Chromite Deposits of Kempirsai, Kazakhstan: a Study of Solid and Fluid Inclusions in Chromite'. *Journal of Petrology*, 38 (10), 1419–58.

Men and Women of Wabash River Energy Limited (2000). 'Wabash River Coal Gasification Repowering Project'. West Terre Haute, Indiana, USA.

Mendelsohn, R. and Dinar, A. (2009). 'Land Use and Climate Change Interactions'. *Annual Review of Resource Economics*, 1, 21.1–21.24.

Méndez, R. (March 17, 2008). 'Repsol cambiará CO_2 por petróleo'. *El País*.

Merour, J. (April 8, 2009). (R. Mills, Interviewer)

Metals Place (March 13, 2009). 'Tellurium production not sufficient to support solar cell industry'. Retrieved 9 May 2009 from Metals Place: http://metalsplace.com/news/articles/26144/tellurium-production-not-sufficient-to-support-solar-cell-industry/

Metz, B., Davidson, O., de Coninck, H., Loos, M. and Meyer, L. (2005). *IPCC Special Report on Carbon Dioxide Capture and Storage*. Cambridge, UK: Cambridge University Press.

Metzger, R. A. and Benford, G. (2001). 'Sequestering of Atmospheric Carbon through Permanent Disposal of Crop Residue'. *Earth and Environmental Science*, 49 (1–2), 11–19.

Meyfroidt, P. and Lambin, E. F. (2009). 'Forest transition in Vietnam and dis-placement of deforestation abroad'. *Proceedings of the National Academy of Sciences*, 106 (38), 16139–44.

Michaels, P. (February 12, 2009). 'Climate scientists blow hot and cold'. *The Guardian*.

Middle East Economic Digest (August 18, 2006). 'PROJECT RISK: Spanners in the works'. Retrieved 4 August 2009 from *Middle East Economic Digest*: http://www.meed.com/news/2006/08/project_risk_spanners_in_the_works.html

Millard, R. (October 13, 2009). 'Carbon storage key to UN climate deal: min-isters'. AFP.

Mills, A. (April 14, 2009). 'Obama's climate control promise is in conflict with his stated goal of energy independence'. *E&P Magazine*.

Mills, R. M. (2008). *The Myth of the Oil Crisis*. Westport, Connecticut, USA: Praeger Publishers.

Mining Exploration News (April 5, 2009). 'Sasol Completed To Develop Basic Engineering Design Coal Mining an Surface Gasification Plant in Secunda'. Retrieved 8 April 2009, from *Mining Exploration News*: http://paguntaka. org/2009/04/05/sasol-completed-to-develop-basic-engineering-design-coal-mining-an-surface-gasification-plant-in-secunda/

MIT (April 29, 2009). 'AEP Alstom Mountaineer Fact Sheet'. Retrieved 15 May 2009 from Carbon Capture and Sequestration Technologies @ MIT: http://sequestration.mit.edu/tools/projects/aep_alstom_mountaineer.html

Modern Power Systems (July 1, 2008). 'Edwardsport project costs increase while AEP plans derailed'. *Modern Power Systems*.

Modern Power Systems (October 1, 2004). 'Horns Rev reveals the real hazards of offshore wind'. Retrieved 21 May 2009 from *ModernPowerSystems*: http://www.modernpowersystems.com/story.asp?storyCode=2030103

Monbiot, G. (2007). *Heat: How to Stop the Planet Burning*. London, UK: Penguin.

Monbiot, G. (March 24, 2009). 'Woodchips with everything. It's the Atkins plan of the low-carbon world'. *The Guardian*.

Mondal, A. (2008). 'Gondwana Basins in India—Vast Geologic Storage Sites for CO_2 Injection' in M. Goel, B. Kumar and S. N. Charan, *Carbon Capture and Storage: R&D Technologies for Sustainable Energy Future* (p. 224). New Delhi, India: Narosa Publishing House.

Mongabay (2009). 'Aluminum price chart'. Retrieved 11 September 2009 from Mongabay: http://www.mongabay.com/images/commodities/charts/alumi-num.html

Mongabay (2009). 'Steel, rebar price chart'. Retrieved 11 September 2009 from Mongabay: http://www.mongabay.com/images/commodities/charts/chart-steel_rebar.html

Moon, T. (February 1, 2008). 'Oil-shale extraction technology has a new owner'. *Journal of Petroleum Technology*.

Morales, A. and van Loon, J. (August 6, 2009). 'China Balks at Global Warming—Gas Capture Costs'. Bloomberg.

Morgan, J. (September 29, 2009). 'Kenya's heart stops pumping'. Retrieved 29 September 2009 from BBC News: http://news.bbc.co.uk/2/hi/africa/8057316.stm

Morriss, A. P., Bogart, W. T., Dorchak, A. and Meiners, R. E. (2009). 'Green Jobs Myths'. *University of Illinois Law & Economics* (LE09–001).

Mountford, R. (August 18, 2009). 'Carbon capture's role in a sustainable future'. *The Guardian*.

Mulkern, A. C. (May 18, 2009). 'Carbon Capture and Storage May Be Key to Climate Bill'. *Scientific American*.

Müller, B. (2006). 'Some Aspects of the Climate Change Issue' in R. Mabro, *Oil in the 21st Century* (p. 351). Oxford, UK: Oxford University Press.

Müller-Fürstenberger, G., Wagner, M. and Müller, B. (2004). *Exploring the Carbon Kuznets Hypothesis*. Oxford Institute for Energy Studies.

Myers, J., Dickinson, W. and Dickinson, W. (August, 2009). 'New system can safely store CO_2'. *E&P Magazine*, pp. 66–7.

Nader, J. (October 15, 2009). Director Middle East, Suntech. (R. Mills, Interviewer)

Nakićenović, N. (2006). 'Dynamics of Global Energy Transitions'. DOE-EPA Workshop 'Modeling the Oil Transition'. Washington DC, USA.

Nakićenović, N. (2002). 'Long-Term Technological Options'. UNFCCC Workshop on the IPCC Third Assessment Report (p. 31). Bonn: Intergovernmental Panel on Climate Change.

Nakićenović, N., Alcamo, J., Davis, G., de Vries, B., Fenhann, J., Gaffin, S. et al. (2000). *IPCC Special Report on Emissions Scenarios*. Cambridge, UK: Cambridge University Press.

Narain, M. (2008). 'MIT Study on the Future of Coal' in M. Goel, B. Kumar & S. N. Charan, *Carbon Capture and Storage: R&D Technologies for Sustainable Energy Future* (p. 224). New Delhi, India: Narosa Publishing House.

National Energy Technology Laboratory (2008). *Carbon Capture Technology Research and Breakthrough Concepts*.

NationalGrid (2009). *Operating the Electricity Transmission Networks in 2020*.

Natural Gas Supply Association (n.d.). 'Unconventional Natural Gas Resources'. Retrieved 8 June 2009 from NaturalGas.org: http://www.naturalgas.org/overview/unconvent_ng_resource.asp

Natural Resources Canada (January 26, 2009). 'Energy Sources'. Retrieved 11 August 2009, from Natural Resources Canada: http://www.nrcan.gc.ca/eneene/sources/eleele/abofai-eng.php#domestic

Nauclér, T., Campbell, W. and Ruijs, J. (2008). *Carbon Capture & Storage: Assessing the Economics*. McKinsey & Company.

Nayak, D. R., Miller, D., Nolan, A., Smith, P. and Smith, J. (2008). *Calculating carbon savings from wind farms on Scottish peat lands—A New Approach*. The Scottish Government.

Nelson, C. R., Evans, J. M., Sorensen, J. A., Steadman, E. N. and Harju, J. A. (2005). *Factors Affecting the Potential for CO_2 Leakage from Geologic Sinks*. Plains CO_2 Reduction Partnership.

Netherlands Environmental Assessment Agency (June 25, 2009). 'Global CO_2 emissions: annual increase halves in 2008'. Retrieved 31 August 2009, from Netherlands Environmental Assessment Agency: http://www.pbl.nl/en/publications/2009/Global-CO2-emissions-annual-increase-halves-in-2008.html

Nevle, R. J. and Bird, D. K. (2008). 'Effects of Syn-pandemic Fire Reduction and Reforestation in the Tropical Americas on Atmospheric Carbon Dioxide During European Conquest'. American Geophysical Union Fall 2008 Meeting, 89(53).

New York Times (September 4, 2009, p. A20). 'Another Astroturf Campaign'.

New Zealand Trade and Enterprise (n.d.). 'Wood, building and interiors'. Retrieved 28 August 2009, from New Zealand Trade and Enterprise: http://www.nzte.govt.nz/access-international-networks/explore-opportunities-in-growth-industries/growth-industries/pages/wood-building-and-interiors.aspx

Nicot, J.-P. (2009). 'A survey of oil and gas wells in the Texas Gulf Coast, USA, and implications for geological sequestration of CO_2'. *Environmental Geology*, 57 (7), 1625–38.

Nippon (n.d.). 'Nippon GtL'. Retrieved 23 May 2009 from http://www.nippon-gtl.or.jp/pdf/nippongtl.pdf

Nord, L. O., Kothandaraman, A., Herzog, H., McRae, G. and Bolland, O. (2008). 'A modeling software linking approach for the analysis of an integrated reforming combined cycle with hot potassium carbonate CO_2 capture'. *Energy Procedia*.

Nordhaus, W. (May 3, 2007). 'The Stern Review on the Economics of Climate Change' . Retrieved 24 April 2009 from http://nordhaus.econ.yale.edu/stern_050307.pdf

Nugent, P. (April 25, 2009). 'Capturing carbon'. Letter to *The Times* .

Odell, P. (2004). *Why Carbon Fuels Will Dominate the 21st Century Energy Economy*. Multi-Science Publishing.

Oil and Gas Journal (July 29, 2009). 'Encore secures carbon dioxide supply for Bell Creek flood'.

Oil and Gas Journal (June 5, 2009). 'Lacq field CCS project is France's first'.

Oilbarrel (August 20, 2008). 'Providence Resources Advances Spanish Point And Announces Kish Bank Gas Storage And Carbon Sequestration Project'. Retrieved 29 August 2009 from Oilbarrel: http://www.oilbarrel.com/nc/news/display_news/article/providence-resources-advances-spanish-point-and-announces-kish-bank-gas-storage-and-carbon-sequestra/856.html?tx_ttnews[swords]=providence%20kish%20bank&cHash=e7d7e9e876

Ordorica-Garcia, G., Wong, S. and Faltinson, J. (2009). *CO_2 Supply from the Fort McMurray Area 2005–2020*. Edmonton, Alberta, Canada: Alberta Research Council Inc.

Osborne, M. (June 25, 2009). 'NanoMarkets report: CIGS growth not limited by indium supply'. Retrieved 26 June 2009 from PV-Tech.org: http://www.pv-tech.org/news/_a/nanomarkets_report_cigs_growth_not_limited_by_indium_supply/

O'Sullivan, E. (February 19, 2008). 'Emal weighs carbon capture at Taweelah'. *Middle East Economic Digest.*

Pacala, S. and Socolow, R. (2004). 'Stabilization Wedges: Solving the Climate Problem for the Next 50 Years with Current Technologies'. *Science*, 305, 968–72.

Page, L. (February 13, 2008). 'Academics propose carbon-capture kit for cars'. Retrieved 31 May 2009 from *The Register*: http://www.theregister.co.uk/2008/02/13/carbon_capture_onboard_cars_water_petrol/

Page, S. C., Williamson, A. G. and Mason, I. G. (2009). 'Carbon capture and storage: Fundamental thermodynamics and current technology'. *Energy Policy*, 37 (9), 3314–24.

Paltsev, S., Reilly, J. M., Jacoby, H. D., Gurgel, A. C., Metcalf, G. E., Sokolov, A. P. et al. (2007). *Assessment of U.S. Cap-and-Trade Proposals.* Massachusetts Institute of Technology.

Pandey, D. (May, 2009). 'Effect of multiple, intercalated sediments on seismic data'. *World Oil*, pp. 58–65.

Parson, E. A. (2006). 'Reflections on Air Capture: The Political Economy of Active Intervention in the Global Environment; An Editorial Comment'. *Climate Change*, 74 (1/3), 5–15.

Pathirana, S. (September 3, 2009). 'Coconuts used to capture carbon'. Retrieved 3 September 2009 from BBC News: http://news.bbc.co.uk/2/hi/science/nature/8232535.stm

Pearce, F. (February 26, 2009). 'Greenwash: Why "clean coal" is the ultimate climate change oxymoron'. *The Guardian.*

Pearce, F. (August 13, 2009). 'How a wind farm could emit more carbon than a coal power station'. *The Guardian.*

Peltier, R. (July 1, 2007). 'Speaking of Coal Power: IGCC Sticker Shock'. Retrieved 15 August 2009 from *Coal Power*: http://www.coalpowermag.com/plant_design/Speaking-of-Coal-Power-IGCC-Sticker-Shock_55.html

Physorg (May 22, 2007). 'Follow the "Green" Brick Road? Retrieved 2 September 2009 from Physorg: http://www.physorg.com/news99067492.html

Pica, E. (2002). 'Running On Empty: How Environmentally Harmful Energy Subsidies Siphon Billions from Taxpayers' . Green Scissors.

Pielke, J. R. (2009). 'An idealized assessment of the economics of air capture of carbon dioxide in mitigation policy'. *Environmental Science and Policy*, 12, 216–25.

Pielke, R. J. (2007). 'Mistreatment of the economic impacts of extreme events in the Stern Review Report on the Economics of Climate Change'. *Global Environmental Change*, 17, 302–10.

Piling, D. (August 7, 2009). 'Lunch with the FT: Jared Diamond'. *Financial Times.*

PointCarbon (June 3, 2009). 'CO2 trading at €30/t in 2013, €40/t in 2016, predicts Point Carbon'. Retrieved 10 July 2009 from PointCarbon: http://www.pointcarbon.com/aboutus/pressroom/pressreleases/1.1130832

Price, R. and Caruso, D. (August 22, 2009). 'AEP seeks grant for burial of CO_2'. *Columbus Dispatch.*

Pronske, K. (June, 2007). 'Kimberlina: A zero-emissions demonstration plant'. Retrieved 7 June 2009 from Clean Energy Systems: http://www.cleanenergysystems.com/news/june_07.html

Prosser, D. (April 15, 2006). 'So, do you really have to get on board that plane?' *Independent on Sunday.*

Quiggin, J. (December 20, 2006). 'Stern and the critics on discounting'. Retrieved 24 April 2009 from http://johnquiggin.com/wp-content/uploads/2006/12/sternreviewed06121.pdf

QuotationsPage (n.d.). 'The great French Marshall Lyautey'. Retrieved 3 August 2009 from QuotationsPage: http://www.quotationspage.com/quote/1928.html

Rametsteiner, E., Obersteiner, M., Kindermann, G. and Sohngen, B. (2008). 'Economics of Avoiding Deforestation' in C. Palmer and S. Engel, *Avoided Deforestation. Prospects for Mitigating Climate Change.* London, UK: Routledge.

Randerson, J. (April 22, 2009). 'Nicholas Stern: We must not give in to pessimism'. *The Guardian.*

Raymo, M., Ruddiman, W. F. and Froelich, P. N. (1988). 'Influence of late Cenozoic mountain building on ocean geochemical cycles'. *Geology*, 16, 649–53.

Read, P. (March 27, 2009). 'This gift of nature is the best way to save us from climate catastrophe'. *The Guardian.*

Read, P. and Lermit, J. (2005). 'Bio-energy with carbon storage (BECS): A sequential decision approach to the threat of abrupt climate change'. *Science Direct*, 30 (14), 2654–71.

REN21 (2009). *Renewables Global Status Report: 2009 Update.*

Renner, M., Sweeney, S. and Kubit, J. (2008). *Green Jobs: Towards decent work in a sustainable, low-carbon world.* Worldwatch Institute/United Nations Environment Programme.

Rennie, A. (2009). 'Carbon Capture and Storage'. *The Energy Institute Lectures* (p. 33). London, UK: Energy Institute.

Reuters (August 31, 2009). 'Green hopes rest in rare earth'. *The National.*

Reuters (October 26, 2009). 'Summit Power and Blue Source Announce Agreement On CO_2 Management for One of the World's Largest Carbon Capture and Storage Projects'. Reuters.

Reuters (May 20, 2005). 'Vattenfall Plans CO_2-Free Power Plant in Germany'. Retrieved 8 May 2009, from Planet Ark: http://www.planetark.com/dailynewsstory.cfm/newsid/30899/story.htm

Rhodes, J. S. and Keith, D. W. (2008). 'Biomass with capture: negative emissions within social and environmental constraints: an editorial comment'. *Climatic Change*, 87, 321–8.

Riahi, K., Rubin, E. S., Taylor, M. R., Schrattenholzer, L. and Hounshell, D. (2004). 'Technological learning for carbon capture and sequestration technologies'. *Energy Economics*, 26, 539–64.

Richards, K. R. and Stokes, C. (2004). 'A Review of Forest Carbon Sequestration Cost Studies: a Dozen Years of Research'. *Climatic Change*, 63, 1–48.

Riduan, S. N., Zhang, Y. and Ying, J. Y. (2009). 'Conversion of Carbon Dioxide into Methanol with Silanes over N-Heterocyclic Carbene Catalysts'. *Angewandte Chemie International Edition*, 48 (18), 3322–5.

Rigzone. (October 2, 2009). 'Petrobras' CO_2 Injection Project to Serve As Test for Pre-Salt'. Retrieved 5 October 2009 from Rigzone: http://www.rigzone.com/news/article.asp?a_id=80962

Riley, P. D. (2004). *Nuclear Waste: Law, Policy, and Pragmatism.* Surrey, UK: Ashgate Publishing.

Robinson, C. (April, 2008). 'Economics, Politics and Climate Change: are the Sceptics Right?' Julian Hodge Bank Lecture.

Rochon, E. (2008). 'False Hope: Why carbon capture and storage won't save the climate'. Greenpeace.

Romm, J. (June 19, 2008). 'Cleaning up on carbon'. *Nature Reports Climate Change.*

Royal Academy of Engineering (May, 2008). 'The Severn Barrage' (p. 76).

Royal Society (2009). 'Geoengineering the climate'.

Royal Society (June 7, 2005). 'Joint science academies' statement: Global response to climate change'. Retrieved 27 May 2009 from the Royal Society: http://royalsociety.org/document.asp?latest=1&id=3222

Rubin, E. S. (2008). 'CO_2 Capture and Transport'. *Elements*, 4, 311–17.

Rughani, D. (2009). 'Cooling the Planet with Biochar?' *SBSTA Meeting* (p. 33). Bonn, Germany.

Rutledge, D. (2007). 'Hubbert's Peak, The Coal Question and Climate Change'. *APSO-USA* (p. 34). Houston, Texas.

Rydberg, S. (May, 2009). 'Delays in Vattenfall's Danish demo project'. Retrieved 13 September 2009 from Vattenfall: http://www.vattenfall.com/www/co2_en/co2_en/399862newsx/1739438vatte/index.jsp#delay

Salameh, M. G. (2Q 2009). 'How Viable is the Hydrogen Economy? The Case of Iceland'. *International Association for Energy Economics Energy Forum*, 11–17.

Salleh, A. (June 16, 2009). 'Australian Forests Best at Locking Up Carbon'. Retrieved 10 September 2009, from *Discovery News*: http://dsc.discovery.com/news/2009/06/16/australia-forest-carbon.html

Salmon, S., Saunders, P. and Borchert, M. (2009). 'Enzyme technology for carbon dioxide separation from mixed gases'. *Earth and Environmental Science*, 6, 1.

Sarkarinejad, K. (2003). 'Structural and microstructural analysis of a palaeotransform fault zone in the Neyriz Ophiolite, Iran' in Y. Dilek and P. T. Robinson, *Ophiolites in Earth History* (p. 717). Geological Society.

BIBLIOGRAPHY

Sathaye, J. A., Dale, L., Makundi, W. and Chan, P. (2008). 'GHG Mitigation Potential in Global Forests' (p. 26). Washington, DC, USA: World Bank.

Saunders, J. and Nussbaum, R. (2008). *Forest Governance and Reduced Emissions from Deforestation and Degradation (REDD)*. London, UK: Chatham House.

Schlamadinger, B. and Marland, G. (2000). *Land use & Global climate change: Forests, Land Management and the Kyoto Protocol*. Arlington, VA, USA: Pew Center on Global Climate Change.

Schmidt, L. (October 7, 2009). '$865 million announced for Quest carbon capture project'. *Calgary Herald*.

Schmidt, S., Rempel, H. and Schwarz-Schampera, U. (2008). *Reserves, Resources and Availability of Energy Resources 2007*. Hanover, Germany: Bundesanstalt für Geowissenschaften und Rohstoffe.

Schoenbrod, D. and Stewart, R. B. (August 24, 2009). 'The Cap-and-Trade Bait and Switch'. *Wall Street Journal*.

Schuiling, R. D. and Krijgsman, P. (2006). 'Enhanced weathering: An effective and cheap tool to sequester CO_2'. *Climate Change*, 74, 349–54.

Schuur, E. A., Bockheim, J., Canadell, J. G., Euskirschen, E., Field, C. B., Goryachkin, S. V. et al. (2008). 'Vulnerability of Permafrost Carbon to Climate Change: Implications for the Global Carbon Cycle'. *BioScience*, 58 (8), 701–14.

ScienceDaily (November 6, 2008). 'Rocks Could Be Harnessed To Sponge Vast Amounts Of Carbon Dioxide from Air'.

Scottish Centre for Carbon Storage (September 17, 2008). 'Oxy-Fuel Combustion Capture'. Retrieved 15 May 2009 from Scottish Centre for Carbon Storage: http://www.geos.ed.ac.uk/sccs/capture/oxyfuel.html

ScottishPower (May 29, 2009). 'UK First At Longannet As ScottishPower Brings Clean Coal Technology One Step Closer To Reality'. Retrieved 13 August 2009 from ScottishPower: http://www.scottishpower.com/PressReleases_1876.htm

Seddon, D. (2006). *Gas Usage & Value: The Technology and Economics of Natural Gas Use in the Process Industries*. PennWell Corp.

Seifritz, W. (1993). 'Partial and total reduction of CO_2 Emissions of Automobiles Using CO_2 Traps'. *International Journal of Hydrogen Energy*, 18, 243–51.

Sengul, M. (2008). 'Reservoir-Well Integrity Aspects of Carbon Capture and Storage' in M. Goel, B. Kumar and S. N. Charan, *Carbon Capture and Storage: R&D Technologies for Sustainable Energy Future* (p. 224). New Delhi, India: Narosa Publishing House.

Shackley, S. and Gough, C. (2006). 'Conclusions and Recommendations' in S. Shackley and C. Gough, *Carbon Capture and its Storage* (p. 313). Surrey, UK: Ashgate Publishing.

Shackley, S., Gough, C. and McLachlan, C. (2006). 'The Public Perception of Carbon Dioxide Capture and Storage in the UK' in S. Shackley and C.

Gough, *Carbon Capture and its Storage* (p. 313). Surrey, UK: Ashgate Publishing.

Sharma, S. and Dodds, K. (2008). 'Challenges in Developing a Monitoring and Verification Scheme for Australia's First Geosequestration Project' in M. Goel, B. Kumar and S. N. Charan, *Carbon Capture and Storage: R&D Technologies for Sustainable Energy Future* (p. 224). New Delhi, India: Narosa Publishing House.

Sharp, J. D., Jaccard, M. K. and Keith, D. W. (2009). 'Anticipating public attitudes toward underground CO_2 storage'. *International Journal of Greenhouse Gas Control*, 3, 641–51.

Sheffield, J., Obenschain, S., Conover, D., Bajura, R., Greene, D., Brown, M. et al. (2004). 'Energy Options for the Future'. *Journal of Fusion Energy*, 23 (2), 63–109.

Shenker, J. (August 21, 2009). 'Nile Delta: "We are going underwater. The sea will conquer our lands"'. *The Guardian*.

Shi, J. Q. and Durucan, S. (2005). 'CO_2 Storage in Caverns and Mines'. *Oil and Gas Science and Technology*, 60 (3), 569–71.

Shukman, D. (August 25, 2009). 'Averting a perfect storm of shortages'. Retrieved 25 August 2009 from BBC News: http://news.bbc.co.uk/2/hi/in_depth/8219184.stm

Shukman, D. (July 9, 2009). 'Project to "grow carbon sinks"'. Retrieved 10 September 2009 from BBC News: http://news.bbc.co.uk/2/hi/science/nature/8139351.stm

Sielhorst, S., Molenaar, J. W. and Offermans, D. (2008). *Biofuels in Africa: An assessment of risks and benefits for African wetlands*. Amsterdam, Netherlands: Wetlands International.

Sierra Club (April, 2008). 'The Basics of Carbon Capture and Sequestration'. Retrieved 7 September 2009, from Sierra Club: http://www.sierraclub.org/energy/factsheets/basics-sequestration.pdf

Simmonds, M., Hurst, P., Wilkinson, M. B., Reddy, S. and Khambaty, S. (2003). 'Amine Based CO_2 Capture from Gas Turbines'. Second Annual Conference on Carbon Sequestration.

Sioshansi, P. (2008). 'If not coal then what?' *International Association for Energy Economics Conference* (p. 39). New Orleans, USA.

Sivertsson, A. (2004). *Study of World Oil Resources with a Comparison to IPCC Emissions Scenarios*. M.Sc. Thesis, Uppsala University, Department of Earth Sciences, Uppsala.

Slaughter, A. J. (2007). *Gas to Liquids: Working Document of the NPC Global Oil and Gas Study*. National Petroleum Council.

Smil, V. (2008). *Energy in nature and society*. Cambridge, MA, USA: MIT Press.

Smith, P. (2009). 'Global greenhouse mitigation in agriculture'. *Climate Change: Global Risks, Challenges and Decisions* (p. 1). IOP Conference Series.

Snow, J. E. (June 21, 2000). 'Economic Theory of Exhaustible Natural Resources: Surprises for the Geologist'. Retrieved 9 May 2009 from http://www.mpch-mainz.mpg.de/~jesnow/MineralEcon/habil/index.html

Snow, N. (October 23, 2009). 'INGAA study takes midstream look at long-term gas supply'. *Oil and Gas Journal*.

Socolow, R. H. and Lam, S. H. (2007). 'Good enough tools for global warming policy making'. *Philosophical Transactions of the Royal Society*, 365, 897–934.

Socolow, R. (2005). 'Stabilization Wedges: Mitigation Tools for the Next Half Century'. *Avoiding Dangerous Climate Change: A Scientific Symposium on Stabilisation of Greenhouse Gases*. Exeter, UK.

Socolow, R. (2006). 'Stabilization Wedges: Mitigation Tools for the Next Half Century' (p. 43). Washington, DC, USA: World Bank.

Sohngen, B. and Brown, S. (2006). 'The Influence of Conversion of Forest Types on Carbon Sequestration and Other Ecosystem Services in the South Central United States'. *Ecological Economics*, 57, 698–708.

Sommer, M. (August 14, 2009). 'Shift in Jamestown "clean coal" plans decried, lauded'. *Buffalo News*.

Sonde, R. R. (n.d.). 'Demonstration of capture, injection and geological Sequestration (storage) in Flood Basalt Formation of India'. Retrieved 13 June 2009 from Carbon Sequestration Leadership Forum: http://www.cslforum.org/publications/documents/Basalt_Formation.pdf

SRI International (November 11, 2005). 'SRI International Presents Novel Direct Carbon Fuel Cell Technology at Industry Event'. Retrieved 3 June 2009 from SRI International: http://www.sri.com/news/releases/11–11–05.html

Stangeland, A. (August 17, 2007). 'A Model for the CO_2 Capture Potential' . Retrieved 8 September 2009 from Bellona: http://www.bellona.no/filearchive/fil_Paper_Stangeland_-_CCS_potential.pdf

Stangeland, A. (October 9, 2007). 'Why CO_2 Capture and Storage (CCS) is an Important Strategy to Reduce Global CO_2 Emissions'. Retrieved 8 September 2009 from Bellona: http://www.bellona.org/position_papers/WhyCCS_1.07

StatoilHydro (2008). *Carbon dioxide: Underground storage of carbon dioxide*.

Stavins, R. N. and Richards, K. R. (2005). *The cost of U.S. forest-based carbon sequestration*. Arlington, VA, USA: Pew Center on Global Climate Change.

Steelonthenet (July 30, 2009). 'MEPS Steel Product Price Levels across 2008—2009'. Retrieved 5 September 2009 from Steelonthenet: http://www.steelonthenet.com/prices.html

Steinberg, M. and Cooper, J. F. (2002). 'High Efficiency Direct Carbon Fuel Cell for CO_2 Emission and Sequestration'. *Fuel Chemistry Division*, 47 (1), 85–7.

Stern, N. (2007). *The Economics of Climate Change*. Cambridge, UK: Cambridge University Press.

Stevens, S. H. (2005). 'Natural CO_2 Fields as Analogs for Geologic CO_2 Storage' in S. M. Benson, C. Oldenburg, M. Hoversten and S. Imbus, *Carbon*

Dioxide Capture for Storage in Deep Geologic Formations—Results from the CO₂ Capture Project (pp. 687–97). Oxford, UK: Elsevier.

Stevens, S. H., Pearce, J. M. and Rigg, A. A. (2001). 'Natural Analogs for Geologic Storage of CO₂: An Integrated Global Research Program'. *First National Conference on Carbon Sequestration* (p. 12). Washington, DC, USA: National Energy Technology Laboratory.

Stolaroff, J. K. (2006). *Capturing CO₂ from ambient air: a feasibility assessment.* D.Phil. Thesis, Carnegie Mellon University, Engineering, Pittsburgh, USA.

Stolaroff, J. K., Keith, D. W. and Lowry, G. V. (2008). 'Carbon Dioxide Capture from Atmospheric Air Using Sodium Hydroxide Spray'. *Environmental Science & Technology*, 42, pp. 2728–35.

Stolaroff, J. K., Lowry, G. V. and Keith, D. W. (2005). 'Using CaO- and MgO-rich industrial waste streams for carbon sequestration'. *Energy Conversion and Management*, 46, 687–99.

Stuber, N., Forster, P., Rädel, G. and Shine, K. (2006). 'The importance of the diurnal and annual cycle of air traffic for contrail radiative forcing'. *Nature*, 441, 864–7.

Sun, S. (2008). 'Power Plant Emissions to Biofuels'. *NREL-AFOSR Workshop on Algal Oil for Jet Fuel Production* (p. 28).

Sundkvist, S. G., Griffin, T. and Thorshaug, N. P. (2001). 'AZEP—Development of an Integrated Air Separation Membrane Gas Turbine'. *Second Nordic Symposium on Carbon Dioxide Capture and Storage* (pp. 52–7). Göteborg, Sweden.

Sun Journal (August 24, 2007). 'Wind turbine shortage continues; costs rising'.

Sustainable Land Development Today (2009). 'SLDI Project Goes Carbon Negative' .

Sweet, W. (n.d.). 'Danish Wind Turbines Take Unfortunate Turn'. Retrieved 21 May 2009 from IEEE Spectrum Online: http://www.spectrum.ieee.org/print/4005

Tacoli, C. (September 29, 2009). 'Climate migration fears "misplaced"'. Retrieved 30 September 2009 from BBC News: http://news.bbc.co.uk/2/hi/science/nature/8278515.stm

Tahil, W. (2006). *The Trouble with Lithium: Implications of Future PHEV Production for Lithium Demand.* Meridian International Research.

Taku Ide, S., Friedmann, S. J. and Herzog, H. J. (2006). 'CO₂ leakage through existing wells: current technology and regulations'. 8th International Conference on Greenhouse Gas Control Technologies. Trondheim, Norway.

Taleb, N. N. (2008). *The Black Swan: The Impact of the Highly Improbable.* London, UK: Penguin.

Tamayo, R. L. (2005). *Geologic Carbon Dioxide Sequestration for the Mexican Oil Industry: An Action Plan.* Massachusetts Institute of Technology.

Tececo (n.d.). '3rd Party Research and Development'. Retrieved 3 September 2009 from Tececo: http://www.tececo.com/rdandd.thirdparty.php

Terracina, M. (2009). 'Tax Incentives for Capturing and Using Carbon—Real or Aspirational?' Houston, Texas, USA: KPMG.

Teske, S. (2008). 'energy [r]evolution: a Sustainable Global Energy Outlook'. Greenpeace International, European Renewable Energy Council.

Think Carbon (July 11, 2009). 'Comparison of Waxman-Markey, EU ETS and CPRS Emissions Trading Schemes'. Retrieved 23 August 2009 from Think Carbon: http://thinkcarbon.wordpress.com/2009/07/11/comparison-of-waxman-markey-eu-ets-and-cprs-emissions-trading-schemes/

Thornley, D. (April, 2009). 'Energy & The Environment: Myths and Facts'. Retrieved 9 May 2009 from Manhattan Institute: http://manhattan-institute.org/energymyths/myth2.htm

Tierney, J. (April 20, 2009). 'Use Energy, Get Rich and Save the Planet'. *New York Times* .

Tol, R. S. (2009). *An Analysis of Mitigation as a Response to Climate Change.* Copenhagen, Denmark: Copenhagen Consensus Center.

Transport Watch UK (December, 2008). 'Facts sheet 5c Electric cars'. Retrieved 5 June 2009 from Transport Watch UK: http://www.transport-watch.co.uk/transport-fact-sheet-5c.htm

Traufetter, G. (December 13, 2007). 'China and India Exploit Icy Energy Reserves'. *Der Spiegel.*

Tripati, A. K., Roberts, C. D. and Eagle, R. A. (2009). 'Coupling of CO_2 and Ice Sheet Stability Over Major Climate Transitions of the Last 20 Million Years'. *Science.*

Trumper, K., Bertzky, M., Dickson, B., van der Heijden, G., Jenkins, M. and Manning, P. (2009). 'The Natural Fix? The role of ecosystems in climate mitigation' . United Nations Environment Programme.

Turta, A., Singhal, A. and Sim, S. (2003). 'Enhanced Gas Recovery (EGR) and CO_2 Storage in Dry Gas Pools'. *CO_2 from Industrial Sources to Commercial EOR Recovery: PTAC Workshop* (p. 29). Calgary, Canada.

Tusiani, M. D. and Shearer, G. (2007). *LNG: A Nontechnical Guide.* PennWell Corp.

Tweeten, L., Sohngen, B. and Hopkins, J. (2000). 'Assessing the Economics of Carbon Sequestration in Agriculture' in R. Lal, J. M. Kimble, R. F. Follett and B. A. Stewart, *Assessment Methods for Soil Carbon Pools.* Boca Raton, Florida, USA: CRC/Lewis Publishers.

Uibu, M., Uus, M. and Kuusik, R. (2009). 'CO_2 mineral sequestration in oil-shale wastes from Estonian power production'. *Journal of Environmental Management*, 90 (2), 1253–60.

UK Department of Trade and Industry (2003). 'Energy white paper 2003: our energy future—creating a low carbon economy'.

UNData (December 17, 2008). 'Solar Electricity'. Retrieved 20 June2009 from UNData: http://data.un.org/Data.aspx?d=EDATA&f=cmID%3AES%3BtrID%3A015

UNFCCC (n.d.). 'Greenhouse Gas Inventory Data—Detailed data by Party'. Retrieved 31 May 2009 from UNFCCC: http://unfccc.int/di/DetailedBy Party.do

United Nations Statistics Division (December 17, 2008). 'Petroleum Coke'. Retrieved 27 June 2009 from UNData: http://data.un.org/Data.aspx?d=ED ATA&f=cmID%3APK%3BtrID%3A013

US Department of Energy (June 12, 2009). 'Secretary Chu Announces Agreement on FutureGen Project in Mattoon, IL'. Retrieved 4 August 2009 from US Department of Energy: http://www.energy.gov/news2009/print2009/7454.htm

US Department of Energy/Wabash River Coal Gasification Project Joint Venture (September, 2000). 'The Wabash River Coal Gasification Repowering Project'.

US Geological Survey (2009). 'Mineral Commodity Surveys: Soda Ash'.

US Geological Survey (May 29, 2009). 'Significant Gas Resource Discovered in US Gulf of Mexico'. Retrieved 12 June 2009 from United States Geological Survey: http://www.usgs.gov/newsroom/article.asp?ID=2227

Upstream Online (August 17, 2009). 'Gorgon LNG gets Oz CO_2 support'. Retrieved 17 August 2009 from Upstream Online: http://www.upstreamonline.com/live/article185930.ece

Upstream Online (April 16, 2009). 'Japan fires up trial GTL plant'. Retrieved 23 May 2009 from Upstream Online: http://www.upstreamonline.com/live/article175828.ece

Upstream Online (August 13, 2009). 'Majors caught in climate change scandal'. Retrieved 3 September 2009 from Upstream Online: http://www.upstreamonline.com/live/article185596.ece

Upstream Online (August 13, 2009). 'Origin calls for carbon agreement'. Retrieved 3 September 2009 from Upstream Online: http://www.upstreamonline.com/live/article185597.ece

Upstream Online (May 18, 2009). 'Oslo delays StatoilHydro oil sands vote'. Retrieved 13 September 2009 from Upstream Online: http://www.upstreamonline.com/live/article178750.ece

Upstream Online (August 20, 2009). 'Oz may sync carbon laws with NZ'. Retrieved 23 August 2009 from Upstream Online: http://www.upstreamonline.com/live/article186110.ece

Upstream Online (October 13, 2009). 'Saudis eye CO_2 injection at Ghawar'. Retrieved 14 October 2009 from Upstream Online: http://www.upstreamonline.com/live/article195770.ece

Upstream Online (August 13, 2009). 'Shell joins UK carbon race'. Retrieved 13 August 2009 from Upstream Online: http://www.upstreamonline.com/live/article185598.ece

Upstream Online (August 24, 2009). 'Town vows to fight Shell's CCS plan'. Retrieved 29 August 2009 from Upstream Online: http://www.upstreamonline.com/live/article186418.ece

van der Schaaf, M. (May–June 2–3, 2009). 'Spectacular developments in Dutch "Energy Valley"'. *Carbon Capture Journal*.

van Loon, J. (June 2, 2008). 'Germany Slashes Solar Subsidies, Threatening Industry'. Bloomberg.

BIBLIOGRAPHY

Vattenfall (2007). 'Climate Map 2030' . Vattenfall AB.

Vattenfall (n.d.). 'Pilot Plant Schwarze Pumpe'. Retrieved 15 May 2009 from Vattenfall: http://www.vattenfall.com/www/co2_en/co2_en/879177tbd/879 211pilot/index.jsp

Vattenfall (n.d.). 'Vattenfall's project on CCS'. Retrieved 29 October 2009 from Vattenfall: http://www.vattenfall.com/www/co2_en/co2_en/index.jsp

Viner, D. and Jones, P. (August, 2000). 'Volcanoes and their effect on climate'. Retrieved 24 April 2009 from Climatic Research Unit, University of East Anglia: http://www.cru.uea.ac.uk/cru/info/volcano/

Walker, G. and King, S. D. (2009). *The Hot Topic*. London, UK: Bloomsbury.

Wang, T. and Watson, J. (2009). 'China's Energy Transition' . Brighton, UK: Tyndall Centre for Climate Change Research.

Wardle, D., Nilsson, M.-C. and Zackrisson, O. (2008). 'Fire-Derived Charcoal Causes Loss of Forest Humus'. *Science*, 320 (1).

Wasas, J. (n.d.). 'practical applications for the stenger-wasas process' . Retrieved 2 September 2009 from Swapsol: http://www.swapsol.com/images/docs/practical-applications.pdf

Watkins, K. (2007). *Human Development Report 2007/2008: Fighting climate change*. United Nations Development Programme.

Watson, M. N., Tingate, P. R., Boreham, C. and Gibson-Poole, C. M. (2006). 'Natural CO_2 Generation, Entrapment and Water-Rock Interaction of the Otway Basin'. 2006 American Association of Petroleum Geologists International Conference. Perth, Australia.

Watts, J. (June 23, 2009). 'China suspends reforestation project over food shortage fears'. *The Guardian*.

Weitzman, M. L. (2008). 'On Modeling and Interpreting the Economics of Catastrophic Climate Change'. Harvard University.

Wetlands International (2006). 'Peatland degradation fuels climate change'.

Wilcox, M. (November 17, 2008). 'Greenpeace ships in Rotterdam coal-power protest'. Retrieved 15 August 2009 from Radio Nederland Wereldomroep: http://static.rnw.nl/migratie/www.radionetherlands.nl/currentaffairs/region/netherlands/081117-Greenpeace-Rotterdam-redirected

Williams, S. (October 12, 2009). 'CO_2 Capture, Storage Should Be Part Of Climate Deal -UK Min'. Retrieved 13 October 2009 from Nasdaq: http://www.nasdaq.com/aspx/stock-market-news-story.aspx?storyid=2009101219 16dowjonesdjonline000297&title=co2-capture-storage-should-be-part-of-climate-deal—uk-min

Wilson, E. J. and Keith, D. W. (2003). 'Geologic Carbon Storage: Understanding the Rules of the Underground'. *Proceedings of the 6th Greenhouse Gas Control Conference* (pp. 229–34). Kyoto, Japan.

Wilson, E. J., Friedmann, S. J. and Pollak, M. F. (2007). 'Research for Deployment: Incorporating Risk, Regulation, and Liability for Carbon Capture and Sequestration'. *Environmental Science and Technology*, 41, 5945–52.

Wilson, M. and Monea, M. (2004). 'IEA GHG Weyburn CO_2 Monitoring and Storage Project Summary Report 2000–2004'. *7th International Conference on Greenhouse Gas Control Technologies* (p. 283). Vancouver, Canada.

Wilson, N. (May 10, 2008). 'Chimneys sweep BP clean coal plan away'. *The Australian*.

Wired (August, 6.08, 1998). 'Building a greater wall'.

Wood, E. (September 7, 2009). 'Energy sprawl: The next worry?' Retrieved 11 September 2009 from Renewable Energy World: http://www.renewableenergyworld.com/rea/blog/post/2009/09/energy-sprawl-the-next-worry

Woods Hole Oceanographic Institution (April 10, 2003). 'Will Ocean Fertilization To Remove Carbon Dioxide From The Atmosphere Work?' *Science Daily*.

Woody, T. (September 29, 2009). 'Alternative Energy Projects Stumble on a Need for Water'. *New York Times*.

Woolf, D. (2008). *Biochar as a Soil Amendment—A review of the Environmental Implications*. Swansea, UK: Swansea University.

World Bank (2006). 'Assessment of the World Bank/GEF Strategy for the Market Development of Concentrating Solar Thermal Power'.

World Bank (2006). 'The Costs of Attaining the Millennium Development Goals' .

World Bank (2007). *The Little Green Data Book*. Washington DC, USA: World Bank.

World Coal Institute (May, 2005). *The Coal Resource: A Comprehensive Overview of Coal*. London, UK.

World Commission on Environment and Development (1987). *Report of the World Commission on Environment and Development: Our Common Future*.

World Economic Forum, Cambridge Energy Research Associates (2009). 'Thirsty Energy: Water and Energy in the 21st Century'.

World Oil (October, 2009). 'ConocoPhillips preparing to field test a process for producing gas from gas hydrates'. *World Oil*.

World Resources Institute (2009). 'Climate Analysis Indicators Tool (CAIT) Version 6.0'. Retrieved 27 October 2009 from World Resources Institute: http://cait.wri.org/cait.php?page=yearly&mode=view&sort=val-desc&pHints=shut&url=form&year=1970&sector=natl&co2=1&lucf=1&update=Update

Worrell, E., Price, L., Hendricks, C., & Meida, L. O. (2001). Carbon Dioxide Emissions from the Global Cement Industry. *Annual Review of Energy and Environment* , 26, 303–329.

WWF (n.d.). 'Carbon capture and storage'. Retrieved 7 September 2009 from WWF: http://www.panda.org/what_we_do/footprint/climate_carbon_energy/energy_solutions/carbon_capture_storage/

Wynn, G., Gardner, T. and Maeda, R. (August 13, 2009). 'FACTBOX-Carbon trading schemes around the world'. Reuters.

Xue, Z., Tanase, D., Saito, H., Nobuoka, D. and Watanabe, J. (2005). 'Time-lapse crosswell seismic tomography and well logging to monitor the injected CO_2 in an onshore aquifer, Nagaoka, Japan'. *Society of Exploration Geophysicists*, 24.

BIBLIOGRAPHY

Yandle, B., Bhattarai, M. and Vijayaraghavan, M. (2004). 'Environmental Kuznets Curves: A Review of Findings, Methods, and Policy Implications'. Bozeman, MT, USA: Property and Environment Research Center.

Zeman, F. (2006). 'Direct Extraction of CO_2 from Air, a Fixed Solution for a Mobile Problem'. *The First Regional Symposium on Carbon Management* (p. 12). Dhahran, Saudi Arabia.

Zeman, F. (2007). 'Energy and Material Balance of CO_2 Capture from Ambient Air'. *Environmental Science and Technology*, 41, 7558–63.

Zeman, F. S. and Keith, D. W. (2008). 'Carbon neutral hydrocarbons'. *Philosophical Transactions of the Royal Society*, 366, 3901–18.

ZINC Research/Dufferin Research (February 25, 2009). 'Canadians support carbon capture and storage'. Retrieved 11 September 2009 from Dufferin Research: http://www.dufferinresearch.com/downloads/News%20Release%20-%20Canadians%20supportive%20of%20Carbon%20Capture%20&%20Storage%20%28Feb%2025%202009%29.pdf

Zittel, D. W. and Schindler, J. (March, 2007). *Coal: Resources and Future Production*. Ottobrunn, Germany: Energy Watch Group.

GLOSSARY

Acid gas
: A mix of carbon dioxide (CO_2) and hydrogen sulphide (H_2S), the waste products from processing sour gas. Often injected underground for disposal.

Albedo
: The proportion of solar radiation reflected by a surface, ranging from 1 (all light reflected) to 0 (all light absorbed). The oceans have an albedo of about 0.1, the land is 0.2–0.4 and snow and ice 0.6–0.8.

Amine
: A class of organic chemicals containing a nitrogen atom; some can be used as solvents to remove carbon dioxide and other pollutants.

Anthracite
: The highest-rank coal, with very low moisture content. Not normally used in power generation.

Anthropogenic climate change
: Climate change caused by human activities, particularly the release of carbon dioxide and other greenhouse gases.

Aquifer
: A water-bearing underground rock formation. Aquifers may contain potable water, or undrinkable saline water. Oil and gas fields are often underlain by aquifers.

ASU
: Air Separation Unit, a piece of equipment used to separate oxygen from the air (also known as an oxygen plant). Most use cryogenic methods (cooling the air until components drop out as a liquid). Used to supply oxygen to gasification plants, oxyfuel boilers, and other applications.

AZEP	Advanced Zero Emission Power, a concept for a gas-fired power plant with combustion in a reaction chamber where oxygen is separated out using a membrane.
bbl	Usual abbreviation for 'barrel', an industry measure for volumes of oil, equal to 42 US gallons or 0.159 cubic metres.
Biochar	A type of fine charcoal created by pyrolysing biomass, which can potentially be used as a soil additive to sequester carbon and improve fertility.
Biofuel	A liquid fuel derived from biomass, and typically used as a substitute for, or additive to, petrol or diesel, the main examples being ethanol and biodiesel.
Biomass	A general term covering all biological material derived from recently living organisms, which can be used for energy (or other applications such as chemicals and fibres).
Biosphere	That part of the Earth which contains living organisms, including the oceans, soil, plants, etc.
Bituminous coal	A high-rank coal, in energy content just below anthracite.
Brown coal	See lignite.
BtL	Biomass-to-liquids, a group of techniques for making liquid hydrocarbons from biomass.
Business-as-usual	A forecast made assuming that no additional climate change policies are enacted.
Cap-rock	Synonym for 'seal'.
Carbonate	A rock mainly constituted of carbonate minerals, predominantly calcium carbonate and calcium magnesium carbonate (dolomite), with the rarer siderite (iron carbonate). An important class of petroleum (and carbon dioxide storage) reservoirs. The weathering of silicate minerals by carbon dioxide yields carbonate minerals.
Carbon credit	A credit (payment) received for preventing the emission of carbon dioxide to the atmosphere, or for removing carbon dioxide from the atmosphere.

Carbon dioxide	A gas, chemical formula CO_2, produced by burning carbon-containing fuels. The main cause of anthropogenic climate change.
Carbon negative	A process that removes carbon dioxide permanently from the atmosphere, such as underground storage of carbon dioxide derived from biomass.
Carbon tax	A tax levied on the emission of carbon dioxide to the atmosphere.
Catalyst	A substance which speeds a chemical reaction without itself being consumed in the reaction.
CBM	Coal-Bed Methane, methane trapped (adsorbed) in coal-beds, from which it can be removed by drilling wells and draining off water. A growing source of unconventional natural gas production. Known as CSM (Coal-Seam Methane) in Australia. CBM extraction can be enhanced by injecting carbon dioxide (ECBM).
CCS	Carbon Capture and Storage (or Sequestration), the process of taking gaseous carbon dioxide (from a source of emissions, or directly from the air), and storing it in a location or form which keeps it out of the atmosphere.
CDR	Carbon Dioxide Removal, a class of geo-engineering techniques for removing carbon dioxide from the atmosphere.
CFC	Chloro-fluoro-carbons, a class of industrial gases used in applications including refrigeration; powerful greenhouse gases, and also responsible for the hole in the ozone layer.
CHP	Combined Heat and Power, a plant which generates electricity and uses its waste heat productively, for instance in industrial processes or to heat buildings. If there is sufficient demand for this heat, the plant efficiency can be very high, around 80%.
Clastic	A type of sedimentary rock formed by the breakdown of pre-existing rocks; main examples include sandstones, siltstones and mudstones/shales.

CLC	Chemical Looping Combustion, a possible future type of combustion for power plants, where a metal is used to carry oxygen rather than burning the fuel in oxygen directly.
Climate change	The change over time in the Earth's climate (temperature, rainfall, cloudiness etc.). The climate changes both naturally, and due to human activities.
CMM	Coal Mine Methane, the extraction of methane from working coal mines, for power and to reduce greenhouse emissions. Similar to, but should not be confused with, CBM.
CNG	Compressed Natural Gas, methane gas under pressure, that can be used, for example, to power vehicles.
Coal	A fossil fuel composed mainly of carbon, derived from the remains of land plants altered by heat and pressure.
Combined Cycle	A modern, highly-efficient power plant configuration where gas is burnt to drive a gas turbine, and the hot combustion products then power a steam turbine. The technology of choice for new natural gas power stations.
CO_2	Carbon dioxide, the main gas responsible for anthropogenic climate change. Emitted when carbon-containing fuels are burned.
CO_2e	CO_2-equivalent, a way of expressing other greenhouse gases in terms of their global warming potential relative to carbon dioxide. The conversion depends on the timescale that is being considered, since the gases have different atmospheric lifetimes; a century is usually taken.
CtL	Coal-to-liquids, a group of techniques for making liquid hydrocarbons from coal. Generally releases large amounts of carbon dioxide.
DC	Direct Current, an electrical current that does not change direction (as opposed to Alternating Current, AC, used in power grids and homes). Can be used to send electricity over long distances with small losses.

Discount rate	An annual factor used to convert costs or benefits occurring in the future into today's money.
DME	Dimethyl ether (C_2H_6O), a possible future transport fuel which can be burned in modified diesel engines.
DSF	Deep Saline Formation, an alternative term for saline aquifers.
ECBM	Enhanced Coal-Bed Methane, the extraction of natural gas from coal by advanced methods (typically, injecting carbon dioxide).
EGR	Enhanced Gas Recovery, an advanced technique for recovering more natural gas from depleted fields by injecting carbon dioxide (or other gases, such as nitrogen).
Enzyme	A biological catalyst.
EOR	Enhanced Oil Recovery, a group of advanced techniques for recovering more oil from fields. Carbon dioxide injection is a major EOR method.
Fault	A break in a rock layer where layers are offset vertically or horizontally. When faults move they cause earthquakes.
FCC	Fluid Catalytic Cracker, a unit of an oil refinery used to convert heavy fractions of the oil to lighter products, particularly petrol (gasoline). A candidate for carbon capture.
Field	A sub-surface accumulation of petroleum (or coal).
Fossil fuel	A fuel composed of the remains of once-living organisms from the geological past, consisting primarily of carbon and hydrogen. Coal, oil, natural gas and peat are the main fossil fuels.
Fracture	A break in a rock layer where there is no visible offset, generally therefore smaller than a fault. Fluids may move preferentially in the sub-surface via fractures.
Geo-engineering	The purposeful large-scale modification of the Earth (typically to moderate climate change).
Global warming	The aspect of anthropogenic climate change reflected in higher average global temperatures.

Greenhouse gas (GHG)	A gas which, in the atmosphere, prevents heat from escaping and so increases global temperatures. The main greenhouse gases are water vapour, carbon dioxide, methane, nitrous oxide, ozone and a number of industrial gases.
Gt	Gigatonne, 1 billion metric tonnes (1,000,000,000 tonnes).
GtL	Gas-to-liquids, a group of techniques for making liquid hydrocarbons from natural gas. Generally releases large amounts of carbon dioxide, but less than CtL.
GW	Gigawatt, a unit of power equal to 1,000 megawatts.
HHV	Higher Heating Value, one of the two ways (the other being Lower Heating Value, LHV) of measuring the heat content of fuels.
Hydrate	A substance consisting of water ice which traps various gases, including methane and carbon dioxide, at low temperatures and/or high pressures. A potential future source of energy, but also of releases of greenhouse gases if hydrates are exposed to global warming. Found particularly in Arctic areas and the deep oceans.
Hydrocarbon	A chemical substance made of hydrogen and carbon. Natural gas and crude oil are examples.
Hydrogen sulphide	A toxic and corrosive gas, chemical formula H_2S, often found as a contaminant of natural gas. Known for its rotten eggs smell. May be co-disposed with carbon dioxide.
IDGCC	Integrated Drying Gasification Combined Cycle, a type of IGCC (see below) which uses waste heat to dry low-rank lignite (brown coal).
IEA	International Energy Agency, a body set up to coordinate energy policy amongst the OECD countries.
IFCC	Indirectly Fired Combined Cycle, a proposed type of power plant where coal combustion is used to heat air to drive a gas turbine and then a steam turbine.

IGCC	Integrated Gasification Combined Cycle, a type of power plant which gasifies a solid fuel (typically coal) before burning it in a combined-cycle system.
Ionic trapping	Trapping of carbon dioxide in water in the form of ions such as carbonate (CO_3^{2-}) and hydrogen carbonate (HCO_3^-).
IPCC	Intergovernmental Panel on Climate Change, a body established by the World Meteorological Organization and the United Nations Environment Programme in 1998 to evaluate the issues and risks of human-caused climate change.
kWh	Kilowatt hour, a unit of energy. The energy consumed by a 1 kilowatt unit (say, an electric fire) running for one hour. A 500 MW power station produces 500,000 kWh every hour.
Leakage	(1) The escape of carbon dioxide from geological storage. (2) Displacement of carbon emissions to another place, driven by policy and economics. For instance, if carbon caps are tight in Europe and lax in China, then carbon-intensive industry may relocate to China, causing a 'leakage' of emissions out of the European carbon cap.
LHV	Lower Heating Value, one of the two ways (the other being Higher Heating Value, HHV) of measuring the heat content of fuels. LHV assumes that the water from combustion is not condensed. LHV is about 4% lower than HHV for coal, and 10% lower for natural gas.
Lignite	Also known as brown coal, the lowest-rank coal, with low energy content and high moisture.
LNG	Liquefied Natural Gas, natural gas cooled to a liquid at -163°C, so that it can be easily transported in special tankers.
LPG	Liquefied Petroleum Gas, a mix of propane and butane which forms a liquid under modest pressures. Extracted from natural gas or formed as a

	by-product of oil refining, LPG is used as a cooking fuel in bottled gas canisters, and as a fuel for modified car engines.
Membrane	A substance with selective permeability, i.e. one that allows certain substances to pass more easily than others, and can therefore be used to separate them.
Methane	A gas, chemical formula CH_4, which is the main constituent of natural gas, and is also released by rotting vegetation, animal digestion and melting gas hydrates. It forms carbon dioxide when burnt. Methane is a much more powerful greenhouse gas than carbon dioxide.
Mt	Megatonne, 1 million (1,000,000) metric tonnes.
MW	Megawatt, the standard unit for power plant sizes. A typical coal-fired power station may be around 500–1,000 MW, with gas-fired stations being somewhat smaller. USA electricity generation averaged 493,000 MW during 2008[1] (i.e. equivalent to almost a thousand 500 MW plants running continuously).
Natural gas	Fossil fuel consisting mainly of methane (CH_4).
NGO	Non-Governmental Organisation, a group legally constituted but not under the control of a government, excluding for-profit companies. Usually campaigns for or advances certain goals, often charitable or policy-related. Examples relevant to CCS include the Royal Society, Bellona and Greenpeace.
OECD	Organisation for Economic Co-operation and Development, a grouping of major developed economies, including, amongst others, most EU countries, the USA, Canada, Mexico, Japan and Australia.
Oil sands	A sedimentary deposit, the best known example of which is in Alberta and Saskatchewan, Canada, containing extra-heavy oil in sandstones. The oil can be extracted, with high carbon footprint and

other environmental impacts, by mining or various *in situ* methods, usually using steam.

Oil shale — A type of sedimentary rock containing large amounts of kerogen, immature organic matter that can be transformed to oil and gas. Oil extraction from oil shales is not yet commercial, but there are various pilot projects around the world. Some proposed processes are very carbon-intensive.

Olivine — A mineral found in ultrabasic rocks, which reacts with carbon dioxide.

OPEC — Organisation of Petroleum Exporting Countries, an international grouping of major oil exporters, comprising Saudi Arabia, Iran, Iraq, Abu Dhabi, Kuwait, Qatar, Venezuela, Ecuador, Algeria, Libya, Nigeria and Angola (but not Russia, Mexico or Norway).

Opex — Operating costs, the ongoing costs to run a power plant or other facility, such as fuel purchase, salaries, maintenance and insurance.

PBR — Photobioreactor, a sealed transparent tube in which algae can be grown.

PC — Pulverised coal, the conventional type of coal-fired power station.

Permeability — A measure of how easily a rock (or other material) allows fluids to flow through it. A good reservoir has high permeability; a good seal has low permeability.

Petroleum — A generic term for oil, natural gas and related hydrocarbons.

Petroleum coke — Or 'petcoke', a coal-like solid carbon residue from oil refining, which can be used as a fuel.

Porosity — A measure of the 'void space' in a rock available to be filled with fluids (water, oil, gas, air, carbon dioxide, etc.). Typical porosities of reservoir rocks are 15–30%.

Positive feedback — A process which amplifies an original disturbance. For instance, melting ice leaves bare rock underneath, which absorbs more heat and so speeds up melting of remaining ice.

ppb	Parts per billion, a measure of very low concentrations of a gas (or other substance).
ppm	Parts per million, a measure of low concentrations of a gas (or other substance).For example, the concentration of carbon dioxide in the atmosphere was about 280 ppm prior to the industrial age.
PV	Photovoltaic, a technology that uses solar cells to generate electricity directly from light.
Pyrolysis	The process of heating a material (typically biomass) in the absence of air to break it down.
Reservoir	A rock with permeability and porosity sufficient to hold fluids (such as oil, gas and carbon dioxide) and for these to be commercially extracted or injected.
Royalty	A payment, calculated as a percentage of production or revenues, levied in cash or in kind on the extraction of oil or gas (or other minerals). In most countries, it is paid to the government, but in the USA it is due to the holder of the mineral rights.
Seal	An impermeable rock formation that does not allow fluids (such as oil, gas, water and carbon dioxide) to flow through it. The main sealing rock types are mudstones and shales, salt, anhydrite and low-permeability carbonates.
Sink	Part of the Earth system (e.g. forests, or the ocean) into which more carbon is flowing in than is coming out. The opposite of source.
Source	Part of the Earth system (e.g. forests, or the ocean) into which more carbon is flowing out than is going in. The opposite of sink.
Sour gas	Natural gas containing the contaminant hydrogen sulphide.
SRM	Solar Radiation Management, a class of geo-engineering techniques designed to cool the Earth by reflecting or intercepting a small amount of the sun's rays.

Sub-bituminous   A coal type ranking between lignite and bituminous.

Supercritical fluid   A fluid under certain pressure and temperature conditions, which cause it to have the properties of both a liquid and a gas, expanding to fill a space, and having the viscosity of a gas, but the density of a liquid. For carbon dioxide, this occurs above 31.1°C temperature and 7.38 MPa pressure.

Trap   A geometric configuration of reservoir and sealing rocks that prevents fluids from rising to the surface. A structural trap relies on faults and/or folds. A stratigraphic trap exists due to lateral changes in rock properties. Combined traps have features of both.

Unconventional oil   Oil that cannot be produced by conventional flow from wells, and requires advanced technology for extraction. Major examples include the 'oil sands' found in Alberta, Canada and elsewhere, oil shales, oil derived from coal and natural gas and, under some definitions, biofuels. Some unconventional oil sources have high carbon intensity.

Well   A hole drilled into the Earth for extracting fluids (oil, gas, water, carbon dioxide) from the sub-surface, or injecting fluids into it. Wells are normally lined with steel casing to prevent them collapsing and to seal off intermediate rock layers.

ZECA   Zero Emission Coal Alliance, a proposed method for gasifying coal with hydrogen to generate power and capture carbon dioxide.

INDEX

Massachusetts, 57; *see also* Bellingham co-generation plant; Brayton Point synthetic fuels plant; USA

Massachusetts Institute of Technology (MIT), 249, 287 *see also* carbon capture and storage, research into

Matter, Jürg, 124

McCain, John, 301

McElmo Dome, 135, 327; *see also* carbon dioxide, from natural sources; United States

McKinsey, 38n, 180, 215, 221–2, 249, 322, 329

Megawatt (MW), 424

Meigs County, 89; *see also* Ohio; postcombustion capture

Membrane, 424; for capture of carbon dioxide, 54–5, 59, 61, 65–6, 68–9, 85, 246, 279, 418; for oxygen separation, 60, 63–4; *see also* capture, of carbon dioxide

Merck, 256

Mercury, 49, 59, 299; *see also* pollution

Meri Pori, 89; *see also* Finland; precombustion capture

Merkel, Angela, 282; *see also* Germany

Metal-organic framework, for carbon dioxide capture, 55

Metamorphic rocks, 99, 110

Methane, 22, 51, 74, 107–8, 127, 135, 157, 171, 240, 323, 422, 424; as greenhouse gas, 6n, 7, 10, 32, 51, 116, 156, 158, 169, 176, 263, 299; in gasification, 62, 67; *see also* aquifer gas; coal-bed methane (CBM); coal-mine methane (CMM); gas; gas hydrates; greenhouse gases; landfill gas; unconventional gas; underground coal gasification (UCG)

Methanol, 97; carbon dioxide capture from manufacture, 195, 333, 338; as alternative fuel, 85, 87, 96, 209; *see also* synthetic fuels

Mexico, agro-forestry in, 215; climate conference, 257; enhanced oil recovery in, 104, 153; forestry management in, 256; natural carbon dioxide in, 135; *see also* Latin America; Organisation for Economic Cooperation and Development (OECD)

Michigan, 272, 282; *see also* Holland, Michigan; United States

Microbes, 146; in the biosphere, 157, 168, 172–3, 176; enhanced oil recovery, 101

Micro-generation, *see* distributed generation

Midale field, 203, 273; *see also* Canada; enhanced oil recovery; Weyburn

Middle East, 306, 325; carbon capture and storage in, 276–7; carbon dioxide emissions from, 48, 188, 327; carbon dioxide storage capacity of, 126, 132, 144; enhanced oil recovery in, 104–5, 204; implications of carbon capture and storage (CCS) for, 268; oil resources, 17, 22, 36, 100–1, 264; solar potential of, 22; wind potential of, 42; *see also* names of individual countries

Miliband, Ed, 34, 257

Millennium Development Goals, 226

Miller field, 90, 104, 204, 238–9; *see also* BP; enhanced oil recovery; North Sea; Peterhead

Mineral carbonation, *see* carbonation; *see also* coal ash; mineralisation

Mineralisation, of carbon dioxide, 2–3, 6, 56, 78, 97–8, 119, 121–5, 133, 147–8, 154, 181–2, 206, 212, 246, 260, 291, 324, 326–7; cost of, 206, 227; *see also* carbonation; olivine; peridotite; ultrabasic rocks

Mines, 108, 148, 170, 238, 278, 323, 420; for carbon dioxide storage, 115; *see also* coal-mine methane (CMM); coal mining

Minev, Denis, 155; *see also* Amazon, Brazil

Miombo, *see* agro-forestry

Miranga field, 276; *see also* Brazil; enhanced oil recovery

Mississippi, natural carbon dioxide in, 135; *see also* United States

Missouri, seismicity in, 144; see also United States

273, 417, 421; *see also* amines; ammonia; ammonium carbonate; flue gas; nitrogen oxides; nitrogen trifluoride; nitrous oxide; oxyfuel; urea
Nitrogen oxides, 49–50, 53, 55, 59–60, 64, 66, 79, 200, 299; *see also* acid rain
Nitrogen trifluoride, 33; *see also* greenhouse gases
Nitrous oxide, 6n, 7, 31, 156, 156n, 158, 172, 263, 422; *see also* acid rain; greenhouse gases
Nordjyllandsværket, 282; *see also* Denmark; Vattenfall
Non-governmental organisation (NGO), *see* environmentalism
Nordhaus, William, 219–20, 322; *see also* carbon tax
North Africa, carbon dioxide storage in, 133; implications of carbon capture and storage for, 268; renewable energy in, 36, 37n; *see also* Africa; Algeria; Egypt; Libya; Middle East; Tunisia
North Carolina, 7; seismicity in, 144; *see also* United States
North Dakota, 62, 75, 89–90, 103; *see also* Beulah synfuels plant; Blue Flint ethanol plant; United States
North Sea, 21, 23, 99; carbon dioxide storage in, 142, 153, 201, 203, 205, 207, 245, 258, 282; carbon dioxide storage capacity, 126, 131; enhanced oil recovery in, 101, 104, 204–5, 239, 280; natural carbon dioxide in, 135; *see also* Beatrice field; Brent field; Cleeton field; Denmark; enhanced oil recovery; Forties field; K12B field; Miller field; Netherlands; Norway; Sleipner field; Statfjord field; UK
Norway, xv, xvii, 265, 268, 269, 284, 336; carbon tax in, 219, 236, 269; gas CCS and, 243; Haltenbanken project, 90, 104, 269; Kårstø gas processing plant, 90, 269; Mongstad refinery, 76, 77, 90, 244, 269, 329, 333; ocean storage of carbon dioxide in, 121; protection of Amazon rainforest, 255; readiness for CCS,

264; Risavika, 90; and Sargas, 54, 90; Sleipner project, 44n, 56, 89, 114, 153, 269; Snøhvit project, 56, 57, 89, 114, 153, 269; Statfjord field, 239; storage capacity of, 131; and Utsira aquifer, 258, 259; *see also* Husnes
Novacem, 97; *see also* cement; Imperial College, London
NowGen, 272; *see also* integrated gasification combined cycle (IGCC); Summit Power; Texas
Nuclear power, 1, 15n, 16, 24, 26–8, 30–1, 37n, 37–8, 41, 43–4, 50, 72, 83, 85, 106, 133, 148, 157, 160, 187, 190, 195, 201, 210–1, 221–3, 232, 234, 240, 247, 247n, 264, 268, 281, 283, 284, 295, 298, 305, 310, 313–4, 330; contribution to emissions reduction, 180; cost overruns, 199, 297; environmental attitude to, 287, 289; in Germany, 30n; for generating hydrogen, 59, 87; and investors, 319; public attitudes to, 300–1; for ships, 84n, 84; subsidies for, 238; use of CO_2 in, 96
Nuclear fusion, 238, 244, 314
Nuclear waste, 19n, 97, 261, 294, 300, 316
Nuon Magnum, 89, 270; *see also* integrated gasification combined cycle (IGCC); Netherlands
NZEP, *see* Near Zero Emissions Power (NZEP)

Obama, Barack, 271; *see also* United States
Ocean acidification by carbon dioxide, 9, 41, 83, 121, 124, 177
Ocean acidity modification, 121, 210, 259; *see also* electrochemical weathering
Ocean alkalinity, *see* ocean acidity modification
Ocean energy, 15n, 23, 30, 32, 37, 190, 284, 314; *see also* renewable energy
Ocean fertilisation, 158, 174–6, 213, 218; *see also* iron; phosphate; urea
Ocean storage, of carbon dioxide, 98, 120–1, 130, 154, 246, 259, 289, 310;